AF856052

THE POLITICS AND RHETORIC OF SCIENTIFIC METHOD

AUSTRALASIAN STUDIES IN HISTORY AND PHILOSOPHY OF SCIENCE

General Editor:

R. W. HOME, *University of Melbourne*

Editorial Advisory Board:

W. R. ALBURY, *University of New South Wales*

B. D. ELLIS, *La Trobe University*

L. A. FARRALL, *Deakin University*

F. R. JEVONS, *Deakin University*

R. JOHNSTON, *University of Wollongong*

H. E. LE GRAND, *University of Melbourne*

A. MUSGRAVE, *University of Otago*

D. R. OLDROYD, *University of New South Wales*

J. RONAYNE, *University of New South Wales*

J. J. C. SMART, *Australian National University*

VOLUME 4

THE POLITICS AND RHETORIC OF SCIENTIFIC METHOD

Historical Studies

Edited by

JOHN A. SCHUSTER

Department of History and Philosophy of Science, University of Wollongong, Australia

and

RICHARD R. YEO

School of Humanities, Griffith University, Australia

D. REIDEL PUBLISHING COMPANY

A MEMBER OF THE KLUWER ACADEMIC PUBLISHERS GROUP

DORDRECHT / BOSTON / LANCASTER / TOKYO

Library of Congress Cataloging-in-Publication Data

CIP

The Politics and rhetoric of scientific method.

(Australasian studies in history and philosophy of science ; 4)
Includes index.
Contents: The Galileo that Feyerabend missed / Alan Chalmers – Cartesian method as mythic speech / John A. Schuster – Steady as a rock – methodology and moving continents / H. E. Le Grand – [etc.]
1. Science–Methodology. 2. Science–History. 3. Science–Philosophy. 4. Science–Social aspects. I. Schuster, John A., 1947– . II. Yeo, Richard R., 1948– . III. Series.
Q175.3.P65 1986 502.8 86–6578

ISBN-13: 978-94-010-8527-4 e-ISBN-13: 978-94-009-4560-9
DOI: 10.1007/978-94-009-4560-9

Published by D. Reidel Publishing Company,
P.O. Box 17, 3300 AA Dordrecht, Holland.

Sold and distributed in the U.S.A. and Canada
by Kluwer Academic Publishers,
190 Old Derby Street, Hingham, MA 02043, U.S.A.

In all other countries, sold and distributed
by Kluwer Academic Publishers Group,
P.O. Box 322, 3300 AH Dordrecht, Holland.

All Rights Reserved
© 1986 by D. Reidel Publishing Company, Dordrecht, Holland
Softcover reprint of the hardcover 1st edition 1986
No part of the material protected by this copyright notice may be reproduced or utilized in any form or by any means, electronic or mechanical, including photocopying, recording or by any information storage and retrieval system, without written permission from the copyright owner.

TABLE OF CONTENTS

FOREWORD

The institutionalization of History and Philosophy of Science as a distinct field of scholarly endeavour began comparatively early — though not always under that name — in the Australasian region. An initial lecturing appointment was made at the University of Melbourne immediately after the Second World War, in 1946, and other appointments followed as the subject underwent an expansion during the 1950s and 1960s similar to that which took place in other parts of the world. Today there are major Departments at the University of Melbourne, the University of New South Wales and the University of Wollongong, and smaller groups active in many other parts of Australia and in New Zealand.

"Australasian Studies in History and Philosophy of Science" aims to provide a distinctive publication outlet for Australian and New Zealand scholars working in the general area of history, philosophy and social studies of science. Each volume comprises a group of essays on a connected theme, edited by an Australian or a New Zealander with special expertise in that particular area. Papers address general issues, however, rather than local ones; parochial topics are avoided. Furthermore, though in each volume a majority of the contributors is from Australia or New Zealand, contributions from elsewhere are by no means ruled out. Quite the reverse, in fact — they are actively encouraged wherever appropriate to the balance of the volume in question.

R. W. HOME
General Editor
Australasian Studies in History and Philosophy of Science

INTRODUCTION

I

Since the seventeenth century, belief in the existence of a single, transferable method responsible for the progress of scientific knowledge has been a major element of the Western intellectual tradition. Bacon and Descartes proposed formal methodologies which claimed to elaborate rules and procedures capable of accounting for the achievements of the 'new science' and facilitating its further growth. Other major natural philosophers of the period, such as Galileo, Gassendi, Boyle, Hooke and Newton also offered models of proper scientific method. Not surprisingly, there was no agreement as to the detailed nature of the method of science and differing views quickly proliferated: from the inductivism of Bacon and the strict deductivism of the young Descartes, to the more probabilist and hypothetico-deductively coloured views of the mature Descartes and the natural philosophers of the next generation, down to the creation of a specifically Newtonian inductivism in the works of the master and his British and Dutch followers.[1]

During the eighteenth century, belief in the idea of a single, efficacious method continued to flourish. Appeals to the supposed use of method became an important means of distinguishing between 'scientific' and 'non-scientific' approaches to the natural world, and, in the Enlightenment, between 'rational' and 'irrational' approaches to the ordering of the human world of politics and social institutions. By the nineteenth century there was wide agreement inside and outside the scientific community about the existence of a scientific method which made natural science the most secure form of knowledge, and considerable effort on the part of both practising scientists and philosophers was devoted to its systematic elaboration. Since at least the early twentieth century, disciplines such as psychology, sociology, anthropology and linguistics have often justified their claims to 'scientific' status by attempting to emulate what was taken to be the method of physics, chemistry or biology. Similarly, in this century, 'methodology',

J. A. Schuster and R. R. Yeo (eds.), The Politics and Rhetoric of Scientific Method, ix–xxxvii.
© 1986 *by D. Reidel Publishing Company.*

the systematic analysis and exposition of the supposed scientific method, has been a major concern in the increasingly specialized fields of philosophy and history of science.

Scientific method, usually construed in terms of assumptions about the rules of procedure in the physical sciences, continues to be widely treated as the epitome of reason and portrayed as an instrument capable of producing objective knowledge wherever it is properly applied. Other forms of inquiry, which themselves admit that they do not follow this method, or which can be portrayed as not following it, are held to be less trustworthy and objective: their lack of method renders them vulnerable to such distorting variables as passion, subjectivity, prejudice, political bias and socio-economic interest.[2] Scientific method is thus identified with the highest standards of intellectual rigour and the most reliable procedures for gaining and assessing knowledge, and it is contrasted with those modes of investigation which prevail in hermeneutic disciplines such as history, philosophy or literary criticism. From this perspective, therefore, to speak of the political and rhetorical aspects of scientific method would be an abuse of terms and perhaps a contradiction, since rhetoric and politics are usually thought to involve just those elements of subjectivity, ideology and irrational persuasion supposedly excluded, or very much minimized, by adherence to scientific method. At best, to study the rhetoric and politics of scientific method would be to inquire into the abuse, misuse and misapplication of that doctrine.

Over the last two decades, however, the traditional belief in the existence of a single, transferable, efficacious scientific method has been challenged. This has opened a range of questions about the actual roles played by methodological doctrines in the development of science and in the social dynamics of scientific communities. Although representing somewhat divergent intellectual programmes, the work of both Alexandre Koyré and Thomas Kuhn cast considerable doubt on the relevance of general doctrines of methodology to any historical understanding of scientific practice and its development. Both writers stressed the particularity of the conceptual, procedural, evaluative and metaphysical elements which constitute any given tradition of scientific inquiry, and they indicated the variety of such traditions in the history of science. No single method could grasp or explain the internal dynamics of any given tradition, or the dynamics of the displacement of traditions over time. The scepticism of Koyré and Kuhn about the existence of a single, universally efficacious method of discovery or evaluation in science was endorsed by other historians, who argued, or at least implicitly accepted, that the formal methodologies constructed by philosophers did not come to grips with the rich complexity of scientific practice in specific historical situations. More recently, from within the philosophy of science, there have been critiques of what Hilary Putnam has termed 'method fetishism', which challenge the possibility of separating a distinctive *method* of science from the substantive *content* of its various branches.[3] More radically, in his controversial book, *Against Method*, Paul Feyerabend has asserted that all systematic doctrines of method have seriously misrepresented the actual practice of science, and that they function as the central myth of the scientific establishment. At the same time, recent research in the sociology of scientific knowledge has examined the ways in which the elements of a scientific tradition, such as norms and rules, are constructed and interpreted in different contexts.[4] This again casts doubt upon the existence of a single, genuinely efficacious and transferable method.

But to reject these previous assumptions about methodology is not to contend that ideas about method are irrelevant to the practice of science. Rather, the issue becomes one of identifying and analyzing the actual socio-cognitive functions of method doctrines within the practice and social organization of the sciences. Broadly speaking, there are three levels of scientific activity at which the deployment of method claims may be studied: (1) the 'internal' level of technical debate and argument where knowledge claims are initially framed, negotiated, and evaluated; (2) the level of institutional and disciplinary organization and politics; and (3) the level of the 'public politics' of the scientific community. A brief glance at these levels will begin to illustrate the range of functions involved and the senses in which it is appropriate to speak of a politics and rhetoric of scientific method.

The first level includes the domains of laboratory practice, the interpretation of results, and the construction and presentation of evidence and theory. Internalist historians of science and sociologists of scientific knowledge have increasingly devoted attention to the phenomena of this level, and it would be fair to say that there is a strong body of opinion which claims that scientific argument, even in published form, does not conform to the dictates of a formal method. On this view, arguments are pieces of practical rather than formal reasoning, more akin in their public, printed manifestations to legal briefs than to solid chains of strictly valid inferences.[5] Typically, the burden of a scientific argument is to promote some novel, or revised, claim about the objects of inquiry within a given discipline or sub-discipline. To that end various resources may be deployed. Appeals are made to data, old and new, which, of course, are thoroughly theory- and standard-laden; to various claims about the objects, tools and techniques currently accepted in the field; and, implicitly at least, to the standards of adequacy and relevance which, in the absence of absolute rigour, guide the assemblage of these resources into a compelling argument. This means that scientific argument is essentially persuasive argument and therefore is rightly termed *rhetorical* in the sense defined by students of 'the new rhetoric', where 'rhetoric' denotes the entire field of discursive structures and strategies used to render arguments persuasive in given situations.[6]

It is in this context that it is possible to appreciate what John Ziman has called 'the tremendous *rhetorical* power' of the concept of scientific method.[7] For even if formal methodologies do not adequately describe the genesis or evaluation of scientific knowledge, the various doctrines of scientific method are routinely available as argumentative resources in scientific practice and debate. To this extent it is correct to say that methodological discourse can be partially constitutive of knowledge claims in science. However, the fact that some scientists believe that appeals to methodology have a privileged status is itself a cultural belief requiring historical and sociological explanation. All of these considerations, then, point toward the first and most internal sense in which there is a rhetoric of scientific method: the corpus of method discourses provides one set of valued and frequently contested resources aiding in the formulation of those essentially rhetorical arguments which characterize scientific discussion.

Still remaining on our first level, recent work in the sociology of science has stressed that the evaluation and circulation of knowledge claims is a political process, and that any and all of the elements bearing upon the process may be contested.[8] The recourse to methodological discourse is simply a possible, though frequent, tactic in this domain. Hence such deployments, and the methodological debates which may be consequent upon them, are part of the micro-politics of knowledge construction in any scientific field. Claims to methodological correctness/incorrectness, purity/impurity, are resources in the knowledge-making game and can become objects of contention within it.

The rhetorical and political aspects of scientific method typical of the first level carry over onto the second – that of the institutional organization of the sciences. Increasingly, historians and sociologists of science have noted the ways in which questions of institutional control and disciplinary hegemony can help to shape the cognitive content of the sciences.[9] On this level one finds essentially political phenomena, such as the formulation and enforcement of institutional ideologies within scientific societies; the demarcation (and internal consolidation) of socio-cognitive groupings within or across disciplines; and the production and advancement of vested interests in particular disciplinary approaches, tools or techniques. These phenomena can be directly relevant to the 'internal' evolution of fields and to the construction of knowledge within them. Such institutional and disciplinary politics are often conducted in part through attempts to impose views of proper methodology in order to account for alleged weaknesses in theories, and to advance preferred research programmes. Here again, therefore, methodological discourse is involved in the politics and social dynamics of the sciences, and can be partly constitutive of the shifting content of knowledge claims. That scientists often taken seriously their methodological claims on this level (as on the first level) adds significance to the phenomenon as viewed by the historian or sociologist of science.

Finally, at the third level, assumptions about scientific method have functioned as a rich cultural resource for the apologetics of the scientific community. At the level of debate which F. M. Turner has labelled 'public science', appeals to notions of proper scientific method have been frequently involved in attempts to advertise the political and cultural value of science to wider audiences.[10] This has been a feature of the public relations of the scientific community since the seventeenth century, and it is important to recognize the ways in which a *consensus* on method, as contrasted with disputes about particular theories, has been presented as the defining characteristic of science. This emphasis on method is the medium through which science has been, in different times and places, both linked with, and distinguished from, other intellectual discourses such as theology or philosophy, and broader social and political ideologies.[11] As mentioned earlier, the power of this rhetorical role of the idea of scientific method is demonstrated by the deference it has elicited from individuals, disciplines and institutions outside the domain of the natural sciences.

When scientists deploy methodological claims, on any of these three levels, they draw upon cultural resources provided by the grand tradition of theorizing about method which extends through a series of philosophers and scientists from Bacon (indeed from Aristotle) to Popper and his followers. It is a corollary of the view of the politics and rhetoric of method advanced here that the tradition of formulating theories of method must be treated in the same contextualist manner which has been advocated in relation to the problem of method in science itself. The history of methodological theories is not a 'whiggish' progression toward ever more clear insights into the 'truth' about method, achieved by a series of gifted methodologists, each working in a narrow internalist way. Each methodologist has operated with (and against) the available formal discourses on method. But each new methodology has been constructed by its author in the light of problems and goals which might relate to the tradition itself, to the perceived state of one or more of the contemporary sciences, or to other discourses believed to be relevant, such as natural theology, political theory, and moral philosophy. The perception and weighting of such concerns by a methodologist is a complex function of his biography, social location, institutional affiliations and perceived interests. The historian's task is to consider these perceptions and weightings when analyzing the resulting disourse, thus abandoning a 'Whig' history of methodological ideas for a social and political history of theorizing about method. In the broadest sense, then, the history of methodology must take as its objects: (1) the tradition of theorizing about method, thus historicized and contextualized; (2) the politics and rhetoric of method on the three levels of scientific discourse; and (3) the relations between these domains.

The chapters in this volume generally share the scepticism recorded above concerning the existence of a single, transferable scientific method, and hence they reject its use as a category of whiggish and generally inductivist historical explanation. However, they affirm the importance of method as a subject of historical inquiry. Each of the eight historical studies in this volume either analyzes the rhetorical and political dimensions of a systematic methodology, or examines the manufacture and deployment of methodological ideas on one or more of the three levels of discourse outlined above. The chapters deal with systematic methodologies associated with writers such as Descartes, Condillac and Whewell; with less formal but significant statements about method in controversies within the physical and biological sciences; and with the place of methodological pronouncements in specific political and institutional contexts. In analyzing a range of historical material from these perspectives, the chapters are intended to exemplify the types of rhetorical strategies and political functions which are served by methodological discourses and which in turn shape their content. As such, the chapters presented here join a small but growing body of revisionist work on the history and sociology of scientific method.[12] Before introducing the chapters, it is therefore worth inquiring how the history of method has reached this new point of departure.

II

Because method has traditionally been seen as the defining feature of the scientific enterprise, historians and philosophers of science have, until quite recently, focused almost entirely upon the career of methodological ideas and their supposed instantiation in scientific practice. This emphasis on method was apparent in the earliest systematic studies of the history and philosophy of science. In reflecting on past scientific achievements, writers such as Jean d'Alembert, Joseph Priestley, William Whewell and Auguste Comte were attempting to distil the method by which previous discoveries were made, on the assumption that its continued use would ensure future success. In this sense, the histories of science offered by Priestley and Whewell were heuristic exercises: both authors believed, although to different degrees, that a study of history could reveal the best method of attaining further scientific truths. Writing in his *History and Present State of Discoveries Relating to Vision, Light and Colours* of 1772, Priestley asserted that the advancement of all branches of science would be facilitated by 'an historical account of their rise, progress and present state'.[13] And, in his earlier *History and Present State of Electricity* he remarked that such an inquiry 'cannot but animate us in our attempts to advance still further, and suggest methods and experiments to assist us in our future progress'.[14] Similar motives informed Whewell's more explicitly programmatic study of past scientific discoveries. In the *History of the Inductive Sciences*, published in 1837, he suggested that such a survey 'may not only remind us of what we have, but may teach us how to improve and increase our store', and 'afford us some indication of the most promising mode of directing our future efforts to add to its extent and completeness'.[15] The title of Whewell's next major work, *The Philosophy of the Inductive Sciences, Founded upon Their History*, indicated that his views on the methodology and epistemology of science depended upon a knowledge of past scientific practice; and he explained that 'we are most likely to learn the best methods of discovering truth, by examining how truths, now universally recognized, have really been discovered'.[16] From this perspective, therefore, one of the key merits of the history of science was that it provided an understanding of the scientific method.

This preoccupation with method was still apparent when the history and philosophy of science began to crystallize as a distinct discipline in the early twentieth century. In 1917, the British historian, Charles Singer, referred to the establishment of journals such as *Isis* as an indication of a 'widely recognized' need to appreciate the method of science. In launching a new scholarly series, *Studies in the History and Method of Science*, he proclaimed that the historian could teach 'the method by which knowledge has been gained'.[17] This assumption was shared by George Sarton, the founder of *Isis*, for whom the history of science was 'the natural basis of the philosophy of science' and the advancement of mankind was 'largely due to the development of positive knowledge and scientific methods'.[18] In Sarton's vision, general education would include a focus upon 'scientific method and knowledge' provided by historians of science, the 'vanguard of the New Humanism'.[19] The influence of this legacy, although in a qualified form, was still apparent in the *Harvard Case Histories in Experimental Science*, published under the editorship of James B. Conant in 1957. Here again, historical study, by allowing an understanding of the method of science, offered a bridge between scientists and the lay public.[20]

During the twentieth century, as the philosophy of science has matured as a distinct discipline, it has largely dominated the study of formal methodology, which accordingly has become increasingly removed from the concerns of practising scientists. Whereas some of the significant contributions to debate on method in the nineteenth century were made by scientists such as Herschel, Bernard, Pearson and Mach, the major writers of this century have been professional philosophers such as Carnap, Reichenbach, Salmon, Popper and Lakatos. This shift has meant that philosophers of science have largely ceased to deal with the method of discovery, an issue which had always held particular attractions for those committed to what they understood as Baconian or Newtonian method. Indeed, even before the advent of logical positivism, the focus of concern had begun to move towards methods of verification, justification and evaluation, and this led to attempts to formulate ideal methods of testing hypotheses and assessing theories and research programmes which should apply to all genuinely scientific endeavours.[21] Consequently, the previously existing link between methodology and the close historical study of the 'best' science has been broken. Since the 1950s, of course, Popper, Lakatos and their followers have tried to restore that link; but, as discussed below, many historians and sociologists of science now find their efforts both unconvincing and productive of further difficulties about the history of method and its relation to scientific practice. Until recently, therefore, those interested in the historical aspects of method have received little assistance from the formal philosophy of science and this empty niche in the historical ecology tended to be filled by historians of philosophy and of ideas, who proceeded to create a thriving sub-discipline concerned with the 'history of methodological ideas'.

The classic studies in this field – such as those by Randall and Crombie – were predicated on the traditional belief that the development and maturation of an efficacious scientific method was a necessary condition for the emergence of modern science.[22] For example, Crombie claimed that 'modern science owes most of its success to the use of . . . what is often called "the experimental method" '. His thesis was that the 'modern, systematic understanding of at least the qualitative aspects of this method was created . . . in the thirteenth century' and then refined, elaborated and eventually handed down to the fathers of modern science in the early seventeenth century.[23] It is characteristic of these books, as well as other later studies in the history of methodological ideas, that the existence of a single, definitive and efficacious method is presumed. Much scholarly effort is then directed not so much toward illustrating the putative method in action as in dissecting formal treatises on method in a search for the real meaning of a method discourse whose legitimacy and efficacy are not disputed. These textual exegeses then lead to the characteristic 'history of ideas' project of tracing the filiation (and improvement) of the supposedly efficacious methodological ideas through a series of classical texts.[24]

In the 1960s and early 1970s the history of methodological ideas approach was expanded in a series of well known and important anthologies.[25] The studies in these volumes continued to trace the historical filiations of ideas about method; but they were not concerned to document further the supposed historical route by which method first created (or helped to create) modern science. Instead, the emphasis shifted to issues dictated by the highly professionalized, largely positivistic philosophy of science of the 1950s and '60s, with much effort being devoted to explicating the relationships between past theories of method and contemporary philosophy of science. Edward Madden, for example, noted that the studies in the volume which he co-edited were supposed to cast light on a series of canonical methodological issues which included 'theory construction, . . . hypothesis, lawfulness . . . the experimental methods, the uniformity of nature and the rule of succession (the problem of induction) . . .'.[26] As an abstract philosophical exercise this project is unexceptionable. However, from an historical point of view it can lead, particularly in its more extreme forms, to a whiggish and teleological history in which a genealogy of ideas on method is traced across key thinkers, the best being claimed as precursors of modern methodologists. Nevertheless, despite the limitations of such work on the history of methodological ideas, it is important to recognize that the sensitive explication of the internal structure of methodological doctrines has been and continues to be a useful enterprise. Indeed, it is propaedeutic to any properly contextual study of the shaping of method discourse and its political and rhetorical functions.

Against this background of a confident, abstract philosophy of method and a well-entrenched history of methodological ideas, the scepticism of Koyré, and later Kuhn, emerged. Koyré belonged to the second generation of a small group of Continental neo-Kantian historians of philosophy who had turned their attention to the conceptual development of science. Partly following the lead of Ernst Cassirer, Leon Brunschvicg and Emile Meyerson, Koyré had found a philosophically unsophisticated discipline of history of science, dominated by inductivist methodology and empiricist epistemology, issuing, more often than not, in what were later to be termed 'whiggish hagiographies'. Koyré's stress on the theory-bound nature of experience and the historical transformations of conceptual frameworks, began to weaken the belief in a single, transferable method.[27]

As befitted his neo-Kantianism, Koyré did believe in an unproblematic notion of scientific progress. But this progress did not rest upon the application of a general scientific method; rather, it depended upon the historically contingent adoption of appropriate metaphysical frameworks and the subsequent rational pursuit of effective science within them. Galileo's constitution of classical mechanics within the framework of a loosely 'Platonic' and mathematical metaphysics was the paradigm case of such a development. According to Koyré, Galileo's success did not depend upon his discovering and using a universally applicable method, but upon the fact that he worked, perceived and argued within the appropriate metaphysical framework.[28] 'No science has ever started with a treatise on method and progressed by the application of such an abstractly derived method', Koyré announced, and generations of professional historians of science have tended to agree.[29]

Since the 1960s the Koyréan approach to the history of science has been buttressed by two significant developments: the publication of Thomas Kuhn's *The Structure of Scientific Revolutions*, and the wider dissemination of Gaston Bachelard's earlier work. From different perspectives both these scholars were concerned with the emergence of effective fields of mathematical-experimental science from earlier, largely qualitative and discursive bodies of thought. The results of their investigations had implications for the notion of a universal scientific method. For Bachelard, each field of science consisted of a set of interlinked, mathematicized concepts which interacted dialectically with the instrumentalities through which those concepts were objectified, or materialized.[30] To paraphrase Bachelard, the meaning of a concept must include the technical conditions of its material realization. In his view, a general scientific method could explain neither the genesis of sciences nor their respective developments.[31] The implications of Kuhn's work are still being assessed, but its relevance to the history of methodology and its role in science is clear. Kuhn's work strengthened the contention, previously made by Koyré and Bachelard, that the structural elements – metaphysical, conceptual, evaluative and instrumental – which make possible a particular tradition of scientific practice are *sui generis* to that field and sufficient for its satisfactory pursuit. It follows from this that systematic doctrines of method play no part in the real dynamics of scientific practice, a proposition accepted, at least implicitly, within many subsequent detailed, 'internalist' reconstructions of past scientific work. Kuhn understood this, for he castigated history written in terms of the discovery and application of some abstract scientific method. In his view, methodological claims were irrelevant for the modern historian of science. Such claims were typically found in prefaces, *eloges*, hagiographies, autobiographies and the 'historical' sections of textbooks, but not in the *real* texts pointing to the actual dynamics of scientific work – research papers, technical memoirs, the parts of textbooks inculcating the paradigm, laboratory notes and manuscripts, and serious correspondence about research.[32]

Koyré, Bachelard and Kuhn therefore presented strong cases for diminishing the role of method in the birth and development of the sciences. They reduced its role to that of mere rhetoric, in the traditional pejorative sense of rhetoric as a mere art of persuasion. Hence, while their work raised questions about the nature of method and the deference to its importance, their adherence to a reason/rhetoric dichotomy effectively froze further inquiry. While accepting that grand doctrines of method do not function in science in the way those doctrines themselves proclaim, the chapters in this volume take up these questions from a perspective which sees methodology as a resource for both argumentative and rhetorical strategies in science. This, however, is to oversimplify the background, for both prior to and immediately following the appearance of Kuhn's book there was a resurgence of methodology in the grand manner in the work of Sir Karl Popper and Imre Lakatos.[33]

Popper, Lakatos and their followers readily admitted the shortcomings of earlier theories of method, especially logical positivism, and they shared with post-Koyréan historians of science a sensitivity to the role of 'metaphysics' in the life of scientific theories. Nevertheless, Popper and Lakatos both aimed to re-establish belief in a unique and efficacious method. Despite notable differences of opinion, each sought to affirm the existence of a normative method, a method of evaluation, consisting in a concise set of criteria and rules for the acceptance or rejection of theories or 'research programmes'. The existence and efficacy of such a method were to be established in part through recourse to historical inquiry of a special sort. As Lakatos formulated the strategy, one proceeds first by constructing historical case studies of theory (research programme) change. The 'rational reconstructions' of the 'actual' history are articulated around the putative patterns of progressive and rational development singled out and explicated by the canons of Lakatosian method. Then,

> ... one tries to compare this rational reconstruction with actual history and to criticize both one's rational reconstruction for lack of historicity and the actual history for lack of rationality.[34]

The superiority of Lakatos' method is supposed to reside in the fact that rational reconstructions based upon it explain more of the 'actual history' than do rational reconstructions based upon any other methodology.[35] This procedure implies that the best science has always been done according to standards approximating to those of the correct (Lakatosian or Popperian) method. If rational reconstructions are not to drift disastrously astray from 'actual history', it must be the case that the great historical actors such as Copernicus, Newton, Darwin and Einstein had fleeting but decisive glimpses of the correct methodology.[36]

There is now a considerable body of such rational reconstructions, if not of full, two-stage, Lakatosian case studies as defined above.[37] This literature has exerted virtually no influence upon social and contextual history of science. One reason for this is that historically sensitive philosophers of science such as Toulmin and Feyerabend regard Lakatosian history as implausible and inadequate, firstly because of its tendency to ensure in advance the vindication of Lakatos' method, and secondly, because of persuasive historical evidence that scientists' standards of rational appraisal have altered over time in complex relations to the history of scientific practice.[38] An even more damaging charge, however, is that Lakatosian (and Popperian) history, rather than being inadequate or implausible, is not history at all. The fundamental reason is that the central categories and rules of Lakatosian and Popperian method, and hence the resulting 'appraisals', are context dependent. In actual cases the meaning, application and weighting of the rules and categories would be subject to continual reinterpretation and negotiation.[39] Indeed the Lakatosian school itself engaged in such critical revisions in an attempt to 'improve' the rational reconstruction of the Copernican and Einsteinian revolutions. Furthermore, work on 'discourse analysis' by Mulkay and Gilbert suggests that in specific contexts the meaning of Popper's rules depends upon scientists' highly variable personal judgments and interpretations of the cognitive state of play. They conclude that 'in situations of cognitive uncertainty Popperian rules ... cannot, therefore, operate as useful prescriptions, but only as evaluative retrodictions.'[40] In other words, because of the contextual variability in their interpretation, the methodologies of Popper and Lakatos cannot work as literally claimed. For the same reason, they can serve as powerful accounting resources, fulfilling rhetorical purposes. Accordingly, any historiography which posits the literal efficacy of these methods must mystify the situational rationality of historical actors, and occlude the actual roles of method discourse in history.

In the 1960s and early 1970s, however, few of these criticisms had emerged. From the standpoint of historians of science, the surge of Popperian and Lakatosian methodological writings compounded the problem about the relations between methodologies and the history of scientific practice which had surfaced in the work of Koyré and Kuhn. I. B. Cohen captured the perplexity of many historians concerning philosophers' impositions of historically insensitve conceptions of scientific method when he commented that: 'It is a common experience for historians of science to be dismayed on encountering their own subject in the writings of philosophers'.[41] Nor was disquiet confined to historians. Within the philosophy of science itself there were criticisms of the tendency to ignore the variety of past scientific practices in favour of the notion of a universally valid or applicable methodology. In the 1960s, Stephen Toulmin pointed to the gap between the actual practice of scientists and the norms prescribed by logical positivism.[42] Since then, this position has gained some wider support amongst philosophers of science, so that now it is possible, as Putnam puts it, to question the search for 'a formal method, capable of being isolated from actual human judgments about the content of science'.[43] Against this background, Feyerabend's methodological anarchism, with its appeal to historical case studies in order to attack both past and present formulations of a single methodology, can be seen as an extreme version of this earlier confrontation between the history and philosophy of science.

It was in the context of this confrontation that Larry Laudan called for a new approach. In his extensive bibliographical survey of work on the history of methodology, published in 1968, Laudan commented on the poor relations between history and philosophy and suggested that a study of methodology should be the bridge between them.[44] He disagreed with Koyré's contention about the irrelevance of general theories of method to the history of science, but acknowledged the inadequacy of existing work on the history of method. The essential point was that most of this work was written by philosophers whose concern was with the internal consistency and logical coherence of past methodologies, or with their relationships to modern theories. This perspective encouraged the construction of a genealogy of ideas about method, some of these being seen as the precursors of sophisticated modern doctrines. Laudan complained that this approach treated methodology as a set of disembodied ideas and failed to ask significant historical questions. In warning of the danger of anachronism, he indicated the need to study the changing meaning of key concepts such as law, induction, experiment and hypothesis, and suggested that this might allow an appreciation of the complex connections between theories of method and actual scientific practice.[45] In canvassing a different approach, Laudan made two important claims: first, that there was no single, historically invariant methodology of science, but a series of historically specific reflections on its practice; and secondly, that methodological doctrines did not guide scientific practice in any simple fashion and were in any case themselves conditioned by substantive scientific developments.

During the 1970s there was a discernible increase in the study of the role of methodological statements and theories in the history of science. Generally following Laudan's recommendations, these studies usually eschewed philosophical assessments and genealogies, and concentrated instead on the relationship between methodology and scientific practice. Michael Ruse sought to demonstrate that methodological commitments, stemming from the philosophical work of Herschel and Whewell, were relevant to the research and argumentation of major nineteenth-century scientists such as Darwin and Lyell. Martin Rudwick offered a detailed analysis of Darwin's methodological anxieties and strategies associated with his hypothesis about the geological formations at Glen Roy. Richard Olson presented the canons of scientific reasoning sanctioned by the Scottish common sense philosophers such as Reid, Stewart and Brown, arguing that their views on induction, analogy, hypothesis and experiment influenced the intellectual style of leading figures in British physics. Geoffrey Cantor showed how methodological prescriptions deriving from this Scottish tradition played an important role in debates about the theory of light in the early nineteenth century.[46]

While this work treated ideas about method in connection with specific examples of scientific practice, it continued to conceive the discourse on method as constituted independently of substantive scientific theories and concepts. It was therefore able to present methodology as an influence on, or explanation of, various examples of scientific thought. At one extreme, Michael Ghiselin ascribed Darwin's scientific success to his correct application of the hypothetico-deductive method. Ruse criticized Ghiselin for what he took to be the claim that Darwin had 'discovered' this method, by indicating that it was already present in the work of Herschel.[47] In doing so, however, Ruse did not challenge the assumption that doctrines of method were efficacious in guiding scientific research; indeed, he argued that the writings of Herschel and Whewell conditioned the development of Darwin's theory in several ways. Similarly, Olson presented the epistemology and methodology of the Scottish moral philosophers as a body of thought, with its own inner logic, which shaped the practice of natural philosophers.

It is possible, however, to take the data about methodology uncovered by these historical studies as the starting point for different conceptions of its role and status in scientific discourse. Responses to some of the work cited above indicated these alternative lines of inquiry. In reviewing Olson's book, John Christie drew attention to the tendency of this historiographic tradition to see ideas on the philosophy and method of science as a body of abstract reflections insulated from other intellectual and social concerns. As a corrective, he argued that the pronouncements of the Scottish philosophers on method and epistemology, on topics such as analogy and causation, were closely related to their need to uphold religious and metaphysical positions such as those associated with natural theology.[48] In this way, statements about method could be interpretated as normative protectors of sensitive ontological positions. Roger Smith made a similar point in commenting on David Hull's analysis of the reception of Darwin's *Origin of Species*: the writings of Herschel, Mill and Whewell on the philosophy of science were informed by moral and political considerations and therefore could not be regarded as a neutral, logical source of reflection about the practice of Victorian science.[49] From these perspectives, methodology does not necessarily constitute an independent variable which directs science, but rather can be approached as a discourse partly constructed by ideological commitments or investments in particular scientific theories.

Laudan's recent collection of essays, *Science and Hypothesis*, is held together by a thesis that entertains this kind of alternative. Focusing on 'the method of hypothesis', Laudan argues that it was prominent in the seventeenth century, condemned in the century of the Enlightenment, but then revived as the centre of methodological concern during the nineteenth century. His key point is that these developments cannot be sufficiently explained without reference to changes in the character of scientific practice: the method of hypothesis returned to favour because it best described the type of theorizing in the most advanced sciences of the day. Laudan draws upon Cantor's work on debates about the nature of light in order to illustrate the close relationship between methodology and substantive scientific issues. Whereas Cantor analyzed the appeals to existing notions of methodological orthodoxy by opponents of the wave theory, Laudan emphasized the way in which the method of hypothesis was elaborated and refined as a rationale for a theory which, by the 1830s, had come to be seen as scientifically progressive. Furthermore, the fact that some of the major theorists of method, such as Whewell and Herschel, were also leading proponents of the undulatory theory of light, illustrated the symbiotic relationship between scientific practice and discourses on method. Thus one of the recurring themes in Laudan's book is the contention that changes in doctrines of method have little to do with developments in philosophy and epistemology but are dependent on shifts in '*scientific* beliefs'; that the acceptance of new methodologies has been largely determined by their capacity 'to legitimate a preferred scientific theory'.[50]

This kind of work clearly reveals the inadequacies of Whiggish and internalist approaches to the history of methodology. It has shown how the formulation and acceptance of specific doctrines of method relate to other areas of discourse such as theology, moral philosophy and, most notably, substantive scientific practice. Secondly, it has exemplified the ways in which views on method have often been constructed or modified in order to achieve *post hoc* legitimation of particular scientific theories. In doing so, such work marks a considerable advance over earlier approaches to the history of methodological ideas.

But it is worth noting the points at which there might be further convergence between the historical study of methodology and other areas of investigation. For example, various lines of inquiry within the sociology of scientific knowledge imply more radical revisions of the traditional concept of an efficacious scientific method. For while acknowledging the legitimating function of doctrines of method, the recent work of Laudan retains a commitment to the view that such doctrines can provide transferable, rational guidelines for scientific practice.[51] In contrast, the work of Barnes, Mulkay, Gilbert, Woolgar and Latour, insofar as it focuses on this issue, stresses the extent to which norms and rules derived from methodological theories are variously interpreted and negotiated in scientific debate. Such a perspective also allows the contention that the rhetorical role of method is not confined to the *post hoc* rationalization of theories but can be involved, together with empirical data, experimental techniques and a range of linguistic resources, in the complex process by which scientific knowledge is produced and ratified.[52] Furthermore, as Schuster argues in this volume, the early work of Feyerabend may provide a model for understanding how general theories of method misrepresent scientific practice and conceal the actual usages of methodological concepts in science. Finally, recent developments in the social history of science suggest that the focus on specific theories and individuals in previous studies of methodology could be expanded to include a consideration of the ways in which appeals to method are significant in debates surrounding the constitution of disciplines and the politics of scientific institutions. The following chapters are informed to different extents by the various lines of approach discussed above.

III

The first two chapters explore perspectives that have emerged with the post-Kuhnian scepticism about general doctrines of method. Alan Chalmers asks how Feyerabend's treatment of Galileo supports his larger case 'against method'. Arguing that Feyerabend's study warrants significant criticism on a number of grounds, he attempts a more accurate historical analysis of Galileo's work which is intended to provide better evidence than Feyerabend himself produced for the thesis that there is no such thing as a unique, ahistorical, efficacious scientific method. Chalmers contends that Feyerabend was misled by concentrating upon Galileo's *Dialogue Concerning the Two Chief World Systems*. The core of Galileo's new approach to science is not contained in this largely polemical work, but rather in his *Two New Sciences*. Hence there is no reason to conclude, on the basis of the *Dialogue*, that propaganda is of the essence of scientific work. In addition, Chalmers takes issue with Feyerabend's charge that a number of Galileo's argumentative procedures in the *Dialogue* essentially consist in 'trickery' and 'ad hocery'. Following Clavelin and others, Chalmers locates Galileo's decisive break with Aristotelian physics, and the institution of the first version of a classical mechanics, in the *Two New Sciences*. In a manner deriving from Bachelard and Kuhn, he argues that the creation of Galileo's new physics, involving as it did a new conceptual framework and form of experimental practice, *ipso facto* brought into being its own new methodological standards. It follows for Chalmers that Feyerabend is wrong to suggest that, methodologically speaking, 'anything goes'; for although there is no single, timeless scientific method, the creation of new modes of theory and practice produces new, local and contingent methodological standards. Chalmers goes on to argue that Feyerabend radically underplayed the extent of Galileo's methodological revolution by confining himself to the essentially empiricist transformation of 'observation language' urged in the *Dialogue*, while ignoring the more fundamental shift in method accompanying the successful elaboration of a mathematico-experimental physics in the *Two New Sciences*.

John Schuster's chapter attempts to fill the historiographical gap left by Koyré's and Kuhn's debunking of the myth of a single, efficacious method. It addresses the question of how it can be that throughout the history of science, methodologists and their audiences have often genuinely believed in the efficacy of general methodologies. Taking Descartes' work as an example, he offers a structural model of how systematic methodologies operate to create for their audiences 'literary effects' of their applicability, unity and efficacy. Schuster's model builds in part on perspectives developed by Feyerabend and Roland Barthes, and it is intended to contribute to the historiography of method in the sense that an understanding of the internal discursive dynamics of method doctrines is arguably central to the problem of explaining their political and rhetorical functions. An extended portion of Schuster's study is devoted to a reconstruction of how Descartes manufactured his method discourse. Schuster analyzes the stages in this process, the discursive resources employed, and the tactics of literary fabrication involved. This approach illustrates the historical and biographical contingency of Descartes' doctrine, as well as the opportunism involved in its production, qualities Cartesian method presumably shares with all set-piece theories of method. A revised historiography of method discourse, Schuster suggests, should involve such structural and developmental perspectives.

H. E. Le Grand's study is the first of three dealing with science in the twentieth century and with scientific communities rather than individuals. It focuses on the role of methodology in a major theoretical dispute, the early stages of the debate in geology over Alfred Wegener's theory of continental drift. Le Grand documents and categorizes the varied deployments of methodological claims in the course of this debate, viewing them as flexible rhetorical resources and not as independent determinants of scientists' actions or beliefs. He produces a suggestive catalogue of the ways in which methodological discourse is deployed, and defends the relevance and value of studying such deployments in formal texts against the claims of the 'discourse analysis' school. Among other findings he highlights the fact that both sides in the debate were in general willing to concede the hegemony of arguments couched in the 'style of physics', and he shows how this further conditioned the evolution of the debate and its accompanying methodological rhetoric. Le Grand also shows that Wegener altered his rhetoric under the pressure of critical attack, arguing that this was not merely a tactical move on Wegener's part but that it also played a partially constitutive role in the articulation of his theory and hence in the subsequent course of the debate.

T. D. Stokes' chapter continues the focus on contemporary science and intra-disciplinary debate. It deals with the role of methodological commitments that led the Indian molecular biologist V. Sasisekharan and his team to propose an alternative to the Watson-Crick double helix model of the DNA molecule. However, rather than classifying and analyzing instances of method rhetoric, Stokes' study, like those of Chalmers and Schuster, addresses the issue of the post-Kuhnian scepticism about method. In particular, it examines Laudan's contention that methodologies can provide more than window dressing and can play constitutive roles in the construction or evaluation of knowledge claims.

Laudan claims that a perceived conflict between a scientific theory and a preferred methodological metatheory can generate one species of those 'external conceptual problems' with which scientists must occasionally deal. Such conflicts may be resolved by adjusting either the theory or the methodology in question. The first alternative would presumably provide examples of methodological commitments partially constituting knowledge claims or their evaluation. However, an ambiguity inherent in Laudan's theory enters at this point: do such cases illustrate the literal efficacy of a method or merely the flexible use of methodological discourse as one resource, amongst several, in the formulation of a claim?

Stokes shows that Sasisekharan insisted upon methodological standards differing from those of the majority of the relevant scientific community and that these commitments did condition his conclusions. Hence the study supports Laudan's stress on the importance of such methodologically generated conceptual problems. Stokes seems to part company with Laudan over the possibility of literal efficacy, however, for he shows how Sasisekharan's path of research was shaped in part by commitment to a methodological ideal which itself was brought into focus and conditioned by technical, institutional, personal and conceptual variables. Sasisekharan's methodological views were not window dressing, but neither were they independently governing and literally applicable rules of procedure. This reading of Stokes' is confirmed by his critique of Laudan's suggested criteria for deciding the relative weighting of problems involving a clash between theory and methodological metatheory. Stokes contends that it is incorrect for Laudan to assume that there ought to be a disciplinary consensus about the sort of warrant offered to the community's working theory by a particular metatheory. Scientists, he contends, assess the weighting of any clash in terms of their own perceived interests, which are complex functions of their placement in the field, preferred or enforced goals of research, and institutional location. All this tends to show that the very existence of a Laudanian methodological problem, as well as its outcome(s), results from contextually grounded interpretations and social negotiations. Hence Stokes' chapter implies that Laudan's views on the role of methodology should include reference to sociological and historical factors.

Jan Sapp's chapter deals with a long-standing debate in twentieth-century biology concerning the relative importance of the role played in heredity by nuclear chromosomal genes as opposed to the cytoplasmic elements of the cell. Sapp takes his bearing from Bourdieu's notion of the struggle for 'scientific authority' within a field of inquiry, and from his own extensive studies of this particular debate. His approach widens the category of 'social accounting', previously employed by Mulkay and Gilbert, to include rhetorical deployments of formal methodological doctrines. Sapp also adds a novel category of resource termed 'ascriptions of the "technique-ladenness" of observation'. This involves scientists' attempts to show that the objects of inquiry and techniques of dominant competitors are 'intellectually constructed objects created through specific modes of scientific production'. Like Le Grand, Sapp demonstrates the presence of these accounting resources in the kinds of formal texts and contexts downplayed by Mulkay and Gilbert, and he documents the increased currency of these resources, especially that of 'technique-ladenness', during the period in which the disciplinary struggle over cytoplasmic inheritance was heightened.

The last three chapters deal with the role of methodology in the politics of scientific institutions and in the wider social relations of science. W. R. Albury discusses the ways in which a formal text on method – Condillac's *Logic* – became a significant resource for the French Ideologues in their attempts to elaborate a scientific approach to social questions during the immediate post-Revolutionary years. At one level, this chapter shows how the association of a doctrine of method with a successful scientific theory – in this case Condillac's method and its favourable reception by Lavoisier – encouraged its deployment in other areas. But furthermore, it reveals the extent to which this choice was conditioned by political and rhetorical considerations. By adulating Lavoisier as a victim of the Terror, the Ideologues were able to present the method of Condillac as both a symbolic and a practical alternative to the purported anti-intellectualism of the Jacobins. Secondly, the linguistic basis of Condillac's method made it more attractive to men trained in law, philosophy and languages than the mathematical emphasis of Condorcet's methodology. In supplying this contextual detail Albury's chapter suggests that the extrapolation of scientific method to other intellectual domains cannot be explained by reference to its general power or appeal, but must be the subject of specific investigation.

David Miller studies the role of methodology in two early nineteenth-century British scientific institutions. By drawing attention to the different audiences for which statements about method were intended, he shows how they could carry different meanings. Thus an emphasis upon Baconian inductivism with its image of scientific inquiry as open and democratic was the dominant public position of the early Geological Society; but Miller's additional analysis suggests that methodological discourse also served other functions. Statements about the value of observation were coupled with defences of theory, and references to the importance of 'facts' often implied allegiance to rival techniques and theoretical frameworks, such as those represented by mineralogy and palaeontological stratigraphy. In these ways, statements on method were rhetorical weapons in arguments concerning the competing research programmes of scientific disciplines.

In the final chapter, Richard Yeo addresses the three key notions which, since the seventeenth century, have been celebrated as the defining characteristics of the scientific method — the assumptions about its accessibility, unity and transferability. The chapter shows how these assumptions were themselves deployed, but also contested, in debates about the nature of science and its relations to society in nineteenth-century Britain. Methodological statements were thus involved in discussions of the role of theory and observation, the division of labour, the hierarchy of disciplines, the unity of science and the cultural implications of scientific modes of thought and practice. Within these debates the three recurring assumptions about method did not necessarily support a particular position or interest, but were exploited rhetorically in both systematic works and less formal texts. Thus the idea of a single method, which at one level served to affirm the unity of science in spite of increasing specialization, was also a focus of dispute in arguments within the scientific community concerning the relationship between different disciplines in the hierarchy of knowledge. Similarly, the claim concerning the transferability of scientific method, although central to the cultural legitimation of the scientific community, could be turned against it.

Thus while the early chapters of the volume treat the place of methodology in the work of individual scientists, in substantive scientific debates and in conflicts between disciplines, the last three chapters consider the institutional and political dimensions which bear upon these appeals to method. In particular, they reveal the ways in which discourse on method can function as a flexible resource in negotiations about the meaning, direction and socio-political uses of science.

Inevitably, this collaborative volume does not pretend to be a comprehensive survey of the topic it addresses. For example, one important area of study — seventeenth- and eighteenth-century British methodological writings — has not been included, although this omission is to some extent compensated by the existence of published work.[53] There are other absences which suggest possibilities for future research. A large portion of the scholarship on the history of method has dealt with major figures or controversies in the physical sciences, such as the debates associated with the work of Galileo, Newton and Descartes. It would therefore be instructive to have detailed, critical studies of the methodological doctrines associated with the life sciences.[54] This kind of investigation might be expected to explore the ways in which discourse on method is involved in the establishment of new disciplines, and its place in the construction of mythologies about heroes in the history of science.[55] More consideration could also be given to the specific social and intellectual contexts in which major methodologists outside the scientific community, such as J. S. Mill and Karl Popper, came to formulate influential doctrines of scientific method.[56] Finally, it might be profitable to examine the significance of the distinction between content and method in theories of science education and in the public relations of science.

JOHN SCHUSTER
University of Wollongong

RICHARD YEO
Griffith University

NOTES

[1] L. Laudan (1981) *Science and Hypothesis*, Dordrecht; I. Lakatos (1978) 'Newton's Effect on Scientific Standards', in J. Worrall and G. Currie (eds), *Imre Lakatos, Philosophical Papers*, vol. 1, Cambridge, pp. 193–222.
[2] W. R. Albury (1983) *The Politics of Objectivity*, Deakin University Press, pp. 5–17.
[3] H. Putnam (1981) *Reason, Truth and History*, Cambridge, pp. 188–200.
[4] P. Feyerabend (1975) *Against Method*, London; B. Barnes (1974) *Scientific Knowledge and Sociological Theory*, London and Boston, chapter 3; M. Mulkay (1979) *Science and the Sociology of Knowledge*, London, chapter 3; B. Barnes and J. Law (1976) 'Whatever Should Be Done with Indexical Expressions', *Theory and Society* **III**, pp. 223–237.
[5] J. R. Ravetz (1971) *Scientific Knowledge and its Social Problems*, Oxford, pp. 75–108, 119–159; K. D. Knorr-Cetina (1981) *The Manufacture of Knowledge*, Oxford, chapter 5; B. Latour and S. Woolgar (1979) *Laboratory Life: The Social Construction of Scientific Facts*, London, chapter 3 and 4; M. Mulkay, *op cit.* (Note 4), pp. 49–59, 73–92; S. Yearley (1981) 'Textual Persuasion: The Role of Social Accounting in the Construction of Scientific Arguments', *Philosophy of the Social Sciences* **XI**, pp. 409–35; W. Weimar (1977) 'Science as a Rhetorical Transaction: Toward a Nonjustificational Conception of Rhetoric', *Philosophy and Rhetoric* **X**, pp. 1–29; J. Gusfield (1976) 'The Literary Rhetoric of Science: Comedy and Pathos in Drinking Driver Research', *American Sociological Review* **XLI**, pp. 16–34; J. Law and R. J. Williams (1982) 'Putting Fasts Together: A Study of Scientific Persuasion', *Social Studies of Science* **XII**, pp. 535–558.
[6] C. Perelman (1979) *The New Rhetoric and the Humanities*, Dordrecht, pp. 18–24; C. Perelman and L. Olbrechts-Tyteca (1971) *The New Rhetoric: A Treatise on Argumentation*, London, pp. 17–23, 67–70, 120–130, 187–192, 460–471, 502–8,

510–12; W. Weimar, *op cit.* (Note 5); J. Ravetz, *op cit.* (Note 5), pp. 120, 151–4; A. Benjamin, G. N. Cantor, J. R. R. Christie (eds), *The Figural and the Literal: Problems of Language in the History of Philosophy, Science and Literature, 1650–1800*, Manchester (forthcoming).

[7] J. Ziman (1968) *Public Knowledge*, Cambridge, p. 31.

[8] P. Bourdieu (1975) 'The Specificity of the Scientific Field and the Social Conditions of the Progress of Reason', *Social Science Information* **VI**, pp. 19–47; W. R. Albury, *op. cit.* (Note 2); B. Latour and S. Woolgar, *op. cit.* (Note 5), chapters 4 and 5; S. Shapin (1979) 'The Politics of Observation: Cerebral Anatomy and Social Interest in the Edinburgh Phrenology Disputes', in R. Wallis (ed.), *On the Margins of Science: The Social Construction of Rejected Knowledge* [*Sociological Review Monograph* **XXVII**], Keele, pp. 139–78; M. Callon (1980) 'Struggles and Negotiations to Define What is Problematical and What is Not: The Sociologic Translation', in K. D. Knorr *et al.* (eds), *The Social Process of Scientific Investigation* [*Sociology of the Sciences Yearbook* **IV**], Dordrecht, pp. 197–219; S. Shapin (1982) 'The History of Science and its Sociological Reconstructions', *History of Science* **XX**, pp. 157–211, esp. pp. 159–169; M. Mulkay and G. N. Gilbert (1981) 'Putting Philosophy to Work: Sir Karl Popper's Influence on Scientific Practice', *Philosophy of the Social Sciences* **XI**, pp. 389–407.

[9] S. Shapin, *op cit.* (Note 8: 1982), pp. 164–175, 204–206; D. Miller (1986) 'The Revival of the Physical Sciences in Britain', *Osiris*, 2nd series, **II**; J. Morrell and A. Thackray (1981) *Gentlemen of Science*, Oxford; E. Frankel (1976) 'Corpuscular Optics and the Wave Theory of Light: The Science and Politics of a Revolution in Physics', *Social Studies of Science* **VI**, pp. 141–84; W. R. Albury (1983) 'The Politics of Truth: A Social Interpretation of Scientific Knowledge, with an Application to the Case of Sociobiology', in M. Ruse (ed.), *Nature Animated*, Dordrecht, pp. 115–129; J. Sapp (1983) 'The Struggle for Authority in the Field of Heredity, 1900–1932: New Perspectives on the Rise of Genetics', *Journal of the History of Biology* **XVI**, pp. 311–42. M. J. S. Rudwick (1985) *The Great Devonian Controversy: The Shaping of Knowledge Among Gentlemanly Specialists*, Chicago and London.

[10] F. M. Turner (1980) 'Public Science in Britain, 1880–1919', *Isis* **LXXI**, pp. 589–608. See also P. Buck (1975) 'Order and Control: The Scientific Method in China and the United States', *Social Studies of Science* **V**, pp. 237–267. For the argument that such public methodologies can endanger scientific progress, see R. Laudan (1977) 'Ideas and Organisations in British Geology', *Isis* **LXVIII**, pp. 527–538, at pp. 527–528.

[11] For different approaches to this issue, see J. Ben-David (1974) *The Scientist's Role in Society: A Comparative Study*, New York, pp. 72–74; Y. Ezrahi (1980) 'Science and the Problem of Authority in Democracy, in R. K. Merton (ed.), *Science and Social Structure, New York Academy of Sciences* **xxxix**, pp. 43–60; P. Wood (1981) 'Methodology and Apologetics: Thomas Sprat's History of the Royal Society', *British Journal for the History of Science* **XIII**, pp. 1–26; R. Yeo (1981) 'Scientific Method and the Image of Science, 1831–1891', in R. M. MacLeod and P. Collins (eds), *The Parliament of Science: The British Association for the Advancement of Science, 1831–1981*, pp. 65–88; P. Feyerabend (1978) *Science in a Free Society*, London.

[12] See *op cit.* (Notes 10 and 11). For a canvassing of approaches, with reference to the debates on Baconianism, see R. Yeo (1985) 'An Idol of the Marketplace: Baconianism in Nineteenth-Century Britain', *History of Science* **XXIII**, pp. 251–298.

[13] Quoted in A. Rupert Hall (1969) 'Can the History of Science be History?', *British Journal for the History of Science* **IV**, pp. 207–220 (at p. 208).

[14] J. Priestley (1767) *The History and Present State of Electricity with Original Experiments*, London, pp. v–vi; see also his remark that: 'Great conquerors . . . have been both animated, and also, in a great measure, formed by reading the exploits of former conquerors. Why may not the same effect be expected from the history of philosophy to philosophers?', *ibid.* For further analysis, see J. G. McEvoy (1979) 'Electricity, Knowledge and the Nature of Progress in Priestley's Thought', *British Journal for the History of Science* **XII**, pp. 1–30.

[15] W. Whewell (1837) *History of the Inductive Sciences* (3 vols), London, vol. 1, p. 5.

[16] W. Whewell (1840) *The Philosophy of the Inductive Sciences, founded upon their history* (2 vols), London, vol. 1, pp. 3–4. However, Whewell did qualify the optimistic notions of a method or art of discovery. See the chapter by R. Yeo in this volume and his *op cit.* (Note 11), pp. 272–277.

[17] C. Singer (ed.) (1971–21) *Studies in the History and Method of Science* (2 vols), Oxford, vol. 1, p. vi. For a reference to the department of 'History and Method of Science' at University College, London, see A. C. Crombie (1984) 'Beginnings at Oxford', *Isis* **LXXV**, pp. 25–28, at p. 26. For other examples of early twentieth-century confidence in scientific method, see F. W. Westaway (1912) *Scientific Method: Its Practice and Philosophy*, London; A. Wolf (1925) *Essentials of Scientific Method*, London; T. B. Strong (ed.) (1906) *Lectures on the Method of Science*, Oxford.

[18] G. Sarton (1921–22) 'Introduction to the History and Philosophy of Science', *Isis* **IV**, pp. 23–31, at p. 25 and (1924) 'The New Humanism', *Isis* **VI**, pp. 9–34, at p. 26.

[19] G. Sarton, (1921–22) 'The Teaching of the History of Science', *Isis* **IV**, pp. 225–249, at p. 249. See also A. Thackray and R. K. Merton (1972) 'On Discipline Building: The Paradoxes of George Sarton', *Isis* **LXIII**, pp. 473–495; A. Thackray (1980) 'The Pre-History of an Academic Discipline: The History of Science in the United States 1891–1941', *Minerva* **XVIII**, pp. 448–473.

[20] J. B. Conant (ed.) (1957) *Harvard Case Histories in Experimental Science*, Harvard, p. vii. But by this date there was also a strong qualification to the notion of a single, simple scientific method. See Conant (1957) *Science and Common Sense*, London, pp. 42–45; I. B. Cohen (1948) *Science, Servant of Man: A Layman's Primer for the Age of Science*, London, p. 15.

[21] L. Laudan, *op cit.* (Note 1); D. Oldroyd (1986) *The Arch of Knowledge: An Introductory Study of the History of the Philosophy and Methodology of Science*, Sydney and London.

[22] J. H. Randall (1961) *The School of Padua and the Emergence of Modern Science*, Padua; A. C. Crombie (1953) *Robert Grosseteste and the Origins of Experimental Science, 1100–1700*, Oxford.

[23] Crombie, *op cit.* (Note 22), p. 1; cf. Randall, *op cit.* (Note 22), p. 16.

[24] For critical discussion of assumptions in the history of ideas, see J. Dunn (1968) 'The Identity of the History of Ideas', *Philosophy* **XLIII**, pp. 85–104; Q. Skinner (1969) 'Meaning and Understanding in the History of Ideas', *History and Theory* **VIII**, pp. 1–53; J. G. A. Pocock (1972) *Politics, Language and Time: Essays on Political Thought and History*, London, 2nd ed., pp. 6–38.

[25] R. Blake, C. Ducasse and E. Madden (1960) *Theories of Scientific Method: The Renaissance through the Nineteenth Century*, Seattle; R. Butts and J. Davis (eds) (1970) *The Methodological Heritage of Newton*, Oxford; R. Giere and R. Westfall (eds) (1970) *The Foundations of Scientific Method: The Nineteenth Century*, Bloomington.

[26] R. Blake *et al.*, *op cit.* (Note 25), p. iii.
[27] T. S. Kuhn (1977) 'The History of Science' in T. S. Kuhn, *The Essential Tension*, London, pp. 105–126; T. S. Kuhn (1970) 'Alexandre Koyré and the History of Science', *Encounter* **XXXIV**, pp. 67–69.
[28] A. Koyré (1939) *Etudes Galiléennes*, Paris; Eng. trans. J. Mepham (1978) *Galileo Studies*, Hassocks, Sussex; A. Koyré (1969) *Metaphysics and Measurement: Essays in Scientific Revolution*, London.
[29] A. Koyré (1956) 'The Origins of Modern Science', *Diogenes* **XVI**, pp. 1–22; reprinted in V. Bullough, (ed.) (1970) *The Scientific Revolution*, New York, pp. 118.
[30] G. Bachelard (1934) *Le nouvel esprit scientifique*, Paris; G. Bachelard (1949) *Le rationalisme appliqué*, Paris; and Dominique Lecourt (1977) *Marxism and Epistemology: Bachelard, Canguilhem, Foucault*, trans. B. Brewster, London, pp. 40–47, 60–70.
[31] G. Bachelard (1965) *La formation de l'esprit scientific*, Paris, 4th ed., p. 61.
[32] T. S. Kuhn (1970) *The Structure of Scientific Revolutions*, Chicago, 2nd ed., pp. 3–4, 8–9; T. S. Kuhn, *op cit.* (Note 27: 1977), pp. 116–117.
[33] K. R. Popper (1959) *The Logic of Scientific Discovery*, London; (1963) *Conjectures and Refutations*, London; (1972) *Objective Knowledge*, Oxford; I. Lakatos and A. Musgrave (eds) (1970) *Criticism and the Growth of Knowledge*, Cambridge; I. Lakatos (1978) in J. Worrall and G. Currie (eds.), *The Methodology of Scientific Research Programmes, Philosophical Papers* vol. I, Cambridge. Also J. Agassi (1962) 'Toward an Historiography of Science', *History and Theory* **II**, Beiheft.
[34] I. Lakatos, *op cit.* (Note 33: 1978), p. 53.
[35] *Ibid.* pp. 102–38.
[36] These would presumably be glimpses of the 'true method' laid up in Popper's and (Lakatos') 'third world'; for demystifications of which, see D. Bloor (1974) 'Popper's Mystification of Objective Knowledge', *Science Studies* **IV**, pp. 65–76; and P. K. Feyerabend (1975) 'Poppers' *Objective Knowledge*', *Inquiry* **XVII**, pp. 475–507.
[37] C. Howson (ed.) (1976) *Method and Appraisal in the Physical Sciences*, Cambridge; H. Frankel (1979) 'The Career of Continental Drift Theory', *Studies in History and Philosophy of Science* **X**, pp. 21–66; E. G. Zahar (1973) 'Why Did Einstein's Research Programme Supersede Lorentz's?', *The British Journal for the Philosophy of Science* **XXIV**, pp. 95–123, 223–262; I. Lakatos (1975) 'Why Did Copernicus' Research Program Supersede Ptolemy's?' in R. S. Westman (ed.), *The Copernican Achievement*, Berkeley, pp. 354–383. Cf. B. Barnes (1979) *T. S. Kuhn and Social Science*, London, pp. 58–63.
[38] P. K. Feyerabend, *op cit.* (Note 4) and (1976) 'On the Critique of Scientific Reason' in R. S. Cohen *et al.*, (eds.), *Essays in Memory of Imre Lakatos* [*Boston Studies in the Philosophy of Science* **XXXIX**], Dordrecht, pp. 109–143. See also the chapters by Noretta Koertge and Stephen Toulmin in the same volume.
[39] J. A. Schuster (1979) 'Kuhn and Lakatos Revisited', *British Journal for the History of Science* **XII**, pp. 301–317, esp. pp. 309ff.
[40] M. Mulkay and G. N. Gilbert, *op cit.* (Note 8), pp. 404, 407.
[41] I. B. Cohen (1974) 'History and the Philosopher of Science', in F. Suppe (ed.), *Structure of Scientific Theories*, Urbana, pp. 308–373 (at p. 311). For more detailed references to the gaps between general methodology and the practice of science see J. Ravetz, *op cit.* (Note 5), chapter 5. For other contributions which are symptomatic of

the tension between history and philosophy, see J. T. Clark (1959) 'The Philosophy of Science and the History of Science', in M. Clagett (ed.), *Critical Problems in the History of Science*, Madison, pp. 103–140; E. McMullin (1970) 'The History and Philosophy of Science: a Taxonomy' in R. H. Stuewer (ed.), *Historical and Philosophical Perspectives of Science*, Toronto, pp. 12–67; J. Smart (1972) 'Science, History and Methodology', *British Journal for the Philosophy of Science* **XXIII**, pp. 266–274; R. N. Giere (1973) 'History and Philosophy of Science: intimate relationship or marriage of convenience?', *idem.* **XXIV**, pp. 282–297; L. P. Williams (1975) 'Should Philosophers of Science be allowed to write History?', *idem.* **XXVI**, pp. 241–253.

[42] S. Toulmin (1966) 'Are the Principles of Logical Empiricism Relevant to the Actual Work of Scientists?', *Scientific American* **214**, pp. 129–133.

[43] H. Putnam, *op cit.* (Note 3), p. 192.

[44] L. Laudan (1968) 'Theories of Scientific Method from Plato to Mach. A Bibliographical Review', *History of Science* **VII**, pp. 1–63 (at pp. 1–2).

[45] *Ibid.* p. 8.

[46] M. Ruse (1975) 'Darwin's Debt to Philosophy: An Examination of the Influence of the Philosophical Ideas of J. F. Herschel and W. Whewell on the Development of Charles Darwin's Theory of Evolution', *Studies in History and Philosophy of Science* **VI**, pp. 159–181 and (1976) 'Charles Lyell and the Philosophers of Science', *British Journal for the History of Science* **IX**, pp. 121–131. See also R. Laudan (1982) 'The Role of Methodology in Lyell's Science', *Studies in History and Philosophy of Science* **XIII**, pp. 215–250; M. J. S. Rudwick (1974–5) 'Darwin and Glen Roy: A "Great Failure" in Scientific Method?', *Studies in History and Philosophy of Science* **V**, pp. 97–185; R. Olson (1975) *Scottish Philosophy and British Physics 1750–1880*, Princeton; G. N. Cantor (1971) 'Henry Brougham and the Scottish Methodological Tradition', *Studies in History and Philosophy of Science* **II**, pp. 69–80 and (1975) 'The Reception of the Wave Theory of Light in Britain: A Case Study Illustrating the Role of Methodology in Scientific Debate', *Historical Studies in the Physical Sciences* **VI**, pp. 109–132.

[47] M. T. Ghiselin (1969) *The Triumph of the Darwinian Method*, Berkeley; M. Ruse (1974) 'The Darwin Industry: A Critical Evaluation', *History of Science* **XII**, pp. 43–58 (at pp. 44–45). See also F. N. Egerton (1971) 'Darwin's Method or Methods?', *Studies in History and Philosophy of Science* **II**, pp. 281–286.

[48] J. R. R. Christie (1976) 'Influencing People', *Annals of Science* **XXXIII**, pp. 311–318.

[49] R. Smith (1974) 'Darwin and His Critics: Essay Review', *British Journal for the History of Science* **V**, pp. 278–285. See also R. Yeo (1979) 'William Whewell, Natural Theology and the Philosophy of Science in Mid-Nineteenth-Century Britain', *Annals of Science* **XXXVI**, pp. 493–516.

[50] L. Laudan, *op cit.* (Note 1), p. 9. For responses to Laudan's general theses, see the following reviews: G. N. Cantor (1983) *Annals of Science* **XL**, pp. 97–98; A. Lugg (1984) *Erkenntnis* **XXI**, pp. 433–483; E. McMullin (1984) *Isis* **LXXV**, pp. 372–377; J. Milton (1985) *The British Journal for Philosophy of Science* **XXXVI**, pp. 89–93.

[51] In the conclusion to another version of his study of methodology and the debate over optics, Laudan stressed that this case, although it refuted the idea that 'the canons of legitimate scientific inference are perennial and unchanging', did not justify the view that standards of evidence are paradigm specific – a view he identified with Kuhn. See

L. Laudan (1981) 'The Medium and its Message: A Study of Some Philosophical Controversies about Ether', in G. N. Cantor and M. J. S. Hodge (eds), *Conceptions of Ether: Studies in the History of Ether Theories 1740–1900*, London, pp. 180–181. The other version is in Laudan, *op cit.* (Note 1), pp. 111–140.

[52] B. Barnes, *op cit.* (Note 37), pp. 22, 27–35, 60–62, 70–83; B. Latour and S. Woolgar, *op cit.* (Note 5), pp. 75–88; M. Mulkay and G. Gilbert, *op cit*, (Note 8). On the 'discourse analysis' approach in general, see G. N. Gilbert and M. Mulkay (1984) *Opening Pandora's Box: A Sociological Analysis of Scientists' Discourse*, Cambridge. Of the chapters in this volume, those by Schuster, Le Grand and Sapp bear the greatest imprint of this approach, while expressing varying degrees of criticism as well. The issues dividing historians of science and the sociological advocates of discourse analysis, particularly Mulkay and Gilbert, are canvassed in S. Shapin (1984) 'Talking History: Reflections on Discourse Analysis', *Isis* **LXXV**, pp. 125–130.

[53] See G. N. Cantor, *op cit.* (Note 46); P. Wood, *op cit.* (Note 11), S. Shapin (1984) 'Pump and Circumstance: Robert Boyle's Literary Technology', *Social Studies of Science* **XIV**, pp. 481–519; S. Shapin and S. Schaffer (1986) *Leviathan and the Air-Pump*, Princeton; P. Dear (1985) 'Totius in Verba: Rhetoric and Authority in the Early Royal Society', *Isis* **LXXVI**, pp. 145–161.

[54] For some suggestive comments on the connection between methodology and the emergence of disciplines, with reference to Bernard, see W. R. Albury (1982) 'Going Beyond the Introduction: Claude Bernard Scholarship in the 1970s', *Bulletin of the History of Medicine* **LVI**, pp. 271–275.

[55] On hagiography (which often involves statements about correct method), see P. G. Abir-Am (1982) 'How Scientists View Their Heroes', *Journal of the History of Biology* **XV**, pp. 281–316; L. S. Jacyna (1983) 'Images of John Hunter', *History of Science* **XXI**, pp. 85–108. Another possible topic, suggested by feminist critiques, is the connection between gender and methodology in the rhetoric of science. See S. Harding and M. B. Hintikka (eds) (1983) *Discovering Reality: Feminist Perspectives on Epistemology, Metaphysics, Methodology, and Philosophy of Science*, Dordrecht.

[56] S. Toulmin (1980) 'The Intellectual Authority and the Social Context of the Scientific Enterprise', *Minerva* **XVIII**, pp. 652–667; J. R. Ravetz (1984) 'Ideological Commitments in the Philosophy of Science', *Radical Philosophy*, pp. 5–11.

ACKNOWLEDGEMENTS

Many individuals have contributed to this project. In the early stages Ian Langham gave us crucial encouragement and advice. His tragic death in 1984 prevents his seeing the fruits of these generous efforts. Rod Home, the General Editor of this series, has exercised his accustomed patience and critical discernment. The following people have read and commented upon chapters or the editors' introduction: Randall Albury, Alan Gabbey, Rod Home, Rachel Laudan, Homer Le Grand, David Miller, Alan Musgrave, Roy Porter, Terry Stokes and Paul Weindling. From time to time Randall Albury and the staff of the School of History and Philosophy of Science at the University of New South Wales have provided facilities for joint editorial work. Rod Home and the staff of the Department of History and Philosophy of Science at the University of Melbourne also made available editorial facilities. Research and editorial work on this volume was supported in part by grants from the University of Wollongong Research Grants Committee and the School of Humanities, Griffith University. We also thank our respective departmental and school colleagues for their interest, help and critical discussions.

JOHN A. SCHUSTER
RICHARD YEO

ALAN CHALMERS

THE GALILEO THAT FEYERABEND MISSED: AN IMPROVED CASE AGAINST METHOD

What you've got to do is you've got to learn how to defend yourself, and you've got to learn how to attack. That's your only trouble as a boxer. You don't know how to defend yourself, and you don't know how to attack.

Harold Pinter, *The Homecoming*

1. INTRODUCTION

Paul Feyerabend has attempted to employ a study of Galileo's science to support his case 'against method'.[1] In this chapter I will criticize Feyerabend's study and the use he makes of it on a number of grounds. I will attempt to show how, nevertheless, an historically broader and more accurate account of Galileo's innovations can be used to counter claims to the effect that there is, or can be, such a thing as a universal, ahistorical scientific method for guiding the work of scientists or for judging the merits of the theories that they produce.

The chief ways in which Feyerabend's Galileo study is intended to constitute a case against method are as follows. Firstly, Galileo's theories involve a change in experience or observation language of a kind that undermines any empiricist or positivist methodology which regards the observational base as being theory-neutral and given to us via the senses in some unproblematic way. Secondly, Galileo's theories are claimed to be *ad hoc*, thus conflicting with those methodologies, such as Popper's falsificationism, which rule out *ad hoc* hypotheses. A third challenge to methodologies involves what Feyerabend, borrowing a notion from Marxism, refers to as the 'uneven development' of theories. The development of a theory is typically out of step with the auxiliary theories that are needed to supplement that theory if it is to be put to the test and protected from apparent refutations. This phenomenon renders the criteria for the rejection of theories as specified by many methodologies too stringent, and liable to lead to the premature

J. A. Schuster and R. R. Yeo (eds.), The Politics and Rhetoric of Scientific Method, 1–31.
© 1986 *by D. Reidel Publishing Company.*

rejection of promising new hypotheses. A fourth element of the case against method involves the claim that propaganda is of the essence of scientific practice, contrary to the view of the methodologists who wish to offer a purely 'rational' account of science and its mode of progress. As far as these four issues are concerned, I accept Feyerabend's version of the third, but will argue that the first can only be accepted if substantially modified, whilst the second and fourth must be largely rejected.

Various aspects of Galileo's science, especially his development of a mathematical theory of motion to be tested by experiment and his use of the telescope in astronomy, undermined the methodological norms of his day and went a long way towards constituting new ones. He undermined old methods and established new ones in practice. This episode not only provides the basis for a case against the various contemporary methodologies of science that have been formulated, but also renders implausible any search for a generally applicable scientific method. However, rejection of a universal method and of generally applicable standards is not equivalent to the rejection of all method and all standards, as is suggested by some of Feyerabend's utterances.

2. NATURAL INTERPRETATIONS AND A CHANGE IN OBSERVATION LANGUAGE

A major part of Feyerabend's study of Galileo's defence of Copernicus involves the tower argument.[2] That argument, which dates from antiquity, poses a difficulty for the assumption that the earth spins on its axis. According to the argument, if the earth did spin on its axis a stone dropped from the top of a tower would fall, not at the base of the tower as it is seen to do, but some hundreds of metres from the base, because during the descent of the stone the tower would be carried away from it by the spinning earth. Further, if the stone were to participate in the presumed motion of the earth and stay with the tower, its motion in space would be a curved line composed of the linear falling and the circular rotation, conflicting, it was alleged, with the observed linearity of the descent. Feyerabend argues that the hypotheses that Galileo employed to diffuse this argument against the Copernican theory involved 'a change of experience'.[3] They amounted to 'a partial revision of our observation language'.[4] His case can be summarized as follows.

When considering the description of a situation by an observer, we can abstractly distinguish between the sensations involved, that is, the mental experiences that an observer undergoes when confronting the situation, and the verbal description of the situation which the observer subscribes to in the light of those sensations. Feyerabend insists that, although for the purpose of analysis we can make the distinction between the sensation and the verbal description, the two stages are not separable in practice. An observer does not first have a sensation, when confronting a falling stone, say, and then subsequently interpret that sensation as being indicative of a stone falling vertically downwards. Rather the observer simply sees the stone falling down, and is subsequently willing to accept the statement 'the stone fell straight down'. Feyerabend acknowledges that the splitting of the two aspects of observation, even for the purposes of analysis, is a simplification which has its limitations since our sensations can be influenced by our linguistic expression of them. This reservation aside, we can abide by the distinction and claim that, when an observer confronts and describes a situation, he or she automatically makes a connection between sensation and description, between the mental experience and the verbal description accepted on the basis of the sensation. Feyerabend gives the name 'natural interpretations' to those 'mental operations which follow so closely upon the senses'[5] and which constitute the link between having a sensation and accepting a description. Natural interpretations are instilled in us from birth. We acquire them during the process of learning a language, for they enable us to connect language with observable situations. What is more, the natural interpretations embodied in a language and culture at some time will typically have entered it and become part of the observational process many generations previously. Consequently, their nature, and even the fact that they are present, are not readily apparent to the individual.

According to Feyerabend, the tower argument involved a number of natural interpretations from the point of view of early seventeenth-century common sense that Galileo needed to challenge. One of them involved the notion of an absolute space defined essentially by the planetary and stellar system with a stationary earth at its centre.[6] Along with this natural interpretation comes the notion of absolute motion in this space. Absolute motion is assumed to have observable effects. In general, the senses faithfully record real motions. An observer imbued with these natural interpretations will automatically take the observed motion of a falling stone to be its 'real' motion in absolute space. The observation of its linear fall clashes with the consequence of the Copernican theory that its motion should be 'mixed straight and circular'. The Copernican theory is refuted from the point of view of early seventeenth-century common sense, and the natural interpretations that are automatically and unconsciously employed by those who have internalised it. After all, 'how could one possibly be unaware of the fact that the falling stone traces a vastly extended trajectory through space!'[7]

The foregoing paragraph was concerned with the question of the appearance of the linear trajectory of the stone as it falls alongside the tower. There remains the problem of why the stone remains with the tower rather than being left behind as the earth rotates. Here, according to Feyerabend, a further natural interpretation forming part of seventeenth-century common sense leads to the tower argument being seen as a refutation of Copernicus, namely 'the *dynamical* principle that objects (such as the falling stone) which are not interfered with assume their natural motion'.[8] A falling stone will have a natural propensity to fall downwards, but no natural propensity to move circularly around the earth's centre.

3. GALILEO'S RELATIVITY PRINCIPLE

From Feyerabend's point of view, if Galileo was successfully to defend the Copernican theory, it was necessary for him to challenge and replace the natural interpretations forming part of seventeenth-century common sense described above. He did this by introducing new natural interpretations. 'In other words, *he introduces a new observation language*.'[9] However, since the natural interpretations that need to be challenged are well-entrenched and perhaps unconsciously adhered to by those who have internalised them and habitually use them, mere arguments will not suffice to establish the superiority of the new interpretations. Indeed, Galileo uses 'propaganda' and 'psychological tricks' in addition to arguments to achieve his end.[10] The new natural interpretations that Galileo introduces involve what Feyerabend refers to as his *relativity principle* and his *principle of circular inertia*.[11] In the remainder of this section I discuss and criticise Feyerabend's account of the former, leaving a discussion of the latter to the following two sections.

As I shall interpret it, Galileo's relativity principle involves the claim that an observer cannot detect by means of the senses a motion in which he or she shares. Feyerabend shows a tendency to interpret Galileo's relativity principle in a more general sense, a sense more in keeping with what has come to be called 'Galilean relativity', namely, that the uniform motion of an isolated system cannot be detected in any way. My reasons for rejecting the more general interpretation involve the broader context of Galileo's new science of motion and are best left until a later section. For the present I will confine the discussion to relativity insofar as it involves an observer's inability to detect by means of the senses a motion in which he or she participates. I shall refer to this restricted sense of relativity as 'perceptual relativity'.[12] To what extent does Feyerabend's claim that Galileo employed trickery to diffuse the tower argument and defend Copernicus apply to Galileo's use of perceptual relativity?

Feyerabend's case runs roughly as follows. Two paradigms concerning the observation of motion are at least implicit in seventeenth-century common sense. Paradigm 1 involves certain reasonably common situations in which shared motion goes unnoticed. An artist drawing on a moving ship notices the motion of his pen relative to the paper, but not the motion of the ship which is shared by him, his pen and the paper.[13] Paradigm 2 involves the motion of an object against a stable background, such as that of a camel trotting through the desert.[14] Galileo's strategy, allegedly, is deliberately to confuse the issue by talking of the motion of the earth, and of the stone in the tower experiment, as if they were examples of motion belonging to paradigm 1 whereas, according to Feyerabend, seventeenth-century common sense, free from Galileo's influence, would have placed them in paradigm 2. 'We now quite automatically start confounding the conditions of the two cases and become relativists.'[15] Thus Galileo's arguments are observed to be 'arguments in appearance only. For he uses *propaganda*. He uses *psychological tricks* in addition to whatever reasons he has to offer'.[16]

Feyerabend's presentation of Galileo's case is unacceptable. In order to clear the matter up it is important to distinguish between two questions that are juxtaposed by Feyerabend, 'How could one possibly be unaware of the swift motion of a large bulk of matter such as the earth is supposed to be!' and 'How could one possibly be unaware of the fact that the falling stone traces a vastly extended trajectory through space!'.[17] The first question is much more general than the second. There are a variety of reasons why those whom Galileo addressed might have expected any motion of the earth to be discernible. Just one of those involves the expectation of seeing the stone in the tower experiment move with a mixed straight and circular motion and similar expectations associated with the flight of birds, the trajectories of cannon shots and so on. It is to offset objections based on the appearance of such phenomena, and the general expectation that the motion of the earth should be noticed by the senses, that Galileo carries out his discussion of shared motion and utilises the principle of perceptual relativity. Other problems associated with the earth's motion remain, such as the problem of why the stone stays with the tower, and why loose objects are not flung from a spinning earth as they would be from the surface of a rotating wheel. The principle of perceptual relativity was not intended to cope with such objections. Galileo deals with them elsewhere in the *Dialogue*, as we shall see.

Let us restrict ourselves to the issue of how things appear to the senses, then: Many of Galileo's contemporaries would have considered the motion of the earth through space to be an example of absolute motion to be included under paradigm 2. However, what is at issue is the appearance of the stone's trajectory in the tower experiment. This is quite appropriately considered under the shared motion of paradigm 1 whether the earth is assumed to move or not. The crucial distinction between the two paradigms is that in the first the observer shares the motion of what is to be observed, whereas in the second, what is to be observed moves relative to the observer. Since the observer shares that supposed rotation of the earth along with the tower and the stone, Galileo is arguing soundly and straightforwardly, and not indulging in trickery, when he compares observations of falling stones to observations on board a moving ship.

I see no reason why Galileo's contemporaries should not have been convinced by his argument involving perceptual relativity. What is more, the arguments were by no means novel in Galileo's day. Feyerabend himself indicates in a footnote that the principle of perceptual relativity appears in Euclid, was repeated by Copernicus and was a commonplace in medieval optics.[18] What Feyerabend does not mention is that Galileo's arguments were anticipated by Buridan and Oresme in the fourteenth century *in the context of discussing a moving earth.*[19] 'I take as a fact that local motion can be perceived only if we can see what assumes a different position relative to another' wrote Oresme, and he then proceeded to illustrate the point by observations made from a moving ship, before applying it to observations of the stars.[20] Since Oresme could adopt perceptual relativity and remain Aristotelian in many respects, it is clear that Galileo did not need to resort to trickery, in principle or in fact, to convince his Aristotelian contemporaries of the intelligibility of his appeal to that principle.

To summarize my analysis of the point, then, Galileo presented straightforward arguments to defend his claim that motion shared by observers, the motion of the earth in particular, is not directly evident to the senses. They were not 'arguments in appearance only'.[21] There are no strong grounds for assuming that they could not have been intelligible to Galileo's Aristotelian contemporaries, and to seventeenth-century common-sense, and acceptable to many. This is not to say that with the arguments based on perceptual relativity alone Galileo could hope to convince Aristotelians or others of his contemporaries of the inadequacy of the objections to the possibility of the earth's motion. There were other reasons to regard such an assumption to be absurd, as Galileo was well aware, and as his ensuing discussion of the issue in the *Dialogue* indicates.

4. GALILEO'S 'PRINCIPLE OF INERTIA'

Having utilised the principle of perceptual relativity to explain why the stone's trajectory appears linear, it remained for Galileo to explain why the stone stays with the tower and is not left behind. For that purpose Galileo utilised what Feyerabend refers to as his 'principle of circular inertia'. Feyerabend's statement of that principle reads, 'an object that moves with a given angular velocity on a frictionless sphere around the centre of the earth will continue moving with the same angular velocity forever'.[22] Galileo's defence of the principle is alleged to be *ad hoc*[23] and involves a surreptitious generalisation of certain common sense notions on a par with those attributed by Feyerabend to Galileo in the latter's defence of perceptual relativity. In the next section I criticize Feyerabend's account of Galileo's 'principle of inertia' from a broader context than that of the *Dialogue*, to which context Feyerabend's discussion is largely confined.[24] In the remainder of the present section I appraise Feyerabend's position from the narrower context.

It might well be conceded to Feyerabend that there is a propagandistic element to Galileo's presentation of his position on inertial motion in the *Dialogue*. Galileo presented it as if it follows straightforwardly from unproblematic assumptions and as if not in need of observational tests. Objects moving downhill accelerate, since they approach the centre of the earth, whilst objects moving uphill decelerate as they recede from that centre. Objects moving on a spherical horizontal surface, since they neither approach nor recede from the centre of the earth, will neither accelerate nor decelerate. In the absence of impediments they will move indefinitely at constant speed. 'Without experiment, I am sure that the effect will happen as I tell you, because it must happen that way.'[25] This arguably propagandistic passage notwithstanding, Galileo's hypothesis was not *ad hoc*. It was amenable to observational and experimental test as can be discerned later in the *Dialogue* itself.[26]

Before considering the issue of whether or not Galileo's hypothesis concerning inertia was *ad hoc* it will be useful to clarify what is intended by that term, something which Feyerabend does not do. For this purpose it will be convenient to identify three senses of '*ad hoc*'.[27] An hypothesis is *ad* hoc_1 with respect to some problem situation if there are no testable consequences of the hypothesis that are not already part of that problem situation. If it is not to be *ad* hoc_1 an hypothesis must be independently testable.[28] An hypothesis will be *ad* hoc_1 with respect to the problem situation involving the Copernican theory if there are no consequences of it that are not already consequences of the Copernican theory and its background. An hypothesis is *ad* hoc_2 if its independently testable consequences have not survived tests, either because the tests have not been carried out or, more seriously, because the results prove to be negative. Finally, an hypothesis is *ad* hoc_3 with respect to some evidence or theory if the evidence or theory plays a heuristic role in the construction of the hypothesis. An hypothesis will be *ad* hoc_3 with respect to the Copernican theory or to some problematic evidence if that theory or evidence played a part in its construction. If a theory is constructed by building into it some data or assumptions, then the validity of those data or assumptions cannot subsequently be taken as evidence for the theory.

The non *ad* hoc_1 character of Galileo's hypothesis concerning inertial motions can be brought out by constrasting it with Copernicus' efforts in this regard. The hypothesis that Copernicus himself introduced to defend his theory against the tower and related arguments was *ad* hoc_1. According to Copernicus, the earth together with all of its parts have a *natural* propensity to execute the diurnal motion.[29] Since there are no independently testable consequences of this hypothesis it was *ad* hoc_1 and *ad* hoc_2. Further, since Copernicus's attribution of a diurnal motion to earthly objects arose in response to problems arising from the motion assigned to the earth by his theory, the move was *ad* hoc_3 with respect to the behaviour of earthly objects such as that of the stone in the tower argument. By contrast, Galileo's hypothesis is more general. It applies to *any* horizontal motion, whatever the speed and direction, and not simply to the diurnal motion. Galileo's hypothesis can be put to the test in contexts involving a variety of horizontal motions which have nothing to do with the Copernican theory, and Galileo mentions a number of such tests in the *Dialogue*. For example, Galileo lists a number of experiments that can be carried out below decks on a uniformly moving ship (pp. 186—7). It is a consequence of his initial hypothesis that the results of these should be the same as when the ship is stationary. He also mentions dropping a stone from the mast of a uniformly moving ship (p. 144) and tests that involve dropping a ball or throwing a javelin in the air from a galloping horse (pp. 156—7). Admittedly, most of these tests, while they might serve as rough tests of Galileo's inertial hypothesis as compared with the rival Aristotelian claims, do not lend themselves to being made precise. After all, how does one arrange for a sailing ship or a horse to move uniformly? Galileo did perform controlled experiments to test his theory[30] but these are not mentioned in the *Dialogue*. A discussion of them and their significance is reserved for Section 7.

Whether or not Galileo himself referred to or performed adequate tests of his hypothesis concerning inertia, the fact remains that his hypothesis does have consequences that go far beyond the evidence bearing directly on the Copernican theory and hence was not *ad* hoc_1 with respect to that theory and its background. Galileo most likely did not carry out the tests mentioned in the *Dialogue*, whilst the results of his own controlled experiments were by no means conclusive. There is a sense, then, in which Galileo's hypothesis was *ad* hoc_2 to a degree. The way in which Galileo's ideas on inertial motion emerged in the context of his physics, to be described in the next section, rules out the possibility that they were *ad* hoc_3 with respect to the Copernican theory.

This is a convenient point to acknowledge and agree with Feyerabend's notion of the uneven development of science as exemplified by Galileo's defence of Copernicus.[31] The defence of the latter required a new dynamics, and also an improved optics once the telescope was introduced. These were not available to Copernicus and at least a century had elapsed before adequate versions of those theories emerged. It was therefore inevitable that the Copernican theory should remain problematic and in need of positive support for a time and it would have been inappropriate to reject it for that reason.

5. GALILEO'S NEW SCIENCE – INERTIAL MOTION

It is time to broaden the context of the discussion and to examine the issues involved in the tower argument from the broader context of Galileo's new science of motion. Galileo's mature theory is presented in Days 3 and 4 of *Two New Sciences*.[32] The work appeared a decade after the *Dialogue* but it has been established that Galileo had arrived at the substance of that theory by 1609.[33] From the perspective of the new theory of motion it can be appreciated that Galileo arrived at his conception of inertial motion, that is, of motion that persists indefinitely without a continuing cause, by a path that had nothing to do with the Copernican theory. It can also be shown that it is inappropriate to refer to those conceptions as a 'principle' of inertia and that Galileo's physics was not of a kind that talks of the motion of matter in general in Euclidean space, so that there is no question of it accommodating a principle of relativity of any generality. Further, the general character, and in particular, the mathematical character, of Galileo's theory of motion made experimentation of a particular kind necessary, a more radical epistemological break than those discussed by Feyerabend. These claims are elaborated in this and the following two sections.

Archimedes was the chief inspiration of Galileo's physics.[34] His first scientific treatise, *The Little Balance*,[35] bears witness to that influence and it is also manifest in his more substantial early works *On Motion* and *On Mechanics*.[36] Although there were certainly other influences, including Aristotelianism, the medieval impetus theories and the medieval science of weights,[37] it is a fair approximation to conceive of Galileo's physics as an attempt to extend the techniques Archimedes had employed in his statics to systems in motion.[38] A mark of Archimedes' statics was his technique of abstracting from the details of real physical situations and dealing with simplified systems that could be characterised in mathematical terms. So, for example, his treatment of the balance involved geometrical shapes endowed with weight suspended by weightless threads from weightless inflexible balance arms supported by a frictionless pivot. In *On Motion* we find Galileo moving from considerations of the balance with a bent arm, described in geometrical terms, to motion on a frictionless curved surface, and thence to motion down a frictionless plane tangential to that surface.[39] We also find Galileo extending Archimedes' hydrostatics to considerations of falling bodies.[40] Here an object falling through a vacuum is taken to be the fundamental case. The presence of a medium introduces an impediment, proportional to its density, preventing the falling body from reaching its natural speed of fall.

Galileo's mathematical treatment of motion down inclined planes in *On Motion* leads him to the observation that, whilst motion down a plane occurs of itself, and is *natural*, and motion up a plane requires some impetus and is *forced*, 'on the horizontal plane itself the motion of the body is neither natural nor forced', and 'can be made to move by the smallest of all possible forces'.[41] Galileo reiterated these considerations a decade later in *On Mechanics*, where he observed that 'on an exactly balanced surface the ball would remain indifferent and questioning between motion and rest'.[42] This result, which Wisan has called the 'protoinertial principle' and which she regards as having evolved 'in a natural way from the medieval science of weights',[43] was certainly not new in Galileo's time. It had been anticipated at least by Cardano and Benedetti. Galileo designated motion in a horizontal plane as 'neutral', as opposed to natural or forced. That notion had precedents also. A problem for Aristotelian theory as it was perceived by some medieval commentators, notably Albertus Magnus, concerned the motion of the element fire. Aristotle himself had supposed that, once it has executed its natural motion upwards to the region near the moon's orbit, the element then participated in the circular motion of the heavens. As W. Wallace has shown, this problem led some late sixteenth century Jesuits lecturing at the Collegio Romano, notably Mutius Vitelleschi and Ludovicus Rugerius, following Albertus, to posit a motion intermediate between natural and forced to accommodate the circular motion of fire.[44] Galileo read and utilised this lecture material and was almost certainly influenced by it when formulating his notion of neutral motions. Galileo explicitly refers to the issue of the motion of fire in Aristotelian theory in the *Dialogue* (p. 141).

In *On Motion* Galileo also considered the hypothetical case of a sphere rotating at the centre of the universe. He observed that it would move 'with a motion that is neither natural nor forced'.[45] Galileo raised the question of whether its motion should be perpetual or not but he could not answer it. 'For if its motion is not contrary to nature, it seems that it should move perpetually; but if its motion is not according to nature, it seems that it should finally come to rest.'[46] Only natural motions can persist without an external force, whilst forced motions cease as soon as the external force is removed or dies away. With motion and rest as the alternatives, Galileo needs, but cannot find, an intermediate state to attribute to his neutral motions.

By 1603 or 1604 the situation had changed in a significant way. For Galileo had come to abandon the idea, expressed earlier in *On Motion*, that a body falls with a natural *speed*, after an initial period of acceleration, and had come to think of bodies falling with a natural acceleration.[47] As far as falling bodies are concerned, Galileo no longer had a problem finding an intermediate state for his neutral motions once this step had been taken. Bodies moving downwards with their natural motion accelerate. Bodies moving on a smooth, horizontal plane, with a neutral motion, neither accelerate nor decelerate. They move with a constant speed which will persist for as long as the horizontal plane persists. That this reconstruction of Galileo's final step to horizontal inertia is a correct one is supported by the fact that the first hint of it appears in Galileo's work in 1603, during his analysis of the natural acceleration of falling bodies.[48]

The argument for horizontal inertia, the emergence of which I have just traced, is one that recurs in Galileo's writing whenever he wishes to defend it. We have already noted, in Section 4, its recurrence in the *Dialogue*. It also appears, for example, in Galileo's *Letter on Sunspots* of 1613 and in *Two New Sciences*.[49] Further, the path to it that I have traced, beginning with the medieval science of weights and Aristotelian considerations of the motion of fire, and ending with considerations of free fall, did not involve and owed nothing to, the Copernican theory. Galileo's hypothesis concerning horizontal inertia was not *ad hoc*$_3$ with respect to the Copernican theory.

Feyerabend's 'guess' is that '*a clear idea of a permanent motion with(out) impetus developed in Galileo only together with his gradual acceptance of the Copernican view*'.[50] The path that I have traced for the emergence of Galileo's notion of the persistence of motion in a horizontal plane by about 1603 strongly suggests that Feyerabend's guess is a bad one. It is certainly the case that Galileo had arrived at his view before he embarked on his serious defence of the Copernican view towards the end of 1609, following his astronomical use of the telescope. Whilst Galileo had expressed interest in, and perhaps even commitment to, the Copernican theory at least as early as 1597,[51] his support was not active, and it may be that any doubts he had about it were strengthened by his failure to detect the parallax he expected in the nova of 1604.[52] As far as the rejection of the impetus theory is concerned, Galileo had employed notions derived from that theory in *On Motion*, in his somewhat messy account of projectile motion, and also to explain the initial acceleration of falling objects, prior to the attainment of what Galileo at that time considered to be their natural speed of falling. Once Galileo had arrived at his conception of a natural downwards acceleration for all earthly bodies and the notion of conserved horizontal velocity that accompanied it, both of his earlier uses of the impetus theory became redundant. As far as Feyerabend's suggestion that it is necessary to reject all ideas of an outer or inner motor for the conserved motion 'if we want to regard motion as a *relative* phenomenon, depending on the choice of a suitable co-ordinate system'[53] is concerned, Galileo's physics was quite incapable of accommodating a relativity principle of this generality, as we shall see.

6. GALILEO'S NEW SCIENCE (CONTINUED)

In this section I digress a little from the task of criticizing and improving on Feyerabend's case against method to comment on some general features of Galileo's mature theory of motion, and to indicate its restricted character from a Newtonian point of view. From this perspective it can be appreciated that attempts by Feyerabend, and many others, to attribute to Galileo the conception of relativity implicit in Newtonian theory and a *principle* of inertia (whether circular or linear) are seriously mistaken. In Section 8 I return to my main theme, and argue that experimentation, as it originates in Galileo's science, constitutes a radical discontinuity in method overlooked by Feyerabend.

The fundamental principle underlying Galileo's physics of motion as represented in Days 3 and 4 of *Two New Sciences* is that all earthly bodies experience a propensity to move vertically downwards towards the centre of the earth with the same uniform acceleration. Galileo's theory was therefore a limited one. It was confined to the motion of terrestrial objects in the vicinity of the earth. This coincides with the view of A. Koyré when he wrote that 'Galileo's physics is a physics of heavy bodies, a physics of bodies which fall downwards' so that, in this respect, Galileo was 'unable to abstract from gravity'.[54] Galileo's physics was not a theory of the motion of bodies in general in Euclidean space, as was to be specified by the fundamental laws of Newton's theory. The fundamentals of Galileo's theory, his laws of fall plus the conservation of horizontal motion that accompanies it, cannot be formulated without explicit reference to a specific body, namely, the earth. Within Galileo's theory, the earth cannot be treated in Newtonian fashion as just another body whose presence at some place in Euclidean space is a contingency. Remove the earth and the law of fall that relates to it from Galileo's theory and there is no theory left.[55] It is in this sense that Galileo was 'unable to abstract from gravity'.

From this perspective, it can be appreciated that there can be no question of a principle of relativity of any generality in Galileo's theory if that principle is interpreted along post-Newtonian lines as specifying the impossibility of detecting the uniform velocity of an isolated system. In Galileo's physics there can be no isolated terrestrial system. Bodies cannot be isolated from the earth insofar as they possess the 'constitutive property of gravity'.[56] However, it must be admitted that there are utterances to be found in the *Dialogue* that can readily be interpreted by a modern reader as a principle of relativity of considerable generality. For instance, Galileo asserts that 'motion, insofar as it acts as motion, to that extent exists relatively to things that lack it; and among things which all share equally in any motion, it does not act, and is as if it did not exist' so that 'motion which is common to many moving things is idle and inconsequential to the relation of these moveables among themselves'.[57] I suggest that these passages should be read as stating the principle of perceptual relativity only, that motion that is shared cannot be detected by means of the senses. Although the quotations taken in isolation appear, at least to the modern reader, to state a principle of greater generality, that the more restricted interpretation is the more appropriate one is clear, I suggest, from the context of Galileo's discussion of the principle and the use to which he puts it. He uses it to explain why the motion of the heavens will look the same whether it is they or the earth that rotates, he applies it to various observations on moving ships, he uses it to explain why the stone in the tower experiment appears to fall straight rather than 'mixed straight and circular' as we have seen, and he discusses it in the context of the Aristotelian 'criterion of science itself' that 'the senses and experience should be our guide in philosophising'.[58]

From a post-Newtonian perspective, the passage in which Galileo describes the null results of experiments carried out below decks on a moving ship designed to detect its uniform velocity[59] can easily be interpreted as illustrating Newtonian relativity. However, from the content of Galileo's discussion it is clear, I suggest, that the experiments are designed to support Galileo's hypothesis concerning the conservation of speed in the horizontal plane. Having argued for the non-observability of shared motion, there remained for Galileo the issue of just what motions are shared. Will a butterfly or a falling drop of water share the motion of the ship? Will the stone stay with the tower? It is a consequence of Galileo's hypothesis concerning horizontal motion that these motions will be shared insofar as they are horizontal and uniform. It is perhaps appropriate to point out here that if the uniform motion of a ship were at an angle to the horizontal, then its motion would not be shared by loose objects within it. Plumb lines would change their direction and objects would slide along the floor. This observation, the validity of which would have been obvious to Galileo, refutes general interpretations of his statement of a relativity principle but does not refute the principle of perceptual relativity. Since the plumb lines and loose objects in our hypothetical example do not share the motion of the ship, the principle of perceptual relativity is not meant to apply. This is a further reason to suppose that Galileo's relativity is appropriately interpreted as perceptual relativity only.

A further problem for those who wish to interpret Galileo's relativity in a general sense is the fact that Galileo himself, in the *Dialogue*, supports positions that are incompatible with any claim to the effect that the motion of the earth is undetectable in any way by observers who participate in the motion. Quite apart from his ill-fated theory of the tides, Galileo acknowledged that, in principle, according to his theory, cannon shots fired at targets situated due North fall to the East of the targets, since the Eastwards motion of the target 'will be somewhat the slower (than that of the shot), being made along a smaller circle', although, admittedly, 'the difference is insensible because of the small distance from the cannon to the mark'.[60] Once again, this conflicts with a general interpretation of Galileo's relativity but not with the principle of perceptual relativity. It is true that Galileo did explicitly state that 'whatever motion comes to be attributed to the earth must necessarily remain imperceptible to us and as if non-existent, so long as we look only at terrestrial objects'[61] and taken at its face value, this appears to be in direct contradiction with his claims about cannon shots fired North or South. However, Galileo puts forward the assertion quoted, not as a principle which he unequivocally adheres to but as 'the beginning of our reflection', as something to be considered. I suggest that if we interpret Galileo's stance on relativity to involve nothing more than perceptual relativity together with the conservation of horizontal motion than we can render his position quite consistent, provided we adopt the principle of charity rather than the 'principle of spite'.[62]

Galileo's treatment of cannon shots fired North or South indicates that it is inappropriate to interpret Galileo's hypothesis concerning horizontal motion as a principle of conservation of *angular* velocity as Feyerabend does. It is speed, not angular velocity, that is conserved according to Galileo.[63] It is also inappropriate to refer to his hypothesis as a *principle*. Horizontal inertia is not a fundamental assumption in Galileo's physics. Rather it emerges as a consequence of his views on natural acceleration downwards, which *is* fundamental to his physics, in the special case of motion in a plane equidistant from the centre of the earth. The point can be strengthened by the observation that there is another special case of inertial motion to be found in Galileo's writings which emerges as a consequence of his views on natural fall in a similar way, although in this instance the conserved motion is linear. The case involves the motion of two equal weights hanging vertically from either end of a string passing over a pulley. Vertical speed imparted to the weights is conserved because it 'meets with no opposition from any weight' and this is 'precisely the same thing' as happens on a horizontal plane.[64] Those who wish to ascribe a principle of circular inertia to Galileo would thus seem to have equally compelling reasons to ascribe a principle of linear inertia to him. The limited character of Galileo's views on inertial motion, the fact that it applies only to the motion of terrestrial objects under certain special sets of conditions, indicates that there is no scope for a general principle of inertia in his theory, and certainly not one applied to celestial as well as terrestrial phenomena as suggested by some authors.[65]

7. GALILEO AND EXPERIMENT

As we have seen, Feyerabend is concerned to undermine empiricist methodologies by demonstrating that Galileo's innovations involved a 'change of experience' and the replacement of one set of 'natural interpretations' by a new set. In this section I will attempt to show that Feyerabend's discussion concedes too much to empiricism and that the latter doctrine can be undermined more radically by focussing on experiment as opposed to mere observation or experience.

Galileo did introduce experimentation in its modern form as a means of testing his theories. Details of such experiments occur not in the *Dialogue* but in *Two New Sciences* and can also be gleaned from his unpublished manuscripts.[66]

Galileo's mathematical theory of motion could be put to the test only by way of experimentation. That theory involved uniform acceleration downwards, uniform horizontal motion and combinations of the two. For a variety of reasons, naturally occuring processes do not exemplify Galileo's theory of motion, as he himself realized. 'I admit', wrote Galileo, 'that the conclusions demonstrated in the abstract are altered in the concrete, and are so falsified that horizontal (motion) is not equable; nor does natural acceleration occur (exactly) in the ratio assumed; nor is the line of the projectile parabolic, and so on'.[67] A basic reason why actual motions do not generally correspond to those described in Galileo's theory is, of course, the existence of a variety of frictional impediments to motion. 'Considering merely the impediment that the air makes to motions in question here, it will be found to disturb them all in an infinitude of ways, according to the infinitely many ways that the shapes of the moveables vary, and their heaviness and their speeds'.[68] In addition to the problems posed by various types of frictional resistance, there is the fact that the motions that Galileo considers as exemplifications and potential tests of his theory are not of the kind that arise of their own accord. For instance, an important sequence of motions considered by Galileo involves a ball rolling down an inclined plane, being deflected onto a horizontal plane, and then leaving the plane to fall freely.[69] In order for him to test the fundamentals of his theory it was necessary for Galileo to construct artificial situations especially designed for the purpose and which, in addition, reduced unwanted impediments to a minimum.[70] Experimentation of this kind involved the introduction of various techniques for removing impediments and for dealing with those that remain.[71] Whilst it must be admitted that there was room for improvement as far as the *results* of Galileo's experimentation were concerned, what is of significance for my discussion is the character of the new kind of practice that Galileo had introduced.[72]

A number of basic epistemological points can be made with reference to this new practice of appealing to experiment.[73] In particular it is incompatible with the empiricist notion that knowledge involves observable regularities. It is incompatible with the Humean conception of scientific laws as constant conjunctions of events.[74] There is a sense in which, in experimental activity, human agents bring about regularities or constant conjunctions, but they do not thereby bring about the laws of nature. It can be argued that modern science involving the experimental practice initiated by Galileo becomes intelligible only if an ontological distinction is drawn between laws of nature and constant conjunctions. Whilst regularities or constant conjunctions are approximated to under experimental conditions, they do not occur in the world at large. However, the laws of nature that experimentation provides the evidence for are not assumed to be operative only under experimental conditions. For Galileo, bodies always possess the tendency to fall with a uniform acceleration but are usually prevented from doing so by air resistance, winds and other impediments.

From the point of view of Galilean science, the phenomena of the world are subject to a variety of modes of behaviour, inertial, gravitational, frictional, thermal, optical and so on. In typical, naturally occuring situations observable events involve a superposition of these modes. This is why observation of such events cannot supply the basic building blocks for our knowledge as empiricists tend to assume.[75] Experimentation of the Galilean type involves the construction of artificial situations designed to eliminate unwanted modes of behaviour and to arrive at a situation which is sufficiently uncomplicated to provide data suitable for testing the mode or modes of behaviour under investigation. Galileo designed or described experiments which involved the elimination of unwanted frictional and hydrodynamic effects and were sufficiently simple to be interpreted as evidence bearing on the interplay of gravity and 'inertia'.

A crucial point at issue in experimental science is not so much the *truth* of observation reports, which is the preoccupation of empiricists, but their *relevance*. The majority of experiences or observable states of affairs become irrelevant for the testing of theories. Only those observations carried out under appropriate experimental conditions are relevant. Experimental results are 'theory-dependent' in obvious ways. Most experiments are designed to test a theory which is already at least roughly formulated, whilst the adequacy of the set-up and the interpretation of the results will rely on the theoretical understanding of the situation. Experimental results come to be superseded once their inadequacy is appreciated in the light of theoretical advances. For instance, the results of Galileo's inclined plane experiments were superseded once the theoretical significance of the distinction between rolling and sliding was appreciated.

It is important to recognise here that the replacement of Galileo's experimental results by more appropriate ones did not involve a criticism of Galileo's experience or perceptions. There is no reason to doubt that Galileo faithfully and accurately recorded what he witnessed through his senses. Be that as it may, the point is that his inclined plane experiments, whatever other faults they may have possessed, involved balls rolling down inclined planes when they should have involved sliding rather than rolling. What is significant about an experiment is the details of the set-up, the precautions taken and the theoretical assumptions informing them. The observation reports of observers, recording pointer readings, heights of mercury columns etc., will usually be fairly unproblematic in a well-designed experiment. However, even if, for the sake of argument, we assume that observation reports can be arrived at with complete veracity by way of the senses, we do not thereby have access to secure foundations for scientific knowledge. It remains the case that the testing ground for scientific theories constantly changes as old experimental results are replaced by new ones, that is, by those that are deemed to have been carried out under more appropriate circumstances or involve more appropriate procedures. In Galilean science, observation is replaced by the results of publicly testable experimental procedures.[76] Empiricists, looking to experience for the source or foundations of scientific knowledge, are on the wrong track completely.

Galileo's experimental practice involved an epistemological revolution, a revolutionary change in method.[77] It involved a change of the 'observational base' of a form that is completely incompatible with a wide range of empiricist methodologies. The results of experimental procedures replaced mere observation as the testing ground for theories. From this perspective, Feyerabend's account of the change in observation language involved in Galileo's innovations can be seen to be inappropriate and to have *underestimated* the change in method and the threat to empiricism that was involved. Feyerabend sees the change as the replacement of one set of natural interpretations by another. 'He (Galileo) insists upon a *critical discussion* to decide which natural interpretations can be kept and which must be replaced.'[78] 'Galileo's first step, in this joint examination of the Copernican doctrine and of a familiar but hidden natural interpretation, consists therefore in *replacing the latter by a different interpretation*. In other words, *he introduces a new observation language*.' He thereby 'restores the senses to their position as instruments of exploration'.[79] From Feyerabend's point of view, then, the testing ground for theories remains observations made by individual observers. Bearing in mind that, for Feyerabend, natural interpretations are 'mental operations that follow so closely upon the senses' and which are 'so firmly connected with their reactions that a separation is difficult to achieve',[80] the replacement of one set of natural interpretations by another involves the replacement of one set of mental operations by another. Prior to Galileo, then, the normal observer, because of his or her cultural background, everyday experiences, language and so on, is programmed in a way that leads to one particular set of observational experiences and corresponding observation language, whereas the observer that has been subject to the medicine of Galileo's *Dialogue* becomes programmed in a new way that leads to a new set of observational experiences and a new observation language. The change in observation language is located in the individual observer. It is basically a psychological change.

According to the account I am advocating, in the new physics the testing ground for theories is no longer located in individual observers. The testing ground is constituted by the results of suitably performed experiments. Of course, experiments are performed by human agents. They do not perform themselves. Frequently, an experiment will involve contributions from a number of individuals each employing their own special skills. (In the experiments Galileo performed during his productive eighteen years at the University of Padua, he availed himself of the technical skills of Marcantonio Mazzoleni.) The skills and techniques employed will not be those of everyday life. They will be the skills and techniques of the physicist, acquired by experience and training. Insofar as experimentalists are required to *observe* the outcome of their experiments, one mark of a good experiment in physics is that the observations in question do not depend crucially on fine psychological judgements by the experimentalist. They typically involve lining up marks on a scale, taking a pointer reading and the like. As M. Clavelin puts it, Galileo showed that 'instruments or special experimental arrangements could be deployed in such a way as to organise experimental findings in a rational manner and hence to remove them from the sphere of mere opinions' and was thus able to 'save experience from the uncertainty of individual impressions'.[81]

Galileo's physics of motion involved replacement of the observational experiences of normal observers by experiment as its testing ground and thereby constituted a break from common sense. In Clavelin's words, 'the objective base of knowledge had changed'.[82] From this point of view, Feyerabend's talk of the clash of Galileo's theory with seventeenth-century common sense, whilst not entirely incorrect, is beside the point. By concentrating on the *Dialogue*, Feyerabend is able to find support for his view that Galileo's revolution hinged on the replacement of one common sense view by a new one. But the *Dialgoue* is an inappropriate place to seek for the essentials of Galileo's new science. That book was essentially a propagandistic one, designed to convert the educated populace at large, Counts and Popes, merchants and clerics, to the Copernican view. It is to Galileo's presentation of his mature theory of *Two New Sciences* that we should turn to discern the character of his innovation and to gauge the extent of the epistemological revolution it entailed.

However, whilst Galileo's new physics involved a break with common sense, it is not the case that the two were rendered completely distinct. For, whilst Galileo's physics could be put to precise tests only under experimental conditions, it was intended to be applicable, with varying degrees of approximation, to the everyday world. Galileo's theory of motion could not be appropriately tested under normal circumstances because, firstly, those situations are too complex, but, secondly and more fundamentally, they involve a superposition of modes of behaviour and not just the mechanism of falling and unimpeded horizontal motion with which Galileo's theory was concerned. Nevertheless, the theory, although it could only be adequately tested in experimental situations, was intended to be applicable outside of experimental situations also. For instance, Galileo's theory can be employed to explain why, or predict that, an object dropped from the top of the mast of a ship moving approximately uniformly will fall near the foot of the mast and not some distance towards the stern, and why various motions below deck will be unaffected by the motion of the ship.

Thus, although Galileo's physics involved a radically new practice, there remained links between it and the world of common sense. Inevitably, there were clashes between the claims of Galileo's physics and the expectations of the seventeenth-century man in the Venetian gondola. When, in the *Dialogue*, Galileo employed the new physics in his endeavours to sell the Copernican theory, he needed to confront, amongst other things, the common sense resistance. In attempting to accomplish that task he used a number of what can appropriately be referred to as propagandistic devices. By treating the *Dialogue*, as opposed to *Two New Sciences*, as containing the essence of Galileo's new science, Feyerabend is able to find support for the view that propaganda is of the essence of science.

Feyerabend also appeals to the use of the telescope to support his claim that Galileo's defence of the Copernican theory involved a transformation of the 'observational base' and a violation of Aristotelian and other methodologies. I have criticized, modified and extended Feyerabend's account elsewhere[83] and will not repeat it here.

8. SCIENCE AND PROPAGANDA

It is undoubtedly the case that Galileo's writings abound in what might well be called propaganda for the new science, and this is especially true of the *Dialogue*.[84] For instance, in the latter work he presents as obvious claims that are novel and problematic (pp. 144–9) and he backs them up by appeal to experiments that he has not performed and which would be difficult to perform (pp. 186–7). He appeals to those affiliated to Plato by employing the latter's theory of anamnesis (p. 191) and he attempts to disarm Aristotelians near the beginning of the *Dialogue* by offering an account of the planetary system employing the notions of natural order and circular motion (pp. 19 and 32).[85] The point at issue, then, is not whether Galileo employed propaganda or not, but the role and status to be attributed to it. Is it the case that there is a sense in which '*propaganda is of the essence*'[86] of Galileo's science?

A distinction needs to be drawn here between Galileo's contributions to a new science, on the one hand, and the question of the social conditions in which that science is to be developed and practiced on the other. [87] To put the matter crudely and briefly, propaganda has no legitimate role to play in the first and an important role to play in the second. The main source for Galileo's contribution to science itself is his *Two New Sciences*, whilst one of the main sources for his attempts to modify the social conditions in order to make room for the new science is the *Dialogue*. W. L. Wisan has noted that in the latter Galileo 'mixed sound scientific reasoning with dubious arguments which he hoped would persuade his readers of the truth of the Copernican system', whilst in the former 'he consistently sought to produce a rigorous mathematical treatise'.[88] As indicated towards the end of the previous section, Feyerabend fails to appreciate those distinctions.

From the point of view of Galileo's contribution to a new science, what are of importance are the theoretical raw materials that Galileo bequeathed to later scientists to work on and the techniques he introduced for working on them. Of significance here is his geometrical characterisation of terrestrial motions, the mathematical techniques employed for exploring its consequences, and the experimental practice utilised to test their adequacy. If these aspects of Galileo's work had involved propagandistic trickery; if for example, the arguments by means of which theorems concerning the motion of balls down inclined planes were derived from the fundamentals of the theory were 'arguments in appearance only',[89] then the work would have been useless as providing raw materials to be worked on and transformed by future workers such as Torricelli or Newton.

As far as the creation of social space for the new science is concerned, it is certainly the case that there was much more involved than rational argument, given the variety of interests at stake. I can agree with Feyerabend that, in that context, 'Galileo's propaganda machinations are often guided by the insight that established institutions, social conditions, prejudices may hinder the acceptance of new ideas' and that Copernicus was 'a symbol for the ideals of a new class'.[90] In this context, I am content to ask, with Feyerabend, 'what is the use of an argument that leaves people unmoved?'.[91] However, I would wish to add, within science itself, what is the use of a bad argument that convinces everyone?; what is the use of a new technique that does not work?

Insofar as Galileo employed methodological or epistemological arguments in support of his theories, these are best interpreted as something like propaganda. Galileo's remarks in this regard are unsystematic, inconsistent and opportunistic. A theme evident in his science of motion is the need for the theorems of that science to be derived by sound deductions from self-evidence principles, after the fashion of Euclid and Archimedes. There is a tension between this view and the role of experimentation and also with the insistence on the importance of the observational facts to be found elsewhere in Galileo's work.[92] When, in earlier sections, I spell out some of the characteristics and epistemological implications of Galileo's new science of motion, the non-empiricist implications of his experimentation for example, I do not mean to suggest that Galileo himself appreciated the situation in the same way. Galileo's defence of his new science involved propaganda and trickery sure enough; but this is not equivalent to admitting that propaganda is of the essence of science itself, although to assert that it is might have some useful shock value with respect to those 'internalists' who wish to deny the relevance of social or 'non-rational' factors for the history of science generally.[93]

9. A CASE AGAINST METHOD

In the light of the foregoing critique, modification and amplification of Feyerabend's study of Galileo's innovations, we are in a position to construct a case against method that differs somewhat from, and is more substantial than, the one Feyerabend has presented. Galileo's introduction of mathematization and experimentation, and also the use of the telescope, into physics are moves that cannot be accommodated by a vast range of empiricist and other methodologies. But quite apart from serving to refute specific methodologies, those episodes offer a basis for challenging any claim to the effect that there is such a thing as a universal, ahistorical account of scientific method that purports to offer criteria for guiding the work of scientists or for judging the merits of the theories that they produce. Aristotle had offered plausible arguments for the view that a mathematical physics is impossible.[94] By developing his geometrical theory of motion and the practice of experimentation that accompanied it Galileo went some way towards demonstrating *in practice* that mathematical physics can be successfully applied to the world. Similarly, he went some way towards demonstrating in practice that in some circumstances, telescopic observations are to be preferred to naked eye data. This, together with aspects of his experimentation and his view that shared motion remains undetectable by the senses, undermined a methodological principle implicit in Aristotelianism and supported by teleological arguments, namely, 'that the senses and experience should be our guide in philosophising'.[95]

Galileo's science was not guided by the methodological norms of his time and its success constituted a challenge to the adequacy of the standards that they embodied. In a similar way, a century later, the generally held view that the notion of action-at-a-distance is an absurdity was undermined, in practice, through the success of Newtonian theory involving gravitational attraction. It is implausible, to say the least, that such moves could be anticipated and legislated for in advance by methodologists. We can expect methodologies to alter in the light of new discoveries, including practical discoveries, and for this reason the notion of a universal, ahistorical account of method that can serve as the standard, not only for present but for all future knowledge, is an absurdity.

In some places Feyerabend embellishes his stand against method with utterances that suggest that he holds the view that in epistemological matters there are no standards whatsoever. He describes his own position as anarchistic, insists that 'all methodologies have their limits and that the only "rule" that survives is "anything goes"'.[96] Additionally he advocates the view that 'Mankind, and even Science, will profit by everyone doing his own thing'.[97] He seems to adopt a highly subjectivist position when he writes that 'what remains (after the refutation of contemporary methodologies) are aesthetic judgements, judgements of taste, metaphysical prejudices, religious desires, in short, *what remains are our subjective wishes*'.[98] The extreme subjectivist, anarchistic, person-relative position that is suggested by such remarks is certainly not the one that I would wish to draw from the case against method offered by the Galilean episode in science. The abolishment of universal, ahistorical standards is not equivalent to the abolishment of all standards.

There will be standards implicit in a body of knowledge at some stage in its development. In general, if one is to make a contribution to that knowledge it will be necessary to adopt those standards. A medieval scholastic wishing to contribute to Aristotelian theory needed to take common-sense perception and the orderly structure of the earth-centred cosmos seriously, needed to couch his arguments in appropriate syllogistic form, and so on. A scientist wishing to contribute to Galilean physics needed to carry out sound geometrical deductions, to follow appropriate procedures and take adequate precautions when experimenting and so forth. An anarchist attempting to contribute to Galilean science by conducting sloppy experiments or by carrying out unsound geometrical deductions simply on the grounds that such activities satisfied his or her whims or tastes could justly be dismissed as a crank.[99] However, it does not follow from this that standards cannot be challenged and changed. This can be done by moves which violate the standards but whose advantages can be demonstrated in practice. For example, conformity to the dictates of classical logic is not sacrosanct. To challenge such conformity it will not be sufficient simply to assert that we no longer care for logical argument. Rather, one must develop, or, more realistically, contribute to the development of, a theory and practice which involve a violation of those dictates and demonstrate in practice the advantages to be gained from such a move. This will not be easy, and if the change is a substantial one there will be social barriers to be overcome as well as those involved in the development of the new knowledge itself, as we have seen in the case of Galileo. There is no universal method. There are no universal standards. But there are historically contingent standards implicit in successful practices. It is not the case that, in epistemological matters, anything goes.[100]

University of Sydney, Australia

NOTES

[1] I will be concerned largely with Feyerabend's case as it appears in his book *Against Method* (London, 1975).

[2] *Against Method*, Chapters 6—8. Galileo's discussion of the tower argument is in his *Dialogue Concerning the Two Chief World Systems*, S. Drake (tr.), (Berkeley, 1967), p. 126ff. Hereafter I will refer to this work simply as *Dialogue*.

[3] *Against Method*, p. 89, italics in original.

[4] *Ibid.*, p. 87.

[5] *Ibid.*, p. 73. The phrase is taken by Feyerabend from the Introduction to Francis Bacon's *Novum Organum*.

[6] Feyerabend (*Against Method*, p. 74) writes of this absolute space as being defined by the solar system, but this is presumably just a slip. The adjective 'solar' is inappropriate and inadequate for the non-Copernican view that is being characterised.
[7] *Against Method*, p. 75
[8] *Ibid.*, p. 90, italics in original.
[9] *Ibid.*, p. 79, italics in original.
[10] *Ibid.*, p. 81.
[11] *Ibid.*, p. 90.
[12] Insofar as the main sense utilised to detect motion is the sense of sight, it would almost be appropriate to use the term 'optical relativity'. However, Galileo does briefly consider the sense of touch in this connection (*Dialogue*, p. 255) so I will stick to the more general term.
[13] *Dialogue*, pp. 171ff. For other examples see *Ibid.*, pp. 243ff.
[14] *Against Method*, pp. 83–4
[15] *Ibid.*, p. 84, italics in original.
[16] *Ibid.*, p. 81, italics in original.
[17] *Ibid.*, p. 75.
[18] *Ibid.*, p. 83.
[19] This point is made, without elaboration by P. K. Machamer, 'Feyerabend and Galileo: The Interaction of Theories, and the Reinterpretation of Experience', *Studies in History and Philosophy of Science* **IV**, 1973, pp. 1–46. In his reply to Machamer, included as Appendix 2 in *Against Method*, Feyerabend does not respond to this particular point.
[20] N. Oresme, *The Book on the Heavens and on the Earth* in *A Source Book in Medieval Science*, E. Grant (ed.) (Cambridge, Mass., 1974), p. 504.
[21] *Against Method*, p. 81.
[22] *Ibid.*, p. 90.
[23] *Ibid.*, p. 97.
[24] Feyerabend does devote a page or two to the evolution of Galileo's ideas on inertia (*ibid.*, pp. 94–5) but his remarks are inadequate as is made evident in the next section of this chapter.
[25] *Dialogue*, p. 145. I suggest that an adequate understanding of the function of thought experiments in Galileo's work would lead to a decrease in the extent to which this passage can be interpreted as propagandistic.
[26] This point is made briefly in J. Curthoys and W. Suchting, 'Feyerabend's Discourse Against Method: A Marxist Critique', *Inquiry* **XX**, 1977, p. 287.
[27] The senses of 'ad hoc' that are sketched here draw on, although do not precisely duplicate, the senses of that term that have been formulated by I. Lakatos and his followers. For Lakatos's original distinctions see his 'Criticism and the Methodology of Scientific Research Programms', *Proceedings of the Aristotelian Society* **LXIX**, 1968, pp. 149–86 and for modifications see E. Zahar, 'Why did Einstein's Programme Supersede Lorentz's?', *British Journal for the Philosophy of Science* **XXIV**, 1973, pp. 95–123 and 223–62.
[28] For a discussion of the notion of independent testability, see K. R. Popper, 'The Aim of Science' in *Objective Knowledge* (Oxford, 1979), pp. 191–205.
[29] See Section 8, Book I of Copernicus's *The Revolution of the Heavenly Spheres*.

[30] See especially S. Drake, 'Galileo's Experimental Confirmations of Horiztonal Inertia: Unpublished Manuscripts', *Isis* **LXIV**, 1973, pp. 291—305.

[31] *Against Method*, pp. 145—8.

[32] Galileo Galilei, *Two New Sciences*, tr. S. Drake (Madison, 1974). In the following references to this work, page numbers are those of *Le Opere di Galileo Galilei* which are indicated in Drake's and other translations.

[33] For documentation of this point see, for example, W. L. Wisan, 'The New Science of Motion: A Study of Galileo's *De Motu Locali*', *Archive for History of Exact Sciences XIII*, 1974, pp. 103—306 and S. Drake, *Galileo at Work* (Chicago, 1978).

[34] On this point see M. Clavelin, *The Natural Philosophy of Galileo* (Cambridge, MA., 1974), Chapter 3; S. Drake, *Galileo Studies* (Ann Arbor, 1970), Chapter 1; and W. R. Shea, *Galileo's Intellectual Revolution* (London, 1972), Chapter 1.

[35] *The Little Balance*, translated by L. Fermi and G. Bernardini, *Galileo and the Scientific Revolution*, (Greenwich, Connecticut, 1965).

[36] *On Motion and On Mechanics* tr. I. E. Drabkin and S. Drake (Madison, 1960).

[37] See Drake, *op. cit.* (Note 34); Wisan, *op. cit.* (Note 33); and also W. A. Wallace, *Prelude to Galileo* (Boston, 1981).

[38] The path from Archimedes' statics to Galileo's kinematics is traced in some detail by Clavelin, *op. cit.* (Note 34).

[39] *On Motion*, Chapter 14. For discussions of this aspect of Galileo's early physics see Clavelin, *op. cit.* (Note 34), pp. 165—72 and Drake, *op. cit.* (Note 34), Chapter 5.

[40] *On Motion*, pp. 23—50, discussed by Clavelin, *op. cit.* (Note 34), pp. 121—38.

[41] *On Motion*, p. 66.

[42] *On Mechanics*, p. 171.

[43] *Wisan, op. cit.* (Note 33), p. 149.

[44] Wallace, *op. cit.* (Note 37), pp. 270—1, 283—4, and 334—5.

[45] *On Motion*, p. 72.

[46] *Ibid.*, p. 73.

[47] See Drake, *op. cit.* (Note 34), Chapter 11, and, for more detail, Wisan, *op. cit.* (Note 33), pp. 199ff. Note that my argument does not depend on the precise definition of uniform acceleration, over which Galileo was uncertain for a period.

[48] See Drake, *op. cit.* (Note 33), p. 128.

[49] S. Drake, *Discoveries and Opinions of Galileo* (New York, 1957), pp. 113—4 and *Two New Sciences*, p. 243.

[50] *Against Method*, p. 96, italics in original.

[51] This is apparent from Galileo's correspondence, discussed by Drake, *op. cit.* (Note 33), pp. 40—1.

[52] 'It therefore appears likely that Galileo, confronted with failure of the first possible confirmation of Copernicanism by direct observation, lost faith in it from 1605 until 1610', Drake, *op. cit.* (Note 33), p. 110.

[53] *Against Method*, p. 95.

[54] A. Koyré, *Galileo Studies*, tr. J. Mepham. (Brighton, Sussex, 1978), pp. 176 and 187 respectively.

[55] These points are discussed more fully in A. F. Chalmers and R. Nicholas, 'Galileo on the Dissipative Effect of a Rotating Earth', *Studies in the History and Philosophy of Science* **XIV**, 1983, pp. 315—40.

[56] Koyré, *op. cit.* (Note 54), p. 187.
[57] *Dialogue*, p. 116. The two quotations, together with a third to be discussed below, are cited by Feyerabend, *Against Method*, pp. 78—9, footnote 10.
[58] See *Dialogue*, pp. 114—6, 171ff and 248ff.
[59] *Ibid.*, pp. 186—7.
[60] *Ibid.*, p. 179.
[61] *Ibid.*, p. 114. This is the third quotation referred to in Note 57 above.
[62] This latter phrase emerged in a conversation between my colleague, Terry Blake and myself about Feyerabend's critics.
[63] This is also apparent from Galileo's treatment of the hypothetical problem of a ball dropped to earth from the moon. See *Dialogue*, p. 233.
[64] *Two New Sciences*, pp. 336—7.
[65] See, for example, Koyré, *op. cit.* (Note 54), especially p. 154, G. De Santillana, Galileo's *Dialogue on the Great World Systems* (Chicago, 1953), pp. 205—6 and E. J. Dijksterhuis, *The Mechanization of the World Picture* (Oxford, 1961), pp. 349—50. The views on Galileo's inertial concept discussed here are dealt with more fully in Chalmers and Nicholas, *op. cit.* (Note 55).
[66] See *Two New Sciences*, pp. 212—3 and, for example, S. Drake, *op. cit.* (Note 30).
[67] *Two New Sciences*, p. 274. The words in brackets appear in Drake's translation.
[68] *Ibid.*, p. 275.
[69] See Drake, *op. cit.* (Note 30).
[70] Artificial experimentation of this kind did not have a central place in Aristotelian physics because it was basically a theory of nature, a theory about how things act and behave when left to themselves. This aspect of Aristotelian theory is brought out by R. G. Collingwood, *The Idea of Nature* (Oxford, 1945), pp. 80—92 and by F. J. E. Woodbridge, *Aristotle's View of Nature* (New York, 1965).
[71] For an excellent account of the evolution of Galileo's attempts to deal with this problem, see N. Koertge, 'Galileo and the Problem of Accidents', *Journal of the History of Ideas* **XXXVIII**, 1977, pp. 389—408.
[72] S. Gaukroger makes a similar point in *Explanatory Structures*, (Hassocks, Sussex, 1978), p. 182.
[73] I have dealt with these issues more fully in 'A Non-Empiricist Account of Experiment', *Methodology and Science* **XVII**, 1984, pp. 95—114.
[74] This, and related points are well made by R. Bhaskar, *A Realist Theory of Science* (Brighton, Sussex, 1978).
[75] A similar point is made by Feyerabend in one of his earlier papers, 'On the Interpretation of Scientific Theories', *Proceedings of the 12th International Congress in Philosophy* **V**, 1960, reprinted in P. K. Feyerabend, *Philosophical Papers, Volume 1, Realism, Rationalism and Scientific Method,* (Cambridge, 1981), pp. 37—43.
[76] Galileo did not grasp the epistemological implications of his experimentation, as is briefly discussed in Section 9. His contemporaries, Descartes and Francis Bacon, offered more in this regard. Bacon explicitly acknowledged the need to replace observation by experiment, and stressed the public character of the latter. (See the Preface to his *Great Instauration.*) For Descartes' critique of sense perception, see C. Larmore, 'Descartes' Empirical Epistemology' in *Descartes: Philosophy, Mathematics and Physics*, S. Gaukroger (ed.), (Brighton, Sussex, 1980), pp. 6—22.

[77] S. Gaukroger has argued that this conception of experiment was a precondition for Galileo's mathematical physics. See *op. cit.* (Note 72), Chapter 6.
[78] *Against Method*, p. 73, italics in original.
[79] *Ibid.*, pp. 78—9, italics in original.
[80] *Ibid.*, p. 73.
[81] M. Clavelin, *op. cit.* (Note 34), p. 401.
[82] *Ibid.*, p. 398.
[83] 'Galileo and the Telescope Observations of Mars and Venus', *British Journal for the Philosophy of Science* **XXXVI**, 1985, pp. 175—191.
[84] M. A. Finocchiaro, in *Galileo and the Art of Reasoning* (Dordrecht, 1980), p. 4, asserts that 'nearly all scholars recognize a polemical and propagandistic dimension in the *Dialogue*', Koyré writes of the 'weapons of the polemicist' employed by Galileo [*op. cit.* (Note 54), p. 158], whilst L. Geymonat refers to the period of the writing of the *Dialogue* as Galileo's propagandistic period [*Galileo Galilei*, (New York, 1965), p. 58].
[85] On the latter point see Drake, *op. cit.*, (Note 34), p. 253.
[86] *Against Method*, p. 157, italics in original.
[87] This distinction is given a precise formulation in L. Althusser's notion of science as the production of knowledge, where it becomes the distinction between the means of production and the social relations of production. See, for example, L. Althusser, *For Marx* (London, 1969) and L. Althusser and E. Balibar, *Reading Capital* (London, 1970). The production view of knowledge is employed by Curthoys and Suchting in their critique of Feyerabend in *op. cit.* (Note 26) and is given particularly clear expression in W. Suchting, 'Knowledge and Practice: Towards a Marxist Critique of Traditional Epistemology', *Science and Society* **XLIV**, 1983, pp. 2—36.
[88] W. L. Wisan, 'Galileo's Scientific Method: A Re-examination', in *New Perspectives on Galileo*, R. E. Butts and J. C. Pitt (eds.), (Dordrecht, 1978), p. 3.
[89] *Against Method*, p. 81.
[90] *Ibid.*, p. 106 f.n. and p. 154 respectively.
[91] *Ibid.*, p. 25. The limits of argument as a tool for creating a space for the new science had been appreciated by Galileo's contemporary, Francis Bacon, See his *Novum Organum*, Book I, aphorisms xxxv, lxi and cxv.
[92] On the diverse aspects of Galileo's methodology see Wisan, *op. cit.* (Note 88) and E. McMullin, 'The Conception of Science in Galileo's Work' in the same volume.
[93] See, for instance, I. Lakatos's comment: 'The whole development [the change from Ptolemaic to Copernican astronomy] is narrowly internal; its progressive part could have taken place at any time, given a Copernican genius, between Aristotle and Ptolemy or in any years after, say the 1175 translation of the *Almagest* into Latin, or for that matter, by an Arab astronomer in the ninth century. External history *in this case* is not only secondary; it is nearly redundant'. 'Why did Copernicus's Research Programme Supersede Ptolemy's?', in *The Methodology of Scientific Research Programmes*, J. Worrall and G. Currie (eds.), (Cambridge, 1978), p. 189.
[94] See, for example, *On the Heavens*, II, 13, 295b, 10ff; III 7, 306a 9—11; III, 8; *Physics*, 1936, 32—5 and *Metaphysics*, Book E. For remarks to similar effect by a contemporary of Galileo see 'Appendix to the Art of Weighing' in *The Principal Works of Simon Stevin I*, E. Dijksterhuis (ed.), (Amsterdam, 1966).
[95] This was 'the criterion of science itself' according to Scipio Chiaramonti (1565—1652), the author of *Anti-Tycho* quoted by Galileo in the *Dialogue*, p. 248.

[96] *Against Method*, p. 296.
[97] *Ibid.*, p. 215.
[98] *Ibid.*, p. 285.
[99] In an earlier work Feyerabend characterised the crank in a similar way. See his 'Realism and Instrumentalism: Comments on the Logic of Factual Support', in *The Critical Approach to Science and Philosophy*, ed. M. Bunge (New York, 1969), p. 305.
[100] It is possible to find in Feyerabend's writings a view that differs from the extreme anarchistic, subjectivist one that is often attributed to him, and to which interpretation some of his remarks lend themselves, as we have seen. See, for example, his comment on the limitations on human freedom due to the character of the objective situation (*Against Method*, p. 187) and his discussion of reason and practice and of the criticism of standards in *Science and a Free Society* (London, 1978), pp. 16ff.

JOHN A. SCHUSTER

CARTESIAN METHOD AS MYTHIC SPEECH: A DIACHRONIC AND STRUCTURAL ANALYSIS

1. METHOD AND THE PROBLEM OF THE HISTORICAL DESCARTES

As far as historians of science are concerned, Descartes has not proven to be a congenial figure. The last comprehensive monograph on his science and mathematics (but not on his natural philosophy) was written by Gaston Milhaud over sixty years ago.[1] Despite early pioneering work by Leon Bruschvicg, Pierre Boutroux, Paul Mouy and Jacob Klein, historians of science have not seriously addressed the evolution of Descartes' scientific and mathematical work, their interactions over time, or their complex relations to his writings on methodology and systematic mechanistic natural philosophy.[2] Recent historical studies, some of exceptional value, have tended to limit themselves to precise analyses of narrow aspects of his technical interests.[3] As a consequence, we still lack a broadly acceptable historical interpretation of Descartes' career as symptomatic and constitutive of a crucial moment in the origin of modern European science.[4]

One of the chief obstacles to historians getting to grips with Descartes' career is posed by his astounding methodological claims, set forth in the *Discours de la méthode* and related texts.[5] In the *Discours*, partie II, Descartes retails a version of his early autobiography, articulated through the development and successful applicaton of his method. He writes that he initially developed the method in the winter of 1619/20, when he was twenty-three, applying it first in the simplest of mathematical fields. This supposedly issued in a general analytical machinery for all mathematical disciplines.[6] He continues that over the next few years he worked to apply and perfect the method, in mathematics and in the mathematical parts of physical inquiry. The results, he insists, are in the *Essais* which follow the *Discours* — the *Géométrie*, the *Météores* and the *Dioptrique*. Presumably, the method therefore significantly facilitated such achievements as: the discovery of the sine law of refraction of light; the discovery of his mechanistic model for the action of light; the perfection of his theory of lenses and telescopes; the

J. A. Schuster and R. R. Yeo (eds.), The Politics and Rhetoric of Scientific Method, 33—95.
© 1986 *by D. Reidel Publishing Company.*

discovery of the explanation of the rainbow; the solution of the major mathematical problems constitutive of the *Géométrie*. And, of course, there is more: the method, he assures the reader, guided his search for metaphysical bases for his systematic mechanistic natural philosophy. However, he had not undertaken this task until 1628, because back in 1619 he had seen (on methodological grounds!) that metaphysics had to await the elaboration and perfection of the method in the subordinate sciences.[7]

Such claims may be easy to refute on a piecemeal basis: No historian of science accustomed to close textual and contextual reconstruction of scientific practice would have much difficulty in disposing of them.[8] But taken as a whole, these claims have proven off-putting and, I would suggest, embarrassing to historians of science, especially those who have been trained in what might loosely be termed the Koyré/Kuhn perspective on method, as discussed in the Introduction to this volume. Such historians of science (and I include myself among them) have often tended to neglect method doctrines. As pointed out in the Introduction, much of this neglect (indeed disdain) of method doctrines has to do with the plausibility of the claims of Koyré, Kuhn (and Bachelard before them) that grand methodological doctrines never really worked in practice; that they are the stuff of 'preface history', of 'whiggish historiography', of textbook hagiography; and that a course of scientific practice is never adequately explained as the outcome of the successful application of some general method.

As it is the aim of this volume to contribute to the breaking of the post-Koyréan silence about methodology as an object of historical study, so this chapter is intended to extend this new perspective to the tortured case of Descartes in particular. Several overlapping aims are involved here, however. At the level of historical interpretation and analysis this chapter examines how Descartes made or constructed his method doctrine. It explores the stages through which the method passed, the discursive resources out of which it was woven, and the strategies of literary fabrication involved in the process. My purpose here is to make a direct contribution to Descartes scholarship and hence to the study of the Scientific Revolution. I stress the role of Descartes' early scientific and mathematical aspirations in the constuction of his method doctrine. In this way I hope to give historians of science a firmer foothold in this area and in the central text, the *Regulae ad directionem ingenii*, than they have had before. Nevertheless, little of this would licence the inclusion of the chapter in this particular volume. The most that could be said is that the reconstruction of the contingencies and discursive strategies which shaped Descartes' doctrine of method contributes somewhat to the demystification of that doctrine by showing that it is an historical artefact, not the blinding revelation of the 'truth' about method (whatever that may be). In this chapter the diachronic reconstruction of Descartes' making of the method is really propaedeutic to my larger aim which *is* of a piece with the theme of this volume.

My central aim is to raise and answer a question which I think is fundamental to the emerging social history of method as politics and rhetoric: As the chapters in this volume show, we can now study the political and rhetorical deployments of method in scientific controversy and in the public politics of science. This prompts me to ask the following deeper question; 'How can it possibly be that throughout the history of science methodologists and their audiences have often genuinely believed in the efficacy of method doctrines which we post-Koyréans 'know' cannot have worked?' No historian wishes to accuse his subjects of being fools or madpersons, just because they appear to disagree with him. Therefore, we are obliged to discover just what it is about systematic method doctrines which creates and sustains their plausibility to believers, past and present. My answer, to be developed through the example of Descartes' method, will be this: All systematic method doctrines belong to a definite species of discourse. The species is characterized by the presence of a certain discursive structure common to all instances of the type. This structure is such that it necessarily defeats the ability of any methodology to accomplish what it literally announces itself to be able to accomplish. At the same time, this same discursive structure sustains or creates a set of illusions (in the form of literary effects) to the effect that the method in question can indeed accomplish what it claims to be able to do. In other words, all set piece method doctrines have the same underlying discursive structure which explains their lack of efficacy as well as their ability to create the literary effect that they are efficacious. I suggest that an understanding of the discursive structure and dynamics of method doctrines will help underwrite the new historiography of the politics and rhetoric of method, while supplying it with new tools and certainly with new questions.[9]

The case study of Cartesian method in this chapter is therefore both a contribution to the historical study of Descartes, and at the same time a working example of how we might address the deeper historiographical problem of method. It seems to me that we have little chance of comprehensively understanding Descartes historically, according to the prevailing standards in the history of science, until we know what to do about his method; that is, until his method discourse is demystified and historicized. By the same token I would suggest our newer social and political historiography of method must confront the phenomenon of 'methodology' head on, and must not concede too much by claiming to treat only the 'external' or merely 'social' deployment (or 'abuse') of method talk. If we are largely agreed that methodology is bunk, we should be prepared to theorize about why that is so and what that means for writing the history of method. Such a theory can be built up in stages in close connection with historical study, and for this theorist Cartesian method is the chosen testing ground and object of study.

The title of this chapter speaks of Cartesian method as 'mythic speech'. These words are not chosen lightly. In part they are intended to have a certain shock value for those readers who do not share in the common, but I think largely tacit, post-Koyréan opinion that 'method is myth' in the usual dictionary sense of the term. More importantly my choice of words intends something more theoretically precise. My use of the terms 'myth' and 'mythopoeic' [myth making] derive, at one or two removes, from Roland Barthes' early essay 'Myth Today'.[10] Barthes claimed to identify a peculiarly modern ('bourgeois') form of myth in which the trick is to naturalize values, interests and socially negotiated outcomes, making them appear to be factual, natural and inevitable. According to Barthes these effects are brought about by the structure of the discourse in question. It seemed to me that one characteristic modern myth, operating in a structural manner, is the post-sixteenth-century myth of scientific method, of which Descartes was one of the original high priests. Although I do not deploy the kind of semiotic techniques which Barthes advocated, I do claim to have identified certain structural levels in any systematic method discourse and I locate the persuasive power and naturalizing force of such discourses in the relations holding generally between the levels. In my account methodologies are discourses so structured that they necessarily lie about their own powers and capabilities in the interest of turning culture into nature. In these precise senses, then, methodologies deserve the (Barthian) title of myths.

My approach also owes much to the anti-methodism of Paul Feyerabend, although in a precise way which requires clarification.

Post-Kuhnian debunkers of method can, I think, perceive two rather distinct initiatives in Feyerabend's work. On the one hand there is Feyerabend's historical critique of methodology, consisting mainly in case-study illustrations of the non-binding character of any and all systematic methodologies. 'Progress' in science, he persuasively argues, has always broken the pat rules laid down by methodologists, and it has always *had* to do so. New standards are constructed and refined in the act, through the very processes of major scientific change; and, in the cases studied, Feyerabend tends to show that rigid adherence to the rules of contemporary (or later) methodologies would have aborted or obstructed the course of development.[11] Feyerabend's efforts in this direction must seem brilliant and historically revealing to any Koyré- or Kuhn-influenced debunker of method. Nevertheless, this work does not really constitute a great advance in our ability to theorize about the nature of methodological *discourse*. On the other hand, in a small and rather neglected corner of his work, Feyerabend, in my opinion, has offered the first sustained demonstration of the structural sources of the mythic character of method discourse. In his important paper, 'Classical Empiricism', Feyerabend lays bare the discursive mechanisms by which Newton's methodological claims in physical optics present a systematically distorted picture of his actual practice, producing a convincing fairly tale about the genesis and status of his claims in that field.[12] Although Feyerabend did not pursue his analysis in explicitly structural terms, his approach, viewed through my semi-Barthian spectacles, helped to catalyze my own structural schema for methodologies. Later, in Section 6.2, I shall attempt to show how Feyerabend's argument maps onto my schema.[13]

The argument will proceed in three broad stages: First, Section 3 will identify and discuss the core of Descartes' method discourse. Section 4 will contain a detailed reconstruction of how and why Descartes manufactured this particular type of method discourse. In Section 5 Descartes' method will be the subject of a structural analysis, articulating the mythic character and functioning of his (or anyone else's) systematic method discourse. Finally, in Section 6, I shall suggest how one can begin to move from the case of Descartes toward a general theory of methodology as mythic speech. However, before proceeding we need to take note of one further obstacle standing in the way of taking Descartes as a test case in the new historiography of method.

2. THE DESCARTES LITERATURE: PART OF THE PROBLEM OR THE SOLUTION?

In the broadest sense this chapter suggests that grand doctrines of method can become objects of proper historical study if we combine the proposed structural study of methodological discourse with the perspectives and tools of the newer social history of methodology as politics and rhetoric. However, in the case of Descartes, the post-Koyréan phobia of method discourse is not the only obstacle to our project. At least as important is the existence of a large orthodox secondary literature about Descartes. This literature takes little notice of modern developments in the historiography of science, and it is dominated by philosophers, historians of philosophy, intellectual historians and scholars of French literature. Much to the dismay of the historian of science, this literature is characterized by a widespread belief, explicit or implied, in the efficacy of Descartes' method. A book could be written about this phenomenon, which is really an epiphenomenon of the cultural worship of Descartes in particular, and of the modern myth of method in general. Believers in the efficacy of Descartes' method can be grouped into three broad categories: (1) 'literalists', (2) 'positivist pseudo-debunkers' and (3) 'exegetical apologists'.

(1) Literalists accept at more or less face value Descartes' epistemological and autobiographical claims for his method. Accordingly, their scholarly task is simply to explicate and clarify the essential truth of Descartes' tale about his own life and method. Lest this seem improbable, especially to historians of science, the reader is referred, for example, to J. R. Vrooman, *René Descartes, A Biography* (New York, 1970) an informative, readily accessible English language treatment. We read on pages 66 and 67 what is actually an amplification of Descartes' own rhetoric in the *Discours*: The method is real and efficacious; it was tested and used in particular technical fields; it dictated the order of study; it is an epoch making cultural achievement ('. . . the method that would be adapted [*sic*] by virtually the entire civilised world, the method that would be accepted as a monument in the history of Western thought'). None of this is questioned as to evidence or meaning. Indeed the text is a continuation of the mythic speech of Descartes himself.

(2) Positivist pseudo-debunkers are a rare breed, possibly because their project demands at least a broad schooling in the traditional 'whig' history of science. A good example is E. J. Dijksterhuis, 'La méthode et les Essais de Descartes' in *Descartes et le Cartesianisme hollandaise* (Paris, 1950), pp. 22–44. He ventures the opinion that Descartes' method is nowhere to be seen in the three *Essais* following the *Discours*, a view which would seem to make him a debunker of method. But, as a redoubtable 'whig' historian, Dijksterhuis did believe in a scientific method, so he went on to claim that Descartes must have glimpsed some part of the method in order to be able to produce the fine results in the *Essais*. This real method, shared by all real scientists, bears no relation to the pap which Descartes' rhetorically spoons out in the *Discours*.

In a similar vein, Elie Denissoff has dismissed the claimed universality of the four rules of the method as a literary sop intended to hold together the disparate parts of the *Discours*.[14] Yet the chief burden of his study is to read the *Discours* as a coded message from Descartes concerning his real method, which is limited to 'mathematical physics', and which Denissoff clearly believes was indeed efficacious. Such 'positivist pseudo-debunking' offers us perhaps the worst of two worlds: On the one hand there is a dismissive scepticism about Descartes' explicit methodological discourse which defuses critical historical inquiry into whatever it could be that Descartes thought he was talking about. On the other hand there is a cloying, now much outdated faith in the existence of a unique, efficacious, rather positivistically conceived scientific method, of which Descartes was one of the first 'discoverers'.

(3) Exegetical apologists are found almost exclusively amongst leading historians of philosophy. Combining careful textual scholarship with at least tacit belief in the method, they include most of the recognizedly great Descartes' scholars of the last hundred years, for example, Gilson, Gouhier, Sirven, Hamelin, Liard and latterly Jean-Luc Marion. These men are great scholars because they pose a serious scholarly question and try to answer it using the highest standards of philosophical and textual criticism. That question is, 'What do the *Discours* and the *Regulae ad directionem ingenii* really say about method, and how much of these texts corresponds to historically recoverable fact?' It goes without saying that you cannot study Descartes without taking their work as a starting point. Exegetical apologists are all united on one

axiom, despite a wide range of other differences. They virtually all believe at some level in the efficacy of the method.[15] True, they do not spend their time reconstructing Descartes' scientific work in terms of the method; they are not 'whig' historians of science. Rather, tacitly taking the method as efficacious, they pursue other goals: explicating the content of the method or reconstructing its development. Like devout Christian biblical hermeneuts, these 'apologists' are engaged in a difficult, scholarly search for *grains of truth* hidden in a selection of canonical texts. But, as we shall see, you do not have to be a believer in method in order to pursue its history; and if you are a sceptic about method, then all sorts of new interesting questions appear, such as, 'what is method discourse', 'how does it affect believers', 'how can we save believers from their illusions'? etc.

3. THE CORE OF DESCARTES' METHOD DISCOURSE

The structural analysis and demystification of Descartes' method depends upon the identification of his central methodological claims, the core of his method discourse. This core is to be found in the *Discours de la méthode* and the *Regulae ad directionem ingenii*, where Descartes offered formal, systematised versions of the method. The texts proclaim themselves to be the pre-eminent vehicles of his method message; and historical research further supports the claim that these texts were central to Descartes' intentions for systematizing and presenting his method.[16] Although, in what follows, we shall make use of the *Discours*, we shall especially stress the *Regulae*. This is because the *Regulae* are more detailed than the *Discours* as well as much earlier, usually being dated from 1626—28. I shall argue that parts of the *Regulae*, essentially the first eleven of its twenty-one rules, date from much earlier than 1626; that they date in fact from 1619—20, the very period when Descartes first elaborated a method discourse. Hence the early *Regulae* represent the earliest and most detailed text on Descartes' method, a text arguably central to his programme, and essential to our picture of the core of his method discourse.[17]

I am well aware that some Cartesian scholars may be sceptical of my identification of the core of Descartes' method discourse, because they know that Descartes produced disparate and varied informal remarks about methodological matters which are scattered in his correspondence and published works. These remarks include, among other topics, meta-reflections on the explanatory structure of his mechanistic natural philosophy, comments on the argumentative tactics in his *Meditations*, and methodologically relevant debate with critics of his optics and mathematics. Commentators have often sifted these texts in search of 'the method'. I would suggest that such projects cannot succeed: one finishes with an account of Descartes' method skewed to the particular selection and weighting made of these texts. For some, the method is an account of how Descartes discovered and/or deployed his arguments in the *Meditations*; for others it is a supposedly definitive account of how he goes about explaining things in his mechanistic natural philosophy or his optics. What appears to happen in such cases is that the commentator produces a more or less compelling account of how Descartes went about glossing and defending a particular text, either the *Meditations*, the *Principia philosophiae* or the *Dioptrique*. But, the fundamental question is whether these glosses make use in any significant way of the discursive resources of the core of the method doctrine, or whether they are not quite distinct pieces of legitimatory and explicative discourse. Sometimes the core of the method discourse *is being deployed* and the characteristic illusions or literary effects *are being produced*; but, sometimes 'the method' has nothing to do with the construction of the supposedly methodological text in question. Clearly, our first task is to analyse the structure and mythological functioning of the core discourse, before addressing the body of putatively 'methodological' remarks and their contexts.

Let us first examine the method as set out in the *Discours*. In partie II one finds four rules of method surrounded by a text which purports to narrate the history of the method. The narrative includes rhetorical elaborations of certain themes, as well as what might be judged more thematically central explications of technical portions of the method.[18] Leaving aside the question of the historical accuracy of the narrative (and hence the further question of the historical relevance of the elaborations and explications), one can, I think, isolate three related propositions, well founded in the text, which form the context of Descartes' presentation of the four rules of method.

1. All rationally obtainable truths subsist in a network of deductive linkages, and this is the meaning of the 'unity of the sciences'.[19] [This will henceforth be termed Descartes' 'latticework' vision of the unity of the sciences.]

2. As rational beings, humans possess two divinely given faculties for the attainment of truth; the power of intuiting individual truths, and the power of deducing valid links between them.[20]
3. A single mind, exercising intuition and deduction, could in principle traverse the entire latticework; but, some help is required in the form of practical hints or suggestions, heuristic rules, to aid in the preparation of inquiries, the ordering of inquiries, and the checking up after inquiries.[21]

Not surprisingly, therefore, there are two complementary moments or aspects within the statement of the rules of the method. Firstly there is a doctrine of truth. On the one hand, it informs us of what we presumably already know — that we can intuit and deduce truths. On the other hand, it adduces some negative heuristic advice from this fact: trust not in a authority, nor in unclear, indistinct belief, will or emotion; avoid precipitation and hasty judgment; go only as far as intuition and deduction reveal the truth. All this is essentially contained in rule 1 of the *Discours*.[22] Secondly, there is an open ended set of heuristic rules, initially gathered from easy excursions around the latticework of knowledge. These are contained in part in rules 2, 3, and 4 of the *Discours*, and they advise the inquirer to divide each problem into as many simpler parts as possible; to resolve each sub-problem in due order, starting with the simplest and rising by degrees to the most complex; to assume a fictitious order in a problem when there is no natural one; to assess ['enumerate'] relevant aspects and materials of a problem beforehand, and afterwards carefully review ['enumerate'] one's steps so that nothing is omitted, no falsehood admitted or truth overlooked.[23] These heuristic rules are not to be applied mechanically. They require sagacity and practical experience. As inquiry proceeds they can be refined and additional rules added.

The two moments or aspects of the rules of method are closely interrelated, especially when they are set against the backdrop of the vision of the latticework of rational truths and unity of the sciences. On the one hand, there is little point in developing heuristic rules in the absence of at least some preliminary picture of the cognitive terrain (latticework) the conquest of which they facilitate; on the other hand, the capacity for intuiting and deducing truth is likely to be insufficient *in practice*; that is, heuristic aids are necessary if we are to move about the latticework in an efficient manner and if we are to learn and profit methodologically from the experience.

Turning now to the *Regulae*, one can say that the opening sections of the text, roughly rules 1 and 11, simply parallel the *Discours*. Rules 1 and 2 present and develop the notion of the unity of the sciences. Rule 3 discusses intuition and deduction, and hence it corresponds to rule 1 of the *Discours* and to the first moment of the method. Rules 5 to 11 contain detailed heuristic advice, and so they correspond to rules 2, 3, and 4 of the *Discours*, the second moment of the method.[24] What is of interest here, however, are the ways in which the *Regulae* provide a clearer and yet more detailed picture than the *Discours* of the core of Descartes' method doctrine. For our purposes two aspects of this greater elaboration must be considered.

First of all, in rule 6, the reader is treated to Descartes' most elaborate (and we shall find the *earliest*) explication of the vision of the latticework of rational truths. The logical chains of truths consist in 'absolute' terms linked to a 'series' of 'relative' terms through a greater or smaller number of rationally specifiable 'relations' (*respectūs*).[25] Absolute terms are the initial terms in particular deductive series, and they are themselves relative to a small set of what might be termed 'absolutely absolute' terms.[26] Relative terms, properly so called, are those occuring further down deductive series. In some degree they 'participate in the same nature' as their antecedents, the absolutes; but, they also involve complex conditioning factors or 'relations'. Relatives are distanced from their absolute 'to the degree that they contain more relations subordinated one to another'.[27] By keeping in mind this explication of the latticework, we shall find it easier to reconstruct the development of Descartes' method discourse and to subject the discourse to a structural critique.

The second important aspect of the early *Regulae* follows from the first. It is the deceptively simple point that in the *Regulae* the articulated vision of the latticework is the template against which Descartes forms the heuristic rules of the method. Even in the *Discours* the three heuristic rules find their rationale in large measure in the latticework vision. But in the *Regulae*, thanks to the elaboration of rule 6, the process of manufacture of the heuristic rules against the backdrop of the latticework is in plain view. Indeed it could be suggested that Descartes' main intention in articulating the latticework was to facilitate the formulation and presentation of the heuristic rules. This may surprise knowledgeable Cartesian scholars who are aware of a large literature concerned with the correct interpretation of the intended ontological and epistemological statuses of Descartes' absolute and relative terms.[28] What must be realized is that Descartes did not stop to wrangle over these issues when composing these early rules. In 1619—20 he was an aspiring methodologist and not yet the metaphysician and builder of a system of natural philosophy of the post 1628 period. One must read his text keeping in view the nature and trajectory of his methodological project (such as we can reconstruct it). Descartes, at this early stage, simply proceeded to use his articulated version of the latticework, along with the doctrine of intuition and deduction, as a basis for constructing some specimen heuristic tips in rules 6 to 11. This point too will be of use in our reconstructing of the making of Descartes' method discourse, and it will be partially confirmed in the process.

Before turning to the heuristic rules in the *Regulae* we must briefly examine rule 5. Strictly speaking, this rule does not contain any detailed heuristic advice. Occuring before the articulation in rule 6 of the latticework vision, rule 5 simply states that any given question ('proposition') is to be reduced to simpler sub-problems; these are to be resolved in due order, and the overall solution should then be assembled by reintegrating, in order, the sub-solutions.[29] In a word, rule 5 tells us that problem solving involves analysis as well as synthesis. Descartes does not elaborate this point, because it is, after all, highly traditional, and because it represents the least innovative portion of his discourse.[30] The real discursive work in Descartes' presentation, and presumably the main locus of its novelty in his view, resides in the way he tries to elicit concrete heuristic rules from the 'analysis/synthesis' couple, *after* it has been articulated onto the details of the latticework vision.[31] We shall find that Descartes presupposes the existence and necessity of analysis and synthesis as loosely sketched in rule 5. His method consists in a further set of suggestions as to how analysis and synthesis are to be carried out. In his view these suggestions and aids are discovered in practice and can be ploughed back into practice to facilitate the solving of ever more 'complex' or higher 'order' problems.

In rule 6 one finds what is arguably the principal heuristic rule: reflection upon already mastered deductive series will lead to the acquisition of skill in pursuing new inquiries. It is the 'chief secret of the art' always to note the absolute term in question and the order of relations binding the relatives to it.[32] Here the entire heuristic method first gets off the ground with the suggestion that one study the contexture of some already known and available segment of the latticework. Many of the specific suggestions in the heuristic method follow from this admonition. For example, one will acquire insight into the classes of problems which can arise about the series under scrutiny, and, from that, one will develop sagacity in ferreting out the simplest routes of solution to given classes of these problems.[33] This sort of heuristic insight will be of use when one is again confronted with these sorts of problems, whether by themselves or as sub-portions of a more complex problem.[34]

Rule 7 is perhaps the most complex portion of the text on heuristic method, and for that reason contains important hints about how Descartes composed and inscribed his rules. Ostensibly it sets out a number of heuristic suggestions, presented as coherent, well motivated variations and extensions of the procedure called 'enumeration'. This surface order is quite specious, however. A close analysis of rule 7 would show that the text was composed as it were 'on the run', and that it contains traces of several shifts and alterations in Descartes' thinking as he came to grips with the necessity of writing down some heuristic rules. A full analysis of the inner dynamics and tensions in the text cannot be undertaken here, where we are mainly concerned to show how the heuristic suggestions, as stated, relate to the background of the latticework vision and doctrine of intuition and deduction. But to discuss the latter issue requires some attention to the internal complexities of the text. What follows is therefore an uneasy, and, I concede, unsatisfactory mixture: A catalogue of the heuristic advice (and its background) mixed with allusions to some of the discursive dynamics of the text.

'Enumeration' first appears when Descartes advises us to conduct numerous continuous reviews of completed deductive series. This process, he hopes, will lead in many cases to some sort of unitary validating intuition of the link between the widely separated first and last terms. This is needed to overcome the fact that since long chains of deduction take time, the certainty of the procedure to some degree relies upon the *memory* of previous deductive steps. Enumeration in this sense will minimize the role of memory, and, as indicated, perhaps eliminate it entirely in some lengthy deductions.[35] What seems to be happening here is that Descartes has discovered some problems about the 'certainty of synthesis' which are conditioned by the very terms of his discourse upon 'intuition', 'deduction' and the 'latticework'. He proposes to 'solve' these discourse-generated puzzles by adducing some additional heuristic advice, in this case the advice to try to telescope a 'series' of 'deductions' into a 'memory-evading' unitary 'intuition'.

Enumeration, now differentiated as 'sufficient enumeration or induction', also appears as the performance of a non-linear inference based upon the grasp of several independent chains of deductions.[36] This might be necessary, for example, in the final assembly of a synthesis (or in the sub-assembly of the solution of a sub-problem). Here, again, a problem of procedure emerges against the horizon of the discourse on latticework, intuition and deduction; and once again the solution is generated from within the discourse in the ostensible form of 'further heuristic advice', this time by adducing a new species of the genus 'enumeration'. Thirdly, from the first sense of enumeration Descartes literally slides into a new sense of enumeration (called 'enumeration or induction') as preliminary categorisation of the materials and means relevant to the solution of a problem.[37] And this usage, taken in the context of later portions of the rule, appears to subdivide further into senses of enumeration as: (1) the specification (during analysis) of the relevant sub-problems (and assessment of their solvability!), and as (2) the solution of such sub-problems.[38]

By this point Descartes is rather neatly entrapped by the momentum of his own discourse: little puzzles and issues about problem solving procedure keep popping up from within his grand discourse on intuition/deducation, analysis/synthesis, and latticework. The puzzles are laid to rest with the triumphant 'discovery' of further heuristic insights, a trick accomplished by multiplying and ramifying senses of 'enumeration' in terms of the resources of the puzzle generating discourse. One should note for future reference just how far these developments take Descartes from any arguably significant and useful concern with the real practice of any field of inquiry. He is happily churning heuristic rules out of his discourse; but, rules which enjoin us to identify and solve sub-problems hardly begin to tell us *how* to identify and solve them in any actual domain of inquiry. Here we begin to sense the illusory, indeed mythic, character of this discursive enterprise, an issue explored in systematic terms in Section 5 below.

Returning to the text of the *Regulae*, one finds more general advice in rules 8 to 11. For example, to train oneself to intuit well, one should start with simple matters (rule 9). Similarly, to learn to discern the orderly deductive texture of series, one should start with simple and to-hand examples (rule 10); one should not wander where deduction cannot lead, and one must learn to recognize when enumerative reviews need to be 'complete' and when merely 'sufficient' (rule 8).[39] Descartes insists at the end of rule 7 that virtually the whole of the [heuristic] method consists in these profundities.[40] Descartes has had little real difficulty in finding and formulating his dazzling collection of rules, because, as has been suggested, the rules are epiphenomena of his discourse on analysis/synthesis, intuition/deduction, latticework, series, absolutes, relatives and relations; that is, they are answers to puzzles the discourse might seem to entail; or, they are straightforward, textually conditioned remarks on what 'analysis/synthesis' involves when articulated onto the wonderful, new, method-relevant discourse.

4. DESCARTES' MANUFACTURE OF HIS METHOD DISCOURSE: LATE 1618 TO EARLY 1620

4.1. *Dating and Strategy*

Despite the unreliable character and literary embellishment of Descartes' autobiography in partie II of the *Discours*, it is widely accepted that his dating of the genesis and early growth of the method probably is accurate. Placing his initial insights about the method in November 1619, he further claims and implies that sustained work on it continued into the winter of 1619/20.[41] The *Olympica*, some early notes of Descartes partially preserved by his biographer Baillet, tend to confirm the dating and part of the content.[42] Therefore, it probably was in the winter of 1619/20 that Descartes first formulated some fundamental ideas about the method, such as, perhaps, the latticework conception, the enlarged doctrine of intuition and deduction and draft versions of some rules. Nevertheless, one need not follow the *Discours* to the next level of detail and accept its version of the order and content of these ideas. It is not necessarily the case that in late 1619 Descartes formulated just those four rules of method found in the *Discours*, nor need his views on the unity of the sciences have followed in the same order and with the same rhetorical elaboration as found in the *Discours*. Our reconstruction commences with the brute fact of the *dating* from winter 1619/20. The *substance* of the reconstruction will primarily depend upon the *Regulae*. In general I intend to show that the earliest portions of the *Regulae*, roughly rules 1 to 11, were composed in that crucial winter of 1619/20; that they constitute Descartes' earliest methodological treatise. In particular my strategy is as follows: Pursuing and partly modifying the important researches of J.-P. Weber, I shall identify two chronologically and conceptually distinct 'strata' in the text of rules 1 to 11, one dating from mid 1619, the other from the winter of 1619/20.[43] These strata, reflecting stages in the construction of Descartes' method discourse, will be placed in the context of what we know about Descartes' scientific and mathematical work and intentions in the period 1618—20. By so contextualizing the strata, we shall be able to confirm their dating (beyond Weber's arguments), and, more importantly, we shall be able to reconstruct in detail how and why Descartes came to manufacture his method discourse.

4.2. *Strata in Rule 4: Universal Mathematics and Universal Method*

The reconstruction begins with rule 4 of the *Regulae*, which was omitted when rules 1 to 11 were discussed in Section 3. Rule 4, entitled 'a method is necessary for finding out the truth', forms a bridge between the doctrine of intuition and deduction in rule 3 and the heuristic method of rules 5 to 11. Descartes writes that method consists in 'certain and simple rules' which if followed will prevent our assuming 'what is false as true' and our 'spending effort to no purpose'.[44] Distinguishing truth from falsity is the business of intuition and deduction, the first moment of the method; preventing the wasting of effort is the business of the heuristic rules, the second moment of the method.

Thus far rule 4 poses no difficulties, and offers no materials for our reconstruction. But when one looks more closely at rule 4 something odd emerges. Rule 4 consists of two autonomous sections which have simply been juxtaposed.[45] The two sections are almost identical in structure and flow of argument; but, they have two different subject matters. One section discusses 'method', the other discusses something called 'universal mathematics'. These two sections are now termed by Cartesian scholars rule 4A (method) and rule 4B (universal mathematics).[46] Neither section discusses the subject of the other. It is not stated that universal mathematics issues from the method, nor is it claimed that the method derives from universal mathematics. Nowhere are the two enterprises identified.[47] Rule 4A on method fits precisely into the flow of rules 1 to 11, as has just been shown. Rule 4B on universal mathematics finds no hint or echo in these same early rules.[48] Whereas rule 4B describes universal mathemaics as a discipline of limited scope, applied to mathematical fields only, rule 4A extolls the method in wildly enthusiastic tones.[49] Descartes writes that the method 'ought to extend to eliciting true results in every subject'; that it embraces 'an understanding of all that does not surpass the powers of man'.[50] Lofted away by his rhetoric, Descartes concludes 4A by proclaiming that,

> To speak freely, I am convinced that it is a more powerful instrument of knowledge than any other that has been bequeathed to us by human agency, as being the source of all the others.[51]

It is especially noteworthy that rules 4A and 4B, so different in content, tone and linkage to the surrounding text, display nearly identical structures of argument. Descartes points to the futility of disorderly studies (mathematical studies in 4B; studies in general in 4A); he alludes to innate seeds of truth from which the discipline in question can grow (method in 4A, universal mathematics in 4B); and he discerns intimations of the discipline in question within the history of mathematics. So, neither discipline is derived from the other; rather, both are derived independently by means of analogous arguments.[52]

All this would seem to signify the following: Rule 4B on universal mathematics was written prior to rule 4A and provided the model for its structure of argument.[53] Rule 4B was part of a projected treatise on universal mathematics, a treatise probably abandoned in favour of the *Regulae*, a treatise on method. Rule 4B is in effect a fossilised relic of that treatise on universal mathematics, encysted within the text of the *Regulae*.[54] When did this happen? Tentative conclusions, to be confirmed below, run along these lines: The conception of universal mathematics in 4B would certainly seem to pre-date November 1619, when, we presume, Descartes arrived at the grander, all encompassing notion of a universal method. If, as now seems likely, he tried to model important passages in his methodological treatise upon crucial sections of the draft of his discussion of universal mathematics, we could perhaps date rule 4A (as well as 1 to 11 generally) from the winter of 1619/20. We therefore need to know more about what this universal mathematics was and whether it indeed dated from before November 1619.

Descartes writes in rule 4B that universal mathematics is supposed to embrace the axioms, principles and methods common to all properly mathematical fields. It is described as the science of 'order' and 'measure' wherever they appear in the various mathematical disciplines.[55]

'Measure', as we shall soon see, plausibly denotes 'quantity in general', the abstract object with which one deals after one has abstracted from the particular mathematical objects of the particular mathematical disciplines. 'Order', we shall see, plausibly connotes a concern with finding general schemas of analysis for problems, once they are stated in such abstract terms. All of this is quite reflective of neo-Platonically inspired ideas about a 'general', 'common', or 'universal' mathematics which were widely canvassed in the late sixteenth and early seventeenth centuries. But Descartes' broad reference to this debate and tradition cannot further clarify the nature of his 'universal mathematics', nor can much else be gleaned from rule 4B.[56] In order to comprehend, and date, the universal mathematics of rule 4B we must consider Descartes' natural philosophical and mathematical researches in 1619. These will provide a convincing context for rule 4B, and further support its dating just prior to November 1619. More importantly, once universal mathematics is dated and defined, it will be possible to show precisely how the method discourse in rules 1 to 11 was constructed on its basis in the winter of 1619/20.

4.3. *The Natural Philosophical and Mathematical Context of Rule 4B: Descartes' Researches in 1619*

While garrisoned in Breda in November 1618, Descartes met Isaac Beeckman, a Dutch physician, schoolmaster and engineer, eight years his senior. With Beeckman he served a second apprenticeship in natural philosophy and mathematics which decisively shaped his subsequent work. With the possible exception of Thomas Harriot, Beeckman was virtually the first man in Europe to conceive of what was to become the new 'mechanical philosophy', a combination of classical atomism with the belief that atoms (or corpuscles) move according to mathematical-mechanical laws.[57] Beeckman never systematized his speculations; he was no builder of systems and he preserved his piecemeal thoughts on mechanism in his *Journal*, which the young Descartes examined. In 1618 Descartes and Beeckman were youthful, enthusiastic and full of swagger, and they had trouble differentiating between their aspirations and their actual achievements. They insisted that their corpuscularian explanations could be rigorously mathematical, and with that in mind, they termed their project not 'mechanism' but 'physico-mathematics':

the combination of true, that is corpuscular, ontology with mathematical analysis and demonstration.[58] Taking physico-mathematics in this strict sense, one finds that they produced no convincing examples of this discipline at the time. Indeed Descartes never achieved a mathematical mechanics in the sense of a Galileo, Huygens or Newton, let alone a mechanics genuinely applicable to the corpuscular realm. When, after 1628, Descartes turned to the construction of a full-blown system of corpuscular mechanism in *Le Monde*, it turned out to be qualitative and discursive. But, back in 1618—19 Descartes and Beeckman apparently believed that they possessed promising intimations of their physico-mathematics in their piecemeal work on music theory, hydrostatics, mechanics and the law of fall.[59] It is their aspirations and intentions which are important for our reconstruction.[60]

An example of their enthusiasm for physico-mathematics is contained in one of Descartes' fragments from the period, a text which we can term 'the hydrostatics manuscript'.[61] Beeckman had requested that Descartes explain in proper physico-mathematical terms Simon Stevin's teaching concerning what is essentially a special case of the hydrostatic paradox. In his *Elements of Hydrostatics* (1586, 1608), Stevin had demonstrated this result by employing his favoured, rigourously 'Archimidean', style of mathematical argument. He applied *reductio ad absurdum* arguments showing that conditions of static equilibrium obtain between specified macroscopic volumes and weights of water (and of a notional solid of equal specific gravity).[62] By contrast, Descartes' attempted 'physico-mathematical' gloss (the 'real' explanation, he thought) is a *bricolage* of *ad hoc* assertions about the corpuscular structure of fluids and the characteristic motions of their constituent particles.[63] His matter and motion discourse is enfolded in allusions to the existence of a purported system of 'mechanics' which supposedly controls his assertions.[64] And the text is further illustrated with purported geometrical representations of the supposedly underlying patterns of corpuscular 'tendencies to motion' (a favoured 'explanatory' technique in his mature system as well).[65] The details (crucial, I think, for understanding the style of his later natural philosophy) need not detain us here. The significance of this sort of exercise for the emergence of 'universal mathematics' was that Descartes apparently believed that 'physico-mathematics' demanded, and in principle could receive, a properly mathematical, general procedure for stating and resolving problems in corpuscularian physics.

In addition to 'physico-mathematics', Descartes also pursued mathematical researches during the year prior to November 1619.[66] He took a rather instrumental tack at first, devoting considerable effort to devising compasses which would generalize and solve geometrical and algebraic problems. One compass in particular deserves notice here (Figure 1). It is fundamentally a device for generating magnitudes in continued geometrical proportions, and it is the same instrument introduced twice into the *Géométrie* nineteen years later.[67]

In the *Géométrie* Descartes was to use the compass to show that the curves generated by the motion of points *D, F* and *H* are of increasing orders of complexity and can be represented and analysed by means of algebraic equations reflective of those orders.[68] In 1619 Descartes did not attend to the curves, though he named some of them, nor did he represent them algebraically. Rather, he focused upon the articulation of the limbs and branches of the physical compass. He saw that many geometrical and algebraic problems could be reduced to problems of determining magnitudes in continued geometrical proportions, and that those so reduced could be modelled to the architecture of the compass

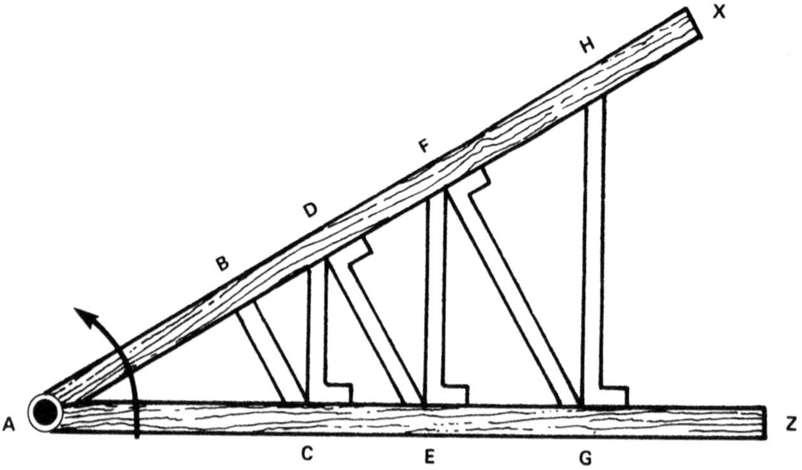

Fig. 1. *Descartes' proportional compass: 1619/1637: AZ* and *AX* pivot at *A*. *BC* is fixed normal to *AX*. All the other rulers are moveable (their extensions beyond *AX, AZ* are not shown. By similar right triangles:

$$\frac{AB}{AC} = \frac{AC}{AD} = \frac{AD}{AE} = \frac{AE}{AF} = \frac{AF}{AG} = \frac{AG}{AH}$$

and solved. I contend that this manner of exploiting the compass was to serve as the veritable 'examplar' for his emerging doctrine of 'universal mathematics'. Let us explore in some detail how this came to pass.

Descartes' use of the compass, and his aspirations for extending and generalizing analytical techniques through it, first emerge clearly in a letter to Beeckman of 26 March 1619.[69] Among other things he announces that with the aid of his compass he has found solutions ('demonstrations') of the thirteen 'permissible' species of cubic equation falling within the three general classes of cubic. That is, equations of the form $x^3 = \pm ax^2 \pm c$; $x^3 = \pm bx \pm c$; and $x^3 = \pm ax^2 \pm bx \pm c$, omitting $x^3 = -ax^2 - c$; $x^3 = -bx - c$ and $x^3 = -ax^2 - bx - c$.[70] Descartes' working notes of the time, preserved in the *Cogitationes privatae*, show how he hoped to use the proportional compass to produce these 'demonstrations'.[71]

The compass first appears in connection with the solution of the equation $x^3 = 7x + 14$. Descartes commits a curious error, arguing that after reducing the equation to the form $(x^3/7) = x + 2$, he will solve $x^3 = x + 2$, and then 'multiply x^3 by 7'.[72] A similar error runs through the notes in this section of the *Cogitationes.* It seems to arise from Descartes' intention to generalize his solutions,[73] for he proceeds to show how $x^3 = x + 2$ can be solved on the compass. *AB* is taken as a unit and the compass is opened until *CE* is equal to two units; then *AC* will be the root.[74] Descartes probably envisioned this procedure as a model for the general 'demonstration' of equations of the form $x^3 = bx + c$, as mentioned in the letter. But he apparently did not notice that the compass cannot be applied to any 'species' of the equation in which a negative term appears on the right-hand side.[75] He seems rather to have been intrigued by the possibility of generalizing his method of solution. This is apparent both from his haste to eliminate the coefficient of x^3 in the example given, and from his claim in the letter to Beeckman to be able to extend his 'demonstrations' to all thirteen permissible cases of the cubic, despite the fact that in this example the cases $x^3 = -bx + c$ and $x^3 = bx - c$ cannot be solved by his instrumental method.

One can conclude, therefore, that by March 1619 Descartes had moved beyond the purely 'geometrical' use of the compass for finding mean proportionals. He had seen, in part mistakenly, that the compass also permitted the solution of certain classes of algebraic equations, provided they could be interpreted as proportions and unfolded on the compass.[76] Descartes' use of the compass in order to transcend the immediate geometrical or algebraical statement of problems, and to reduce them to common forms of relations among proportional magnitudes, was soon to constitute the technical core of universal mathematics (in so far as it had one) and to provide the exemplars for its discursive elaboration. This, it will now be shown, must have occurred between late March and November of 1619.

4.4. *The Construction of Universal Mathematics in 1619*[77]

In March 1619 Descartes did not yet possess the concept of universal mathematics, as expressed in rule 4B. In his letter to Beeckman of 26 March Descartes did not discuss a unified analytical discipline, but rather a loose compendium of analytical tools.[78] In addition, physico-mathematical problems are not mentioned as belonging to the scope of the programme. Universal mathematics could not have crystallized until Descartes envisioned a unified discipline embracing all mathematically articulated fields, physico-mathematics included. What we know of Descartes' researches and aspirations in 1619 begins show how universal mathematics might have emerged between March and November of that year.

In the first place, it is not difficult to see how Descartes could have moved to subsume physico-mathematical problems within the intended scope of his compendium of analysis. Even before March 1619 Descartes was convinced that physico-mathematics should in principle depend upon properly mathematical modes of representation, analysis and demonstration. He need only have realized that his compendium had to include tools suitable for physico-mathematics. Although we have no evidence on this score between March 1619 and rule 4B, the small but logical move to include physico-mathematical problems would have constituted the key step between the letter to Beeckman and rule 4B, as far as physico-mathematics is concerned.

The more important issue is how did Descartes transform, between March and November 1619, the plan for a compendium of analytical techniques into the conception of a unified analytical discipline, universal mathematics? Here we do have hard evidence. The answer would seem to reside in Descartes' enthusiastic pursuit of what to him would have been the 'natural' implications of his researches with the compass.

The early work with the proportional compass convinced Descartes that certain sorts of algebraic and geometrical problems were subject to identical types of analytical treatment, provided that they could be rendered in terms of relations between proportional magnitudes: The proportional magnitudes could be represented or instantiated on the limbs and branches of the compass, and general solutions could be discovered in the form of procedures, literally manipulations, for the unraveling of the relations between known and unknown line lengths. To extrapolate further from such observations would be to speculate that perhaps problems about all types of quantities could and should be rendered in the abstract in terms of relations between some 'magnitudes in general', so that, as a consequence, very general schemas for unpacking classes of relations might be derived and applied. Such a speculation is, of course, very close to the meaning of rule 4B and to our suggested readings of 'order' and 'measure'.[79] If Descartes read such meanings into his compass researches, he would have been well on the way to displacing the earlier idea of a compendium of analytical techniques with the idea of a universal mathematics. It is just such a reading which could have taken place under the influence of the pre-existing debates about the nature of a 'common', 'general' or 'universal' mathematics. Descartes' personal vision of universal mathematics most likely crystallized when, reflecting on the culturally available but vague theme of a 'universal mathematics', he saw expressed in the compass researches concrete exemplars for the central concepts of that discipline: (1) an exemplar for the concept of 'magnitude in general', the abstract, generalized quantity capable of subsuming and representing the objects of any mathematical field, whether 'pure' or 'physico-mathematical'; and (2) an exemplar for the conception of generalized analytical procedures consisting in the identification of the classes of structure (or 'order') of 'relations' holding among such generalized magnitudes when they represent various classes of problem. Rule 4B would therefore seem to be the culmination of a course of research, speculation and discursive elaboration which led Descartes from physico-mathematics and his early compass researches, through the letter of 26 March 1619, down to the fabrication of universal mathematics prior to November 1619.

The universal mathematics of rule 4B was not a viable discipline. It had only the most tenuous links to the curious explanatory exercises which Descartes apparently believed to be genuine instances of it. The researches with the compass were, on the plane of mathematical practice, really quite limited and laced with errors recognizable by other contemporary algebraicists. To create universal mathematics Descartes combined the tenuously grounded discourse on physico-mathematics with partial, tendentious glosses of his compass researches. The amalgam was perfected under the influence of the culturally available discourse on 'common' or 'universal' mathematics, and in turn gave a new, welcome but entirely specious density to that discourse. Thus did he create his version of universal mathematics; or, to be precise, thus did he cobble together a text on the theme 'universal mathematics'. In a word, the universal mathematics and rule 4B is a textual artifact, and, unlike other things sometimes discussed in texts, only a textual artifact.[80]

From the perspective of the young Descartes, however, the situation was quite different. Universal mathematics was no doubt a promising, realizable project, an undertaking emerging naturally from his previous course of research and reflection, and holding out the opportunity to subsume and rationalize those researches while setting them in their proper disciplinary matrix, a matrix which would demonstrate, as against previous inconclusive speculation, just what the prized universal mathematics was to be.

4.5. *The Making of Cartesian Method Discourse: Winter 1619/20*

In 1619 Descartes apparently did no further work on universal mathematics. This was because by November 1619 universal mathematics was overtaken and subsumed by the grander project of the method. Descartes' method discourse was constructed in its main lines by means of a precisely identifiable extrapolation of the central elements of his discourse on universal mathematics, elements which were themselves wildly optimistic metaphorical extensions of aspects of his earlier work, as we have just seen. Recall that at the core of Descartes' teaching on method in the *Regulae* one finds a discourse about intuition and deduction, about 'absolute' and 'relative' terms, about 'series' and 'relations', and about the latticework. Rules 6 to 11 announce heuristic rules, forged against the template of that discourse. It will now be suggested that Descartes' core discourse on method, as well as a number of the heuristic rules, resulted from the analogical extension of terms and rules constitutive of Descartes' discourse on universal mathematics.

In rule 6 there is a remarkable remnant of the sort of process of analogical extension which seems to have produced the method discourse. Near the end of rule 6, the rule which articulates the latticework vision, there is a little mathematical example concerning a series of numbers in a continued geometrical proportion:

$$\frac{3}{6} = \frac{6}{12} = \frac{12}{24} = \frac{24}{48}$$

Such a series is, of course, typical of the sort of object to be treated in universal mathematics. Indeed, it would seem that in universal mathematics all problems were supposed to be reducible to this type of structure. In rule 6 Descartes uses the series to illustrate some of the general heuristic rules he is in the process of unveiling. So, one is advised to intuit the basic defining ratio of the series, and to inspect the order of numbers sequentially generated by reiterated application of the ratio. From this procedure will follow, predictably, insight into how problems about the series may be classified, and hence how to choose the simplest routes of solution.[81] By implication, one could go on to apply other heuristic rules to the tasks of posing and solving problems about this series, and increasingly complex series of which it forms a part.

Descartes offers this example as an illustration of the method, and indeed it serves this purpose well. The mathematical structure of the series illustrates the more profound latticework of truths. The intuition of the defining ratio and the inspection of the structure of the resulting proportions mirror the suggestion to inspect the structure of an already mastered deductive series, that is, to mark the absolute and the order of relatives subordinate to it. From inspection of the mathematical series there flow methodological insights which parallel some rules of the heuristic method. But the very perfection of the example raises the issue of whether it (or another similar case) was not the very model upon which Descartes erected the core of his method discourse. Consider these translations between 'methodological' terms and 'mathematical' terms involved in the example (or in any putative instance of universal mathematics): For the methodological concept of the 'absolute term', *read* 'defining ratio applied to an initial number'; for 'relative terms' *read* 'subsequently generated numbers in continued geometrical proportion'; for the grandiose 'latticework of rational truths', *read* 'orderly interlinked series of numbers in continued geometrical proportions'; finally, for the edifying heuristic rules of the method, *read* 'concrete but fairly trivial pieces of advice concerning the solution of problems arising about such series in continued proportion'.[82] Everything we have previously seen about the genesis and structure of universal mathematics reinforces the conclusion that the 'methodological' terms are the analogues of the 'universal mathematical' terms. Descartes, moving beyond the exciting idea of universal mathematics in rule 4B (itself the product of an overhasty crossing of his mathematical and physico-mathematical interests), analogically extended universal mathematics and happily concluded that *all* knowledge consists in structures of logically related elements; that all such structures have basic and (orderly generated) derivative terms; and that all the heuristic insights adhering to typical examples in universal mathematics can thereby be transformed into generally applicable heuristic rules of method.[83] The method discourse was not directly abstracted from successful practice in some area of mathematics.[84] It was produced by a megalomaniacal performance of operations of analogical extension upon the terms of a discourse, universal mathematics, which itself could not do what it purported to do.[85]

All this permits us to look again at the question of dating. In Section 4.2 it was shown that rule 4A is closely related to rule 4B, being in fact modelled upon it. In Sections 4.3 and 4.4 we saw that rule 4B fits precisely into the context of Descartes' work and aspiration in mid 1619. It has just been argued that rules 1—3 and 5—11 are closely linked to the universal mathematics which emerged during the course of 1619. Just as 4A was modeled upon 4B, so the other rules appear to be based upon the analogical extension of core concepts of the very universal mathematics which 4B discusses. It therefore appears likely that rule 4A and the text of which it is the pivot, rules 1—3, 5—11, date from the winter of 1619/20 and constitute Descartes' first detailed version of his method as it grew out of, and beyond, the project of universal mathematics.[86]

In summary, then, we have seen that during 1619 Descartes was swept forward on a mounting wave of enthusiasm concerning the method. He started with Beeckmanian physico-mathematics, itself an overblown project, and with some interesting moves in mathematical analysis. He was driven on, first to the dream of a universal mathematics, and then, in November, to the founding insights of the method. It is little wonder, therefore, that his famous three dreams of 11 November 1619 display a nearly mystical state of enthusiasm over his recent insights, and that in interpreting the dreams he so eagerly read divine sanction into his project. (Given what we have uncovered about how he produced and transformed this series of discourses, we have here a particularly fine case of false consciousness concerning the provenance and character of theoretical discourses.)

5. THE STRUCTURE OF DESCARTES' METHOD AS MYTHIC SPEECH

5.1. *Introduction*

We can now return to the question posed at the beginning of this chapter: 'How can it be that Descartes and others apparently believed in the reality and efficacy of a method which most post-Koyréan historians of science are convinced cannot have worked'? My fundamental claim is that the vacuity and sterility of the method *and* its appearance of effi[illegible]y are, as it were, opposite sides of the same coin. That 'coin' is the way Descartes' method discourse is structured onto several interacting levels. Descartes' method cannot possibly do what it claims to be able to do, because, as discourse, it has a particular structure; and yet it is that very structure which can create and sustain illusions or literary effects about the efficacy, applicability and unity of the method.

The analysis begins from a naive but fundamental premiss: In order for the rules of the method to be considered *efficacious* in the practice of a given field of research, the rules have to be applied and deployed within inquiry in the target field in ways adequate to the proclaimed goals and foci of the method.[87] Granting this point, it would seem that Descartes must give some arguably adequate account of the contents and workings of the target field and that this account must be couched in terms supplied by his methodological discourse, his talk of 'absolutes' and 'relatives', of 'relations' and 'series', which are aspects of the latticework. This is necessary because the heuristic rules are claimed to apply to entities of this type. The heuristic rules, after all, were formulated directly in terms of, and are clearly 'relevant' and 'applicable' to the discourse about 'absolutes', 'relatives', 'series' and 'relations'. Approaching a target field with his method, Descartes must be able to construe that field in terms relevant to the use and application of his rules. Descartes, as we shall soon see, accepted these conditions and worked within them. He thought his results exemplified the efficacy of the method. I intend to show that his method discourse produces only literary effects of its own efficacy, unity and applicability, and so traps Descartes (and other believers) in illusions about its own efficacy.

5.2. *A Kuhnian Presupposition: Methods Cannot Account for Scientific Practice*

Before looking at Descartes' attempt to apply his method, I have to make clear exactly where I stand on the possibility of rendering, by means of a method discourse, an 'arguably adequate account of the contents and workings of a target field'. I hold that such methodological accounts, adequate in the senses required, cannot be achieved. This I maintain broadly on the basis of the Koyré-Kuhn dispensation in the historiography of science, as buttressed by the relevant writings of Bachelard, as well as by recent developments in the 'interpretive' or 'cognitive' sociology of scientific knowledge.[88] The key here is Kuhn's insight that the elements — metaphysical, conceptual, evaluative and instrumental — which make possible a particular tradition of paradigm based normal research are *sui generis* to that field and sufficient for its pursuit. And since the conceptual structures and modes of practice of living fields are *sui generis*, to redescribe, gloss or translate them into some other framework is simply to reinterpret them for some 'outside' purpose. Practice proceeds from within each field's proper framework and not in terms of glosses provided by putative single, transferable methods.[89]

These points are illustrated in Figure 2. Any given field of science has its own paradigm, its own versions of the generic elements displayed in the matrix: (a) basic concepts and law sketches; (b) metaphysical commitments (that is 'relevant' aspects of 'relevant' 'external' discourses of any sort); (c) tools and instrumentalities (including the theories and standards thereof); (d) standards of relevance and of

CONCEPTS (C)	METAPHYSICS (M)	TOOLS (T)
STANDARDS (S)	AIMS (A)	EXEMPLARS (E)

Fig. 2.

adequacy for the selection of projects and for the formulation and evaluation of knowledge claims; (e) disciplinary goals of any accountably 'internally' or 'externally' generated sort; (f) concrete law-, concept-, and standard-instantiating achievements, Kuhnian 'exemplars' or Bachelardian *phénoméno-techniques*, if you like.[90] At any give moment the scientific domain may then be further represented as in Figure 3, where we have *n sui generis* fields, each with its own particular constellation of matrix elements, constituting for the time being its own paradigm. The sciences are thus many, not one. True, neighboring and cognate fields may share certain elements in common; findings or concepts in one field may be taken up (under translation) as tools or instruments in another; or, groups of fields may have emerged or begun to develop under a more or less common metaphysical umbrella. But, none of this argues the identity or even long term convergence between or among paradigms. Similarites between some elements of two or more matrices always fall short of producing identical paradigms.

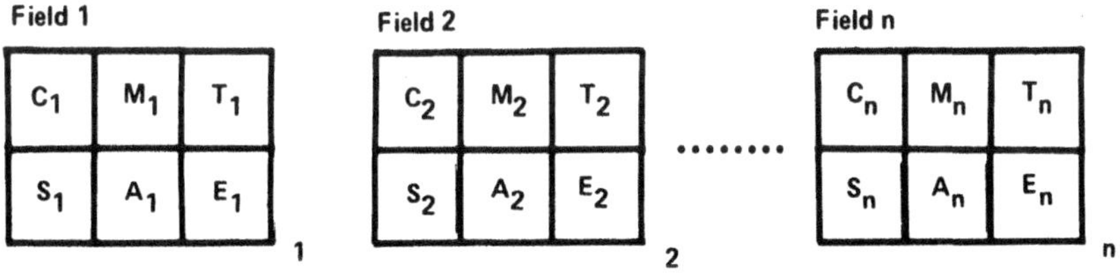

Fig. 3.

One may therefore speak, in Koyréan-Kuhnian terms, of each field having its own method(s), inextricably embedded in its own matrix. To speak of a common, single transferable scientific method is to substitute abstract and necessarily ineffective talk for the lived, thought and practised life of each of the fields. Indeed, the generalizations we may make about the content and practice of *Science* have all the limitations of the generic reference matrix in Figure 2: It abstracts from concrete instances of content, and intra-field relations between matrix elements, and so loses touch with the particular cognitive-practical density of each of the *n* fields. In the remainder of this chapter we are not concerned with further supporting the contention that method disourses, such as Descartes', must fail adequately to gloss or redescribe target fields; rather, we explore the mechanisms which explain that failure and which also explain the creation of the illusion or literary effect that no failure has occurred.

5.3. *An Example of Descartes Methodologizing a Field of Inquiry*

Let us now consider the fate of one field of inquiry when Descartes redescribed it in the terms of his method. Descartes provides a nice example with what may be termed the 'science of magnets' which he discusses twice in the *Regulae*.[91] The example is especially interesting because what Descartes says about 'doing' magnet science will apply to any area in which he attempted to provide corpuscular-mechanical explanations. Methodologically speaking, Descartes instructs us first to isolate a fixed set of experimental data about magnets, in practice the experiments reported in Gilbert's *De magnete* (1600). We are then to inquire into the 'intermixture' of 'simple natures' which will explain the magnet.[92] Here the absolute natures or terms surely are primitive geometrico-mechanical elements, corpuscles, with their properties of size, shape, hardness and state of motion or rest. What, then, are the 'relatives' in this case? Descartes terms them 'intermixtures' of absolutes. That conveys the image of some set of complex corpuscular-mechanical models, models for the structure of lodestones, magnetic 'effluvia', magnetisable bodies etc.[93] Hence in this case Descartes' methodological gloss of 'magnet science' does depend in a loose sense upon the required specification of 'absolutes' and their studied 'complexification' into sets of 'relatives'. And these absolutes and relatives are arguably the sorts of entities which did eventually enter into his detailed explanation of magnetism, set forth later in his *Principia philosophiae* of 1644. The mythologist, however, must ask an embarrassing question: 'Has Descartes provided here an adequate redescription and gloss of what it took to construct corpuscular-mechanical explanations of magnetism (or of anything else for that matter)?'

Modern Cartesian scholarship has begun to give us some answers to the question of what really was involved in Descartes' formulating and inscribing of corpuscular-mechanical explanations, and something approaching a consensus now exists in the literature:[94] In non-methodological contexts and later in his career, Descartes seems to have come to hold that although the details of the corpuscular-mechanical world system cannot be fully deduced from 'first principles', there are some absolutely certain metaphysical principles, for example, that the essence of matter is extension, which place constraints upon what can and cannot be asserted of any detailed corpuscular model designed to explain a particular class of phenomena. Available evidence, and in particular, the 'facts' to be explained, also have an important bearing on the formulation of the detailed explanatory models. Hence the models have a necessarily hypothetical character, and at best can achieve only 'moral certainty'. When, in his later works, Descartes spoke of 'deducing' phenomena from his principles, he did not mean the strictly mathematical deduction envisioned in his central methodological texts, but rather 'deduction' in the looser contemporary acceptation of 'plausibly explain'.[95]

Such, then, were Descartes' own later and more considered views on the production of corpuscular-mechanical discourse. Although they show that his strict methodological views bore little relation to the procedure, they do not quite do full justice to what we might now term the interpretational complexity and fluidity of his project and the indexical character of virtually every move within it. Imagine an ethnomethodologist of science transported back in time to observe Descartes as he attempt to produce and inscribe a piece of corpuscular-mechanical discourse. Our temporal interloper would probably have identified three interacting moments in Descartes' performance. His field notes might read as follows:

(1) Logically and temporally prior to the construction of any particular explanation, Descartes tried to devise and legitimate his basic metaphysical principles which would constrain and condition the formulation of specific corpuscular-mechanical models. Such principles included his fundamental definitions of matter and mind, and his basic laws of motion, collision and the behaviour of directional tendencies to motion.[96] Needless to say, the production of Cartesian conclusions in metaphysics and dynamics, and their legitimation, are not amenable to clear, consistent, rule-bound procedural glossing. One might parody his own account of his procedures and say that 'God only knows how he did it'. Nor is it clear how and in precisely what sense the models should be 'constrained' in any given case. That, too, can only be a matter of on the spot interpretation and 'negotiation', if only with himself!

(2) 'Relevant' empirical evidence has to be selected, weighed and 'appropriately' deployed and described. Evidence can include 'facts' needing explanation, or 'facts' lending credibility to the explanatory model offered (including 'facts' purporting to weaken the credibility of competing explanations). It is not clear that Descartes has procedures for accomplishing these tasks which are any more rule-bound than those ongoing negotiations and 'constructions' of facts and arguments revealingly studied in the twentieth century by Woolgar and Latour, and Collins, Pickering and Pinch.[97]

(3) In the light of the 'evidence' and the metaphysical 'constraints' a specific corpuscular-mechanical model for the phenomena in question has to be constructed. Given the un-methodological character of the proceedings under (1) and (2), it is not to be expected that Descartes' inscribing of characterizations of particular models is a method-bound activity. Consider, additionally, that in Descartes' usages the meaning of 'deduce' in the phrase 'deduce the phenomena from the model' is fluid and reinterpretable. As if this were not enough, there is also the point that each specific model has ultimately to 'fit' into a 'system' of natural philosophy. This raises a host of additional interpretive challenges which reflect back upon the way in which a model is to be constructed. For example, Descartes, the 'systematizer' always asks, 'To what degree does a particular model 'comport' [itself a fraught word] with other specific models within the system in respect of (a) consistency of mode of metaphysical constraint;[98] (b) similarity (or difference) of explananda in view,[99] and (c) the degree of structural, cosmological 'interplay' intended to hold between these models in the overall system of the world machine.[100]

Not to put too fine a point on it, in the field notes of our ethnomethodologist, (1), (2) and (3) constitute a complex undertaking. Each moment involves discursive practices, interpretations, weightings and selections for which no rules were ever given, and from which it is not plausible to imagine any workable and consistent full set of rules could ever be elicited by *post-facto* glossing.[101] *A fortiori* these moments elude Descartes' own recommendation in effect to 'find the absolutes and the structure of relatives and use my heuristic rules in so doing'. Descartes' methodological discourse on how to 'do' magnet science bears no significant relation to whatever it is he must have been 'doing' in order to accomplish magnet science; that is, to inscribe his typical corpuscular-mechanical texts on the subject. The recommendations embedded in his method talk could never have led to the construction of this corpuscular-mechanical discourse about magnets. Here, the methodological recommendations which he gives in the *Regulae* simply batten upon the prior accomplishment of (1), (2) and (3), and upon the uncodified and arguably uncodifiable body of discursive practices which underpin them at every step. The methodological version of what it is to do magnet science drains the practice of magnet science of its actual *sui generis* procedural density (whatever that might have been); and, having drained that density, the method discourse poses as the real basis of Descartes' practice.

Nobody should really be surprised by this. Very few serious students of the history of scientific practice would expect that Cartesian method discourse (or any systematic method discourse) could adequately gloss the content and practice of a field of research. And yet, an important fact presents itself at this point. Descartes' methodological story about doing magnet science has a curious property: Certainly it fails adequately to redescribe its target; but, inadequate as the story is, the rules of the method do seem to apply to it and mesh within it. Consider that in the *Regulae* Descartes discusses the problem of explaining magnetism, and his discourse proceeds by using the terms 'absolute', 'relatives', 'relations' and 'series' ['intermixture']. So, as far as 'just discoursing' goes, one can gear the rules of the method right in to facilitate the investigation; that is, to facilitate the continuation of our little methodological story about how we really should 'do' magnet science. Descartes' redescription of magnet science, his little account of what we should do, is just the sort of text into which additional talk about the rules and their use can be inserted 'convincingly' and 'coherently'.

What then should we make of the fact that the rules of the method can be applied to and deployed within Descartes' account of doing magnet science? The answer depends upon whether we are believers in the method, or are post-Kuhnian mythologists trying to demystify it. Believer and mythologist can agree that the rules 'go together' with the methodological account. But, the mythologist sees that the redescription is a phantom, a parody of what must have gone on when Descartes 'did' corpuscular-mechanical natural philosophy. The situation of the believer, however, is very different, especially if the believer is an early modern methodologist, like Descartes, who did not have the benefit of *our* contemporary critical tools through which to articulate what we mean by saying that the dynamics of the inscribing corpuscular-mechanical explanations were very complex, and quite probably ultimately elusive. Our early modern methodologist (or our surviving contemporary methodologists of various ilks) might have easily overlooked the slide between the practice of corpuscular-mechanism and the methodological glossing of that practice. Or, better expressed, he is unlikely to have overlooked it, because, he probably had no discursive tools through which to thematise it in the first place. He had virtually no resources for explicating his practice other than those offered by a *methodological* discourse of one sort of another, a generic type of discourse, moreover, which enjoyed cultural prestige and was accorded epistemological precedence in the attempt to explicate or 'account for' intellectual practices. Now, if you overlook the structural slide; or if it is just unthematisable for you, then you are bound to be impressed by Descartes' method story, and in particular, you are bound to be impressed by the wonderful way the rules of his method apply beautifully within the story.

Any believer who reaches such a level of conviction would then be in a position to engage in a pair of characteristiç behaviours, quite reasonable from his viewpoint, which reflect his belief in method and would serve to reinforce it. Having constructed the method story about magnet science, or having consumed it from an 'authoritative' source, the believer could then do the following:

(1) He could practise corpuscular-mechanism, more or less proceeding in the way discovered by our time travelling ethnomethodologist, while monitoring his actions to himself and/or to others in terms of the little method story. This would lend credence to the method by seemingly attaching it to segments of the practice as they occur in the flow described by the ethnomethodologist. Such 'voicing over' or real time monitoring of practice could be institutionalized in pedagogy to entrench both the method in question and the theories embedded in the routines so glossed.[102]

(2) He could practise corpuscular-mechanism, and then after the fact claim in good faith that, although what he had done might not fully reflect the dictates of the method, he could in principle have accomplished the same things by strictly adhering to it.[103]

Both (1) and (2) are very easy to accomplish. The method story which the methodologist recites while working, or after the fact (and 'in principle' or 'in fact'), is simple to construct, because it mainly utilizes the core of the method discourse; and, it is easy to accept, because the tame methodologist (and his tame audiences), almost by definition, either have no alternative critical frameworks available for theorizing about practice, or, if they have them to hand, they are for one reason or another unwilling to deploy them.

5.4. *The Structural Levels in Descartes' Method Discourse*

By means of the example of Descartes' magnet science, we have uncovered the basic structural units of his method discourse, indeed the basic structural units of any systematic method discourse. The discussion in the previous section shows that we are dealing with three levels of discourse: The systematic 'core' of the method discourse, the *sui generis* target field, and the methodological accounts, glosses, redescriptions and stories which can be manufactured about the latter, using the resources of the former. Figure 4 is a map of these levels and it is also a guide to the places where the characteristic illusions or literary effects of Descartes' (or anyone else's) method discourse are generated, between and across levels, as we shall soon see.

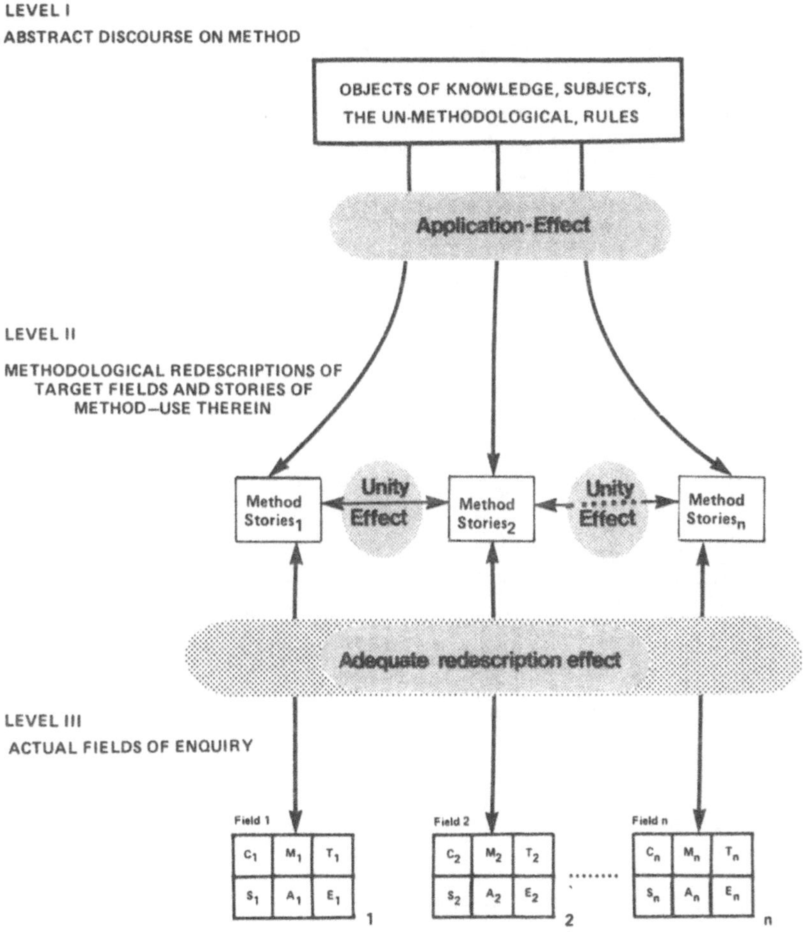

Fig. 4. The structure of Cartesian method-discourse.

Level I is the level of systematic discourse about the method, what has been termed the 'core' of Descartes' method doctrine. It includes, in Descartes' case, the teaching concerning intuition and deduction, the concept of the latticework, the interconnected conceptions of 'absolute' and 'relative' terms, 'series' and 'relations', and, in addition, the statements of the rules of the method.

Level III has been met previously; it is that representation of the *sui generis* character of scientific fields, viewed in Koyréan-Kuhnian perspective which was presented in Figure 3. Here what is represented is the domain of living fields which are candidates for being target fields of the method in question. Magnet science would be one of these fields, and we shall soon meet another of the fields which Descartes believed could be commanded by and through his method. So, once again, we have the set of matrices, each one schematically representing the existence of a discrete, efficacious field of inquiry. Inside each matrix there are symbolised what one might take to be the contents and structure of such a field — its own basic concepts, metaphysics, tools and instruments, standards of relevance and adequacy, goals and exemplars. The point, again, is to represent the neo-Kuhnian thesis that the 'method(s)' of any such field are *sui generis*, for they are inextricably bound up with the contents and structure of the relevant matrix. The aim of the representation is not to establish these points; but rather to pose them as the framework for exploring the characteristic discursive phenomena produced by the overall structure of the method discourse.[104]

Level II consists of a set of methodological redescriptions or accounts of the corresponding fields represented on Level III. The example of magnet science shows that the Level II redescription is formulated using the core methodological terms available on and constitutive of Level I, as well as the rules of the method, which are also part of Level I and which are woven into the resulting story or account.

5.5. *The First Two Structural 'Effects': 'Adequate Redescription' and 'Application'*

If we return to the case of magnet science, bearing in mind the structure in Figure 4, we can begin to locate the sites at which a method discourse generates illusions or creates literary effects concerning its own efficacy, and we can specify its *modus operandi* at those sites. That is, we can examine how the structure of a method discourse contributes to its dynamics as a mode of mythic speech.

The key to the mythological operation of Descartes' method discourse resides in getting the reader/listener to operate on Level II, where he basks in the methodological version(s) of the target field(s) with which he is concerned. On Level II the rules of the method to 'apply to the redescriptions offered'; ongoing work on Level III can be glossed in terms of Level II, or *post-facto* 'accounting' for practice on Level III can be offered in terms of Level II stories. It is crucial that the Level II story both embodies the rules of the method and disastrously eviscerates actual Level III practice. Indeed these two characteristics of Level II stories may be seen as interacting in the very constitution of such stories: Level II stories eviscerate actual Level III practice because they are episodic, fabular versions of the core methodological terms available on Level I. They are fairy tales of methodological comportment spun out of materials and scenarios available on Level I. But, of course, the rules of the method only mesh into such stories because the stories derive from Level I.

One may therefore state that when a reader or listener is confronted with a Level II redescription or story of rule-following, he is in danger of succumbing to two structural illusions or literary effects characteristic of systematic method discourses. Firstly, he may be taken in by the 'adequate redescription effect', producing the illusion that Level II redescriptions are in any sense adequate to Level III. Secondly, he may be taken in by the 'application effect', producing the illusion that the application of the rules in the Level II story is (or could be) the application of the rules to the practice of the target field (Level III). These effects are *structural* in the sense that they are made possible and sustained by the relationships between the three levels of discourse. That structural arrangement also simultaneously explains why a method discourse, such as Descartes', must be inadequate and ineffective in real practice. Its upper two levels work together to suppress from view any given Level III field, and to pose in its place a dessicated phantom of its actual structure and practice. That phantom (the Level II redescription or account) is then solidified and underwritten by its 'obvious' congruence with the grandiose, self-proclaimedly authoritative core discourse on Level I. Since the very process of manufacturing, solidifying and underwriting phantoms on Level II is the source of the effects, we may well reassert the central thesis of this chapter: Descartes' method discourse is a species of mythic speech whose discursive structure renders it vacuous whilst simultaneously sustaining powerful illusions that it is not.

It may well be asked how a reader/listener ever gets to Level II in the first place; what types of reader/listeners are at risk of falling victim to these effects; and what exactly are the dynamics of their acquiescent reading/listening? These are crucial problems which I have begun to address elsewhere.[105] Here it must suffice to say that two broad lines of inquiry are involved. Firstly, one can ask in general how stories or accounts of putative matters of fact come to be heard as true even prior to, and possibly completely independently of any attempt at further inquiry. I suggest that ethnomethodological investigations of the strategies used in making sense of everyday accounts and narratives are relevant here.[106] Secondly, the general theory of construing of accounts needs to be supplemented in any given historical case by an inquiry into the conditions and determinations which turned the historical actor in question into a particular sort of reader/listener. A believer in a method is someone who has chosen or been led to choose (in some wide social-historical senses of the term) to have no other resources for explicating scientific practice than those offered by the method of his 'choice'. We must try to supply an account of the historical context and biographical trajectory which made that method available to him and 'preferred' by him. Analytically speaking there is the historical problem of explaining the election of a method by an actor, and then there is the general problem of explaining how a method discourse functions upon an actor, once he is 'inside' it.[107] In the case of Descartes, how and why he formulated his particular method is an historical problem; how his method could be sterile and yet appear not to be is a structural problem it shares with other method doctrines. To explain how and why Descartes could believe in such a method is a function of both inquiries taken together.

5.6. *The Third 'Effect': the 'Unity' of a Method Discourse*

Thus far the analysis has concerned the relationship between any given target field on Level III and its phantom redescription on Level II. We have yet to consider the consequences of the fact that a method discourse such as Descartes' can generate across Level II a range of redescriptions, each one corresponding to a different target field (cf.

Figure 4). A general method, after all, has to be able to command more than one area of research. In any given systematic method discourse, each and every Level II redescription will be couched in terms of Level I elements. In the case of Descartes these involve reference to 'absolutes', 'relatives', 'series' etc. Hence all the Level II redescriptions in a given method discourse will appear to be similar, although we shall see that serious equivocations are introduced as Level I elements are deployed in several Level II redescriptions. In addition, any Level II story then produced about the use of the method will involve an account of the application of the rules of the method to the redescribed field. Unsurprisingly, the believer in Descartes' or some other method will find that the rules of his favourite method gear into each and every redescription offered across its Level II. With that realization, the application effect gives birth to the unity effect — the illusion that the rules of the method are applicable to some set of discrete fields of inquiry.

In the *Regulae* Descartes provides splendid illustrations of these points. He not only offers his methodological version of magnet science, but also his methodological version of mathematical analysis. He writes that in this case the 'absolute' is the straight line, the coordinate, the 'relative' is a curve.[108] Presumably we are to refer a curve to straight lines by means of equations expressing the nature and degree of its 'relativity' to the coordinates. These clear and distinct algebraic representations of curves then become the vehicles for further inquiry into the properties and relations of curves. The entire discipline is pursued under the aegis of an emerging algebraic theory of equations. Before exploring the unity effect by comparing Level II magnet science with Level II mathematical analysis, we must first consider how this *Regulae* version of mathematical analysis fares in relation to the adequate redescription effect (and hence in relation to the application effect). This will further illustrate points made above in Section 5.5, and in addition it will show the relevance of our earlier reconstruction of the development of Descartes' method discourse.

Recall Descartes' heuristic rules, such as 'note always the absolute in question and the order of relations binding the relatives to it'; or, 'break down questions into simpler parts and resolve them in due order proceeding from the simpler to the more complex'. These rules not only *seem* to be transcriptions of maxims used in resolving equations, they were indeed elicited from parts of Descartes' mathematical practice. As we have seen, the grandiose concepts of the 'latticework', of 'absolute' and 'relative' terms, of 'relations' and 'series', were all derived by analogy from the elements of the sort of little mathematical problems informing 'universal mathematics'; that is, from problems about quantities in continued geometrical proportions, such as the one at the end of rule 6, or the ones which so fascinated Descartes in 1619.[109] Descartes' heuristic rules of method were likewise constituted as methodological analogues of maxims which were useful and to-hand in dealing with such little problems. So mythologists of method can agree with believers in Descartes' method: (1) that the heuristic rules of the method do give a *post-facto* summary of some aspects of the domain of mathematical analysis as Descartes knew it, and (2) that Descartes' entire Level I discourse is at least minimally adequate and applicable to the small domain of mathematical problems which formed the basis of 'universal mathematics', because the Level I discourse and the rules are simply further analogical developments of that base.

However, any believer in Descartes' method would surely wish to go much further than this. Echoing Descartes' text, he would issue a stronger claim: *If* coordinates are 'absolutes' and curves are 'relatives' describable by equations which can be solved using the rules of the method, *then* the method adequately constitutes in principle the entire field of geometrical analysis, by, of course, rendering it part of the domain of application of the method. Descartes would have us believe that if one were to set up such a version of geometrical analysis, he could then use the rules of the method to expand the field further (and to account for its previous achievements as well). This, in historical and mathematical point of fact, is not the case; and, as our structural analysis shows, cannot be the case.

The heuristic rules of Descartes' method were simply too vague and too limited to be of use in the further expansion of the domain of geometrical analysis. The heuristic rules did not, and could not help Descartes or anyone else move beyond the resolution of just the trivial sorts of problems which formed the basis of 'universal mathematics' and of Level I of the method discourse. Nor can they explain the prior development of the field in any way other than through a tendentious and implausible Level II account, a methodologist's 'rational reconstruction'. Consider, for example, that in Book Three of the *Géométrie* (1637) we have Descartes' own statement of the procedures and techniques of contemporary algebra and theory of equations. He summarized, rationalized and extended the rapidly growing early seventeenth century 'tool kit' of algebraic techniques. He showed how to evaluate the number and signs of the roots of an equation; discussed real and imaginary roots of an equation; showed how to alter roots by common factors; how to reduce varieties of equations to canonical forms etc.[110] It is obvious that the heuristic rules of the method are mildly reflective of a small amount of this material. But in themselves they cannot account for the full richness of the theory of equations or its development to that contemporary point of perfection. Algebra, which was increasingly seen as a general theory of equations, was developing within a largely *sui generis* and rapidly changing realm of properly algebraic conception, expression and technique (it was a Level III discourse and practice).[111] The heuristic rules redundantly echo some few of the findings. However, they furnish no significant insight into how the art had developed, nor how it could be further extended.[112]

In short, Descarters did not formulate an adequate Level II version of mathematical analysis. This was despite the fact that his Level I rules were indeed applicable to a small portion of that domain, and the fact that his entire Level I discourse was an analogical transcription of that small portion. So, although Descartes' method-discourse was based upon a sub-domain of mathematical practice, this did not prevent his Level II version of mathematical analysis from suffering from the adequate redescription and application effects, and it has led to much misunderstanding of the real scope and efficacy of his method.

Now we can compare Descartes' Level II version of mathematical analysis to his Level II version of magnet science. Everyone immediately sees that the rules of the method 'apply' to both Level II stories, and that the stories have a certain resemblance: They are both about the 'relations' of 'absolutes' to 'relatives' etc. etc. Mythologists realize, however, that this resemblance is superficial and misleading. It is based on disastrous equivocations about the terms 'absolute', 'relatives' and 'series'. The words are the same, but they are badly abused. Depending upon how one uses the term 'absolute', as coordinate or corpuscule, we then suitably frame our 'relatives' as equations or models. But coordinates bear to curves or equations vastly different 'relations' than basic corpuscles bear to physical conjunctions thereof.[113] *Within* each story there is some unity of expression about absolutes, relatives and relations, whilst across versions there is nothing but an almost comical equivocation.[114]

How can it possibly be that a believer does not see these equivocations? How is it that he falls for the unity effect?

In Section 5.5 it was suggested that a believer in a method tends not to see, or at least not to thematise, the slippage between Level III and Level II, what was termed the evisceration of Level III in favour of of its Level II redescription. Once located on Level II, and lacking any motive or machinery for removing himself from that comfortable perch, the believer may be confronted with two (or more) Level II redescriptions, corresponding respectively to two (or more) target fields. He is now likely to be impressed by the observation that these Level II versions are 'rather similar' if not identical. To the extent that he can think of the target fields in independent, Level III, terms, he is likely to say that the methodological versions of the fields render them 'more alike' than the non-methodological accounts (which accounts may be further stigmatized for their lack of 'methodological articulation').

Now, it is worth noting that even for mythologists there is some truth in this. The Level II versions are indeed more similar than any serious accounts which might be given of their Level III matrices.[115] Level II accounts do not lie about each other in the way they each lie about their respective Level III targets. The Level II accounts are honest (that is both leading and misleading) analogues of each other. They are of the same discursive fabric, Level I, and there are legitimate things to be said about their analogies. All this, of course, greatly aids the unity effect, which in itself is no better, but certainly no worse than any other categorical grouping of instances by 'similarities'. It is the structural location of this effect which makes it an accomplice of the really delusionary two prior effects.

Returning to the believer, we find that he is much impressed by the discoveries that the Level II versions have a wondrous similarity and that the rules of his method gear into each and every one of these accounts. The method truly is 'unitary': For him it reveals the underlying similarity of the fields in its domain, and it empowers him, *via* its widely applicable rules, to command and pursue efficaciously any and all of these fields. To the extent, if any, that such a believer becomes aware of any contrasts or tensions between any Level II version and its independently characterizable target field, he can mobilize a discourse on the 'practical problems of applying the method'; for, as Descartes and other methodologists are wont to say, 'the method consists more in practice than in abstract statement after all'.

As in the case of the first two structural effects, the reason for this third effect is the obverse of the reason why the method cannot actually work in the ways it claims to work. There is a similarity between Level II versions and a gearing of the rules of the method to them *because* those versions are woven out of the cloth of the Level I discourse; but, for that very reason these Level II versions cannot hope to be adequate glosses of the structure and dynamics of living, Level III fields; they eviscerate those fields in the interests of Level I and still necessarily equivocate amongst themselves.

In general, then, the literary effects of Cartesian (or any other) methodology relate to each other in this manner: In any systematic method discourse the adequate redescription effect is fundamental, and it ultimately depends upon the plausibility of Level II stories within a context of cultural precedence accorded to the Level I discourse. The application effect depends upon the adequate redescription effect, for it is the illusion that the application of the rules on Level II is the application of the rules in actual practice. The unity effect results from the iteration of the application effect across the spectrum of fields thought to be commanded by the method in question, and it is facilitated by the fact that Level II entities bear 'legitimate' analogical relations to each other. Note also that there is a 'progress effect': Methodologists can proudly point to 'progress' as new Level II accounts of new domains are added, as the method is 'extended'. They can also label as 'progress' the revising of Level II accounts of old domains in order to 'grasp' and 'explain' new developments in those already 'methodized' fields. Finally, in conclusion, we can also see that any believer who gets this far into the hall of discursive effects will probably see himself as in a legitimate business, the business of prescriptive methodology. He can then devote countless hours to explicating everything that has happened so far. He can expatiate on the unity, applicability, efficacy and progressiveness of his method. He can refine and explicate Level I, his Level II stories and the rules; he can castigate other methodologists and those who do not believe in methodology; and he can comment upon all these matters at ever higher levels of meta-discourse.

6. TOWARD A GENERAL THEORY OF METHODOLOGY AS MYTHIC SPEECH[116]

The structural study of Descartes' method presented in this chapter is intended to illustrate a more general claim, hinted at several times in the text: That all systematic doctrines of method possess the same discursive structure and hence can perpetrate upon their adherents the same set of literary effects concerning their application, efficacy, progress and unity. In this section I want to comment briefly on this general claim and give an illustration of its application.

6.1. *The Generic Structure of Method Discourses*

In moving away from the particular case of Descartes' method, we must first of all note a point which was conveniently neglected earlier in the discussion. It is that all method doctrines encountered in the Western tradition from Aristotle to Popper and beyond are structured around two intertwined core metaphors: (1) to acquire knowledge is a matter establishing a correct subjective grasp, or more typically, *vision* of independently existing, objective objects of knowledge; (2) method, drawing on the literal Greek meaning of the term, is the subject's 'way through' to the objects of knowledge, an ordered set of prescriptions as to the path to be followed by the subject in pursuit of knowledge, where the rules pertain to the initial discovery or grasp of the objects and/or to the task of validation, confirmation or testing of the pedigree of the objects initially grasped.

All particular method doctrines are attempts to fill in this core metaphorical picture, and all are deeply conditioned by the heavy cultural loading of the terms of the base metaphors. The subjects encountered in method doctrines are still largely the implied subject of Aristotelian logic (and physics): Sober, honest, sense organs and brain in good working order, unbiased, especially by nastly social, economic or political contingencies which can only spoil the project. Whether this subject takes on a Cartesian, Popperian or Lakatosian coloring, it is still the classical subject of true and rational knowledge, variously clothed. Similarly, the objects of knowledge in method discourses are curiously reminiscent of the furniture of Aristotle's cosmos. They form a unique, objective, independently existing collection of types, in principle knowable by the reasoning subjectivity, provided all is in good moral and physiological order, the proper philosophical objectivity is maintained and the rules followed. Whether the objects of knowledge are Cartesian 'simple natures', Millian facts or Popperian 'basic statements', laws, theories and the logical relations holding amongst them, these objects share these happy epistemic properties and are in principle always convivially available to the well behaved, rule obeying subject.

The history of method doctrines is in large measure the history of varying and often competing attempts to dress these core notions in conceptual vestments deemed appropriate to the methodologist's perception of the context of debate and structure of socio-cognitive relevances holding in his time and place. Typically, bits of older method doctrines, as well as pieces of neighboring varieties of discourse — theological, natural philosophical, ethical, mathematical, psychological — are the stuff out of which a new doctrine is fabricated.

Leaving the core metaphors to one side, because they are always present and operating upon the elaboration of the discourse, we can turn again to the three levels of discourse which we discovered in Descartes' method doctrine, and which I now suggest are manifested in every concrete instance of a systematic methodology. Hence, in any particular example of the species 'systematic method discourse' we encounter, first, a Level I which will consist in (1) generalized [non-discipline specific] statements of the rules [or some of the rules] of that method, and (2) explicit, more or less systematized, abstract and generalized discourse concerning the canonical themes, 'knowing subjects' and 'objects of knowledge'. The main thrust of the Level I discourse is to establish that the aim of the method is the acquisition, discovery and/or evaluation of these 'objects' by means of the subject's correct deployment of the rules. There is typically present also (3) some discourse on the 'pitfalls', 'obstacles' and 'sources of error' which can deflect a subject, mask or distort the objects, or lead to misapplication of the rules. The sorts of problems thematised in (3) are very much functions of the contents and hidden assumptions of (1) and (2). Sometimes, in addition, all of Level I is packaged within a metaphysical (or even theological) framework.

Level III in any systematic method doctrine consists in just those living *sui generis* fields of inquiry to which the method supposedly applies. In any such method doctrine Level II will consist in a set of stories of methodological practice, of discovery and/or evaluation, corresponding to the set of fields on Level III. It matters not at all whether we are dealing with what contemporary philosophy of science would term a method of discovery or a method of evaluation, the former issuing in tales of discovery on Level II, the latter generating tales of successful, objective and progressive evaluation and appraisal of hypotheses, theories or research programmes. In either case the stories or accounts analytically proceed as follows: First the corresponding Level III field is redescribed or glossed in terms of the elements provided by Level I of the particular method in question; next, the account or story of practice (of discovery and/or evaluation) is woven by reference to an actor or actors (conceived in Level I terms) applying the rules within the redescribed field. That is, actors conceived in terms of Level I 'subjects' are portrayed applying the rules to Level I 'objects of knowledge' embodied in the Level II gloss of the 'essence' of the target field.

Hence Level II stories can only exist in so far as the use of the method's Level I conceptual resources gives them shape. In any given method the stories are specific, episodic unfoldings of the conceptual resources present on Level I as elements in a generalized discourse on method. From these conditions follows the possibility of the method generating its characteristic literary effects.

The foregoing points also explain an aspect of Level II stories which is highly symptomatic of their mythological character. In such stories there is never any real difficulty in the 'subject' applying the rules of the method to the 'objects' in question. The interpretation of the rules is always clear, uncontroversial, unambiguous, unchallenged and un-negotiated — most unlike those significant rules we know about in the real world which do gear into real material practices. In the Level II accounts the rules are so easy to follow, apply and interpret because, of course, nobody is actually doing anything with them except inscribing or hearing a story about them, using the usual (largely tacit) common-sensical procedures for speaking/hearing coherent stories.[117] But the believer in a method is trapped in a structure, a hall of semantic mirrors which produces misleading literary effects; and although he behaves quite rationally in constructing or in hearing the accounts, he must be misled about the nature of the practice in question. His 'possibly true' Level II account is in fact a fairy tale; its truth and the truth about the efficacy of the rules are precisely definable literary effects, which ironically (but necessarily) have their basis in quite ordinary and reasonable everyday practices of speaking/hearing stories.

The critical questioning of any Level II story is very difficult for a believer, because 'acceptable' non-destructive lines of potential criticism are already mapped out by the typifications of the 'un-methodological' on Level I.[118] In general, in the absence of access to resources for radically questioning a Level II story, or for interpolating methodologically significant alternative accounts, the account will be heard as quite possibly true. By 'radically questioning', moreover, is meant more than a denial of the circumstantial facts of the matter recounted in the story which leaves the method doctrine unscathed as such. Method stories allow for the circumstantial facts to be questioned but not the method doctrine itself. Hence the common ploy which says of a Level II account, 'Yes, perhaps things did not quite happen that way, but the point is they could in principle have happened in that manner, and *that* constitutes the grounds for accepting the claim in question'. Any residual, thematised gap between story and counter-account is filled up by discourse on the theme: 'Methods are always difficult to apply in practice, the "real" [read Level II story as criticized] can never fully match up to the ideal [read Level I]'.[119]

6.2. *Feyerabend's Pioneering Study of Methodology as Mythic Speech*

Near the end of Section 1, I mentioned that my approach to methodology as mythic speech has been heavily influenced by Feyerabend's dissection of the mechanics of Newton's methodological claims in his paper 'Classical Empiricism'. We can now examine Feyerabend's argument in an attempt to bring out the implicitly structural character of his analysis, and thereby also to show how his analysis can be mapped onto my general schema.

Feyerabend is concerned with the ways in which Newton's proclaimed empiricist method enabled him to package and defend his theory of physical optics. In Feyerabend's version of Newton's empiricism, the essential methodological rule is: 'The sole basis for the derivation (induction) of theory is "experience"',[120] 'Experience' is the stable, objective, repeatable, public foundation from which theories may be rigorously induced (hence this is part of Newton's Level I). Experience, moreover, requires no interpretation or justification. A methodological corollary, expressed by Newton in versions of his 'fourth rule of philosophising', states that theories derived from experience are not to be challenged on the basis of 'speculative' alternative 'hypotheses'.[121] (This is a Level I counter-discourse on the 'unmethodological').[122]

What interests Feyerabend is the essential vacuity of Newton's central methodological rule (the parallel of our first two 'effects'), for the rule itself states nothing about what experience is; what is to count as experience; how it is to be located; how, if located, it is to be interpreted, etc. Newton's (Level I) pronunciamentos name a privileged basis for theory without being able to specify how that basis is to be produced, recognized, evaluated or inferred from. This produces a situation in which any institutionalized teaching about what is to count as 'experience' can be used by believers to fill the criteriological void.[123] Feyerabend then shows how Newton summons 'experience' into existence by a two stage semantic slight of hand. First we have the paradigmatic experimental illustrations of the theory which are, of course, thoroughly theory- and standard-laden. The 'results' of these experiments, what Newton or we might perceive in them, are first identified with what Newton calls 'phenomena'. But 'phenomena' in his usages are to us highly generalized, idealized and differentially weighted observation reports. Then Newton claims that such 'phenomena' are the 'experience' which we seek as the sole basis of theory.[124] That a theory does seem to follow from such 'experience' is hardly surprising, Feyerabend observes, since such experiences already have the logical form of laws,[125] being general propositions well loaded with the terms of the very theory to be 'induced'. (This, I would suggest, can be seen as a very neat instance of the application effect, the convincing 'realization' that the rules of one's method do apply in the realm of the 'authoritative' method story which one has spun about one's practice using the resources of Level I.)[126]

6.3. *Some Loose Ends*

I have suggested that the structural study of Descartes' method can be extended to any example of a systematic method discourse. In particular I have pointed out how Feyerabend's demystifying treatment of Newton's proclaimed method fits the schema developed in Section 5. Elsewhere I have analyzed methods of evaluation (Popper, Lakatos) in similar terms.[127] But none of this is meant to imply that the structural study of method discourse has reached a mature stage. My free borrowings from Barthes, Feyerabend, Bachelard and Sacks reveal that my models and its materials are very tentative and exploratory. Many problems, and opportunities, remain.

Level II of method discourses seems to be the site of a number of these problems and opportunities. The machinery used by believers in the construction and construal of Level II stories needs to be explored more fully, and perhaps from perspectives beyond the ethnomethodological ones to which I have alluded. Any method's system of such stories and their inner articulation may well be amenable to the tools of structural anthropology and textual criticism. Indeed the entire vast edifice of methodological discourse stemming from the seventeenth century might form an object of study in this regard, the edifice being taken as something akin to a Levi-Straussian myth 'system'.

The need for an ethnomethodological or interpretive sociological approach to Level II stories is further brought out by a point made above in passing (in Note 89): Level II stories can and do appear on Level III as resources deployed in the debates and negotiations which actually make and unmake knowledge claims. For the believer in a method, appeals on Level III to Level II stories will be highly privileged and impressive. These stories will be the only discursive elements in play on Level III which for the believer properly label themselves as 'methodological'. If there is a large measure of agreement in principle about which method is the correct one, then the actors will see the deployment of its Level II stories as further evidence of the rightness and efficacy of their method (although 'allowable' accounts will have to be given of how one's opponents can support erroneous claims despite their apparent commitment to the correct method). If there is no wide agreement about the correctness of one given method discourse, or if two or more are competing, then the debate should be marked by strenuous attempts to establish the superiority and correctness of each party's pet method. In either case, one expects method discourse(s) to appear as resources and as objects of discussion in much the way other chapters in this volume have found them to be.[128]

In sum, then, what I should now like to term the structural/sociological study of method discourse is only just beginning. It invites historians of science and sociologists of scientific knowledge to explore, adapt, revise and deploy the resources of a number of cognate disciplines, in order to get to grips with the nature and history of method discourse. In this chapter I have tried to show how some rough and ready tools, hastily assembled, may help to crack that 'hard case' in the history and historiography of method constituted by René Descartes and his loyal scholars.

ACKNOWLEDGEMENT

For their insightful comments on earlier drafts of this chapter I would like to thank Richard Yeo, Alan Gabbey, Homer Le Grand and Rod Home. Research undertaken for this chapter was supported in part by a grant from the University of Wollongong Research Grants Committee. I would also like to thank Randall Albury, Geoff Sharpe, Don Miller and Henry Krips for their critical comments on oral presentations of the arguments in this chapter.

University of Wollongong, Australia

NOTES

[1] G. Milhaud, *Descartes Savant* (Paris, 1921).

[2] L. Brunschvicg, *Les étapes de la philosophie mathématique* (Paris, 1922); P. Boutroux, *L'Imagination et les mathématiques selon Descartes* (Paris, 1900); P. Mouy, *Le développement de la physique cartésienne 1646–1712* (Paris, 1934); J. Klein, *Greek Mathematical Thought and the Origin of Algebra* (London, 1968). These works share a serious concern with relating Descartes' mathematical and scientific projects to his methodological pronouncements, and they generally display a realistic perspective on the tensions which existed between various of Descartes' intellectual pursuits.

[3] For example, A Gabbey, 'Force and Inertia in the Seventeenth Century', in *Descartes: Philosophy, Mathematics and Physics*, ed. S. Gaukroger (Brighton, Sussex, 1980), pp. 230–320; A. I. Sabra, *Theories of Light from Descartes to Newton* (London, 1967); A. E. Shapiro, 'Light, Pressure and Rectilinear Propagation: Descartes' Celestial Optics and Newton's Hydrostatics', *Studies in History and Philosophy of Science* **V**, 1974, pp. 239–96; E. J. Aiton, *The Vortex Theory of Planetary Motion* (New York, 1972); H. J. M. Bos, 'On the Representation of Curves in Descartes' *Géométrie*', *Archive for History of Exact Science* **XXIV**, 1981, pp. 295–338; and Bruce Eastwood, 'Descartes on Refraction: Scientific vs Rhetorical Method', *Isis* **LXXV**, 1984, pp. 481–502.

[4] In general histories of science Descartes' place in the scientific revolution is, of course, acknowledged, but his work is too often summarily treated under the rubrics of 'methodology' and/or 'the new mechanical philosophy'. Elementary textbook accounts, such as H. F. Kearney, *Science and Change 1500–1700* (New York, 1971), R. S. Westfall, *The Construction of Modern Science* (London, 1971) and B. Easlea, *Witch Hunting, Magic and the New Philosophy* (Brighton, Sussex, 1980), each in their own different ways place a salutory stress on the sudden and contextually conditioned emergence of Cartesian mechanism.

[5] Descartes' *Discours de la méthode* appears in *Oeuvres de Descartes*, C. Adam and P. Tannery (ed.), XIII vols., Nouvelle Presentation en co-edition avec C.N.R.S. (Paris, 1974), vol. VI (hereafter cited as AT, VI, with page reference). The outstanding critical edition is E. Gilson, *René Descartes, Discours de la méthode, Texte et commentaire* (Paris, 1930) in which the pagination of the text of the *Discours* corresponds to that in AT. The standard English translation is in *The Philosophical Works of Descartes*, E. Haldane and G. Ross (ed.), II vols. (Cambridge, 1970), vol. I. (hereafter cited as HR, I, with page reference).
[6] This (fictional) discipline is not to be identified with Descartes' much later analytical geometry, nor with his slightly earlier 'universal mathematics'. See below Note 17 and Sections 4.4 and 5.6.
[7] AT, VI, pp. 20—2; HR, I, pp. 92—4.
[8] For example, elsewhere I have reconstructed Descartes' discovery of the law of refraction and his devising of a construction for all the 'solid' problems of the ancients (a constructive solution to all cubic and quartic equations). (J. A. Schuster, 'Descartes and the Scientific Revolution 1618—1634: An Interpretation', 2 vols. (Ph.D. diss., Princeton University, 1977), pp. 124—49, 299—352.) In neither case is any light shed on the course of research by appeal to the supposed application of Descartes' method. Not surprisingly, analogous remarks apply to that supposedly paradigmatic case of the application of the method, the discovery of the explanation of the formation of the primary and secondary arcs of the rainbow. (Descartes, *Les météores*, AT, VI, pp. 325—44). This was an exceedingly good piece of 'normal' science, the solution to an age old puzzle in geometrical optics, a solution conditioned by the tools, standards and concepts of the discipline. Descartes' sole advantage over others was his possession of an exact law of refraction, which now served, as laws often do, as a tool of further research.
[9] I have begun to generalise from Descartes' case to systematic method discourses of any type in 'Methodologies as Mythic Structures: A Preface to the Future Historiography of Method', *Metascience: Annual Review of the Australasian Association for the History, Philosophy and Social Studies of Science* **I—II**, 1984, pp. 15—36.
[10] In Roland Barthes, *Mythologies*, A. Lavers (tr.), (St. Albans, 1973), pp. 109—59.
[11] Paul K. Feyerabend, *Against Method* (London, 1975) and *Science in a Free Society* (London, 1978).
[12] P. K. Feyerabend, 'Classical Empiricism', in R. Butts and J. Davis (eds.), *The Methodological Heritage of Newton* (London, 1970), pp. 150—70.
[13] My earliest suspicion that there was a specific mechanism of mystification involved in methodological accounts was aroused by reading Gaston Bachelard's *Le rationalism appliqué* (Paris, 1949), where he deals with the systematic role of traditional philosophical perspectives (not particularly methodologies) such as empiricism, rationalism, conventionalism, in producing a structured series of illustory pictures of how theory and practice (or 'applied rationalism' and 'technical materialism') interrelate in the constitution of mature mathematico-experimental sciences. Cf. D. LeCourt, *Marxism and Epistemology — Bachelard, Canguilhem, Foucault* (London, 1975), pp. 41ff.
[14] Elie Denissoff, *Descartes, Premier théoricien de la physique mathématique* (Louvain, 1970), pp. 28, 30, 96—8, and *passim*.

[15] The sceptical reader should consult the following: L. Liard, 'La méthod de Descartes et le mathématique universelle', *Révue philosophique* **X**, 1880 pp. 569—600 (p. 573); J. Sirven, *Les années d'apprentissage de Descartes* (Paris, 1928), pp. 349—53; Gilson, *op. cit.* (Note 5), pp. 196, 222, 180—1; H. Gouhier, *Les premières pensées de Decartes* (Paris, 1958) p. 62; G. Gadoffre, 'Introduction historique' to *Descartes' Discours de la méthode* (Manchester, 2nd ed., 1961) p. xxxviii; R. Lefèvre, *La vocation de Descartes* (Paris, 1956) pp. 145, 149, 152; J. -L. Allard, *Le mathématicisme de Descartes* (Ottawa, 1963) pp. 28, 30, 143; J. Chevalier, '*Le Discours de la méthode*', *Archives de philosophie* **XIII**, 1937, pp. 1—13 (pp. 6—7); L. J. Beck, *The Method of Descartes: A Study of the Regulae* (Oxford, 1952), p. 198; W. Röd, *Descartes' Erste Philosophie* (Bonn, 1971), p. 18, note 8.

[16] On this point see any competent developmental and contextual study of these texts, for example, on the *Discours*, Gadoffre, *op. cit.* (Note 15), or Denissoff, *op. cit.* (Note 14); on the *Regulae*, J.-P. Weber, *La constitution due texte des Regulae* (Paris, 1964).

[17] Cartesian scholars will, of course, appreciate that the *Regulae* are unfinished, having been abandoned at rule twenty-one in 1628. This, however, does not entail that the *Regulae* contain only part of the core of the method. I have argued in detail elsewhere that the *Regulae* consist in a number of textual strata, the earliest of which (rules 1 to 11 dating from 1619—20) contain the essence of the method doctrine. The later strata deal not so much with method as with what Descartes termed 'universal mathematics', a discipline he attempted to articulate in 1626—28 on the template, as it were, of his method doctrine of 1619—20. The situation is further complicated by the fact that the method was itself an analogical elaboration of Descartes' initial conception of 'universal mathematics', dating from 1618—19, as is discussed below in Sections 4.4 and 4.5. See J. A. Schuster, 'Descartes' *Mathesis Universalis: 1619—1628*', in S. Gaukroger (ed.), *Descartes: Philosophy, Mathematics and Physics* (Brighton, Sussex, 1980), pp. 41—96, which develops the line of analysis first advanced by J.-P. Weber, *op. cit.* (Note 16).

[18] AT, VI, pp. 11—2. Gilson's commentary on these passages runs to seventy pages, bringing out their metaphorical and rhetorical elaboration *op. cit.* (Note 5), pp. 155—228.

[19] AT, VI, pp. 11 1.13 — p. 12 1.16 and Gilson's commentary, *op. cit.* (Note 5), pp. 157—62; AT, VI, p. 19 1.6—16.

[20] AT, VI, p. 17 1.11 — p. 18 1.23 (the appeal to logic and mathematics as models and the statement of the first rule of the method, cf. below Note 22).

[21] AT, VI, p. 18 1.24 — p. 19 l. 5 (rules 2, 3 and 4); p. 20 1.25 — p. 21 1.6 (elaboration of precepts in the development of the method through mathematical applications).

[22] AT, VI, p. 18 1.16—1.23: 'Le premier était de ne recevoir jamais aucune chose pour vraie, que je ne la connusse évidememment être telle: c'est-à-dire, d'éviter soigneusement la précipitation et la prévention; et de ne comprendre rien de plus en mes jugements, que ce qui se présenterait si clairement et si distinctement à mon esprit, que je n'eusse aucune occasion de le mettre en doute.' Cf. Gilson's commentary, *op. cit.* (Note 5), pp. 197—204.

[23] On 'enumeration' and 'fictitious order' see below pp. 45—6 and Note 26.

[24] Without wishing to foment pedantic controversy and spliting of hairs, one might set out the following parallels: *Discours*, rule 1, corresponds to *Regulae*, rule 3; *Discours*, rule 2, corresponds to *Regulae*, rules 5 and 7; *Discours*, rule 3, corresponds to *Regulae*, rules 5 and 6; *Discours*, rule 4 corresponds to *Regulae*, rule 7. Cf. F. Alquié, *Descartes,*

Oeuvres Philosophiques (Paris, 1963) t.I, p. 587 Note 1; and J.-P. Weber, *op. cit.* (Note 16), p. 64.

[25] AT, X, pp. 181–182; HR, I, p. 15.

[26] The grounds for introducing the non-Cartesian term 'absolutely absolute' are as follows: Later in rule 6 Descartes indicates that what is absolute and what is relative may vary depending upon the task in question: 'Herein lies the secret of this whole art, that in all things we should diligently mark that which is most absolute. For some things are from one point of view more absolute than others, but from a different standpoint are more relative.' (AT, X, p. 382; HR, I, p. 16)

This implies that Descartes intends a notion of 'methodological' order, created by the subject in furtherance of his cognitive interests, and apparently independent of any 'ontological' order there may be of absolute and relative terms. J.-L. Marion and others take this as Descartes' definitive view of the matter. (Cf. Marion, *Sur l'ontologie grise de Descartes* (Paris, 1975) section 'La fiction de l'ordre', pp. 71–8.) But one should note that Descartes clearly credits the existence of an objective, let us say 'cosmic' order of truths, given for the subject to explore. In the very next passage he discusses the existence of some terms which are absolute in respect of any series in which they play a part (AT, X, p. 382 l.25–26). They are, as one might say, 'absolutely absolute'. Consider, furthermore, that Descartes has already said that 'absolute' applies to such things as 'causes' and 'essences' (text cited above in Note 25).

[27] Relatives are 'whatever is said to be dependent, or an effect, composite, particular, many, unequal, unlike, oblique etc. (AT, X, p. 382 l.8–11; HR, I, p. 16)

[28] There is an immense literature concerned with the interpretation of Descartes' absolute and relative terms. See, for example, J. Hartland-Swan, 'Descartes' Simple Natures', *Philosophy* **XXII**, 1947, pp. 139–52; S. V. Keeling, 'Le réalisme de Descartes et le rôle des natures simples', *Révue de métaphysique et de morale* **XLIV**, 1937, pp. 63–99; J. Le Blonde, 'Les natures simples chez Descartes', *Archive de philosophie* **XIII**, 1937, pp. 163–80; L. J. Beck, *op. cit.* (Note 15), Chapters IV–VI; A. Boyce Gibson, *The Philosophy of Descartes* (London, 1932), Chapter V. For recent debate between 'realist' and 'neo-Kantian' interpretations of 'absolute' and 'relative' terms and 'simple' and 'compound' natures, see Marion *op. cit.* (Note 26) and B. E. O'Neil 'Epistemological Direct Realism in Descartes' Philosophy' (Ph.D. diss., University of California, Berkeley, 1967). Cf. also J. Ree, *Descartes* (New York, 1974), pp. 24–5, 36–8.

In the main, this literature sees the problem of interpretation as one of specifying the ontological and epistemological statuses of these entities and their relations to Descartes' later conceptions of 'innate' and 'simple' 'ideas'. Such attempts almost inevitably construe 'absolute' and 'relative' terms in relation to some picture of Descartes' overall 'system', as supposedly expounded in the *Meditations* (1641) and *Principia Philosophiae* (1644). Although this approach can have some merit in elucidating Descartes' later position in metaphysics, it is ill-suited to the project of viewing the text of the *Regulae* in a strict biographical and diachronic framework for the purpose of analysing the genesis and structure of the method.

[29] AT, X, pp. 379–80; HR, I, pp. 14–5.

[30] On Medieval and Renaissance concepts of analysis/synthesis, resolution/composition see A. C. Crombie, *Robert Grosseteste and the Origins of Experimental Science 1100–1700* (Oxford, 1953); J. H. Randall, 'The Development of Scientific Method in

the School of Padua', *Journal of the History of Ideas* **I**, 1940, pp. 177—206; N. W. Gilbert, *Renaissance Concepts of Method* (New York, 1960); L. Jardine, *Francis Bacon. Discovery and the Art of Discourse* (Cambridge, 1974). On the relation of these concepts to geometrical and algebraic problem solving and theorem proving procedures see J. Klein, *op. cit.* (Note 2) and M. S. Mahoney, 'The Beginnings of Algebraic Thought in the Seventeenth Century' in S. Gaukroger, *op. cit.* (Note 17), pp. 141—55.

[31] This illustrates again the greater elaboration of the heuristic rules in the *Regulae* over those in the *Discourse*, and it also hints at the vacuity of taking off from rule 5 to trace the 'influences' upon it of the traditional methodological discussions of analysis/synthesis, resolution/composition.

[32] AT, X, p. 381, l.7; HR, I, p. 15.

[33] AT, X, p. 382 l.17—l.18, p. 383 l.16—l.26, and the extended example which Descartes gives at AT, X, p. 384 l.20—p. 387 l.7, and which is discussed below in Section 4.5. In this connection it is worth noting that even when Descartes first veers toward a notion of 'methodological order' and then subordinates it to a notion of 'given', 'cosmic' absolutes (see above Note 26), he still is mainly interested in wresting heuristic capital from this version of the latticework, rather than in exploring the precise ontological statuses of the absolute and relative terms: 'These ['few pure and simple natures'] we say should be carefully noted, for they are just those facts which we have called the simplest in any series. All the others can only be perceived as deductions from these, either immediate and proximate, or not to be attained save by two or three more acts of inference. *The number of these acts should be noted in order that we may perceive whether the facts are separated from the primary and simplest proposition by a greater of smaller number of steps.*' (AT, X, p. 383; HR, I. pp. 16—7) (emphasis on Descartes' heuristic advice added).

[34] A more exhaustive and textually meticulous analysis of rules 6 and 7 would produce the following picture of how Descartes' perceived the 'flow' of methodical procedure, leaving specific heuristic suggestions aside:

(1) reflection upon finished series leads to new heuristic insight;
(2) analysis and solution of slightly higher order problems mobilizes those insights;
(3) synthesis of the problem in (2) is followed by
(4) reflection upon the newly finished series, leading to more heuristic insight.

Much of the complexity of the text can be explained as arising from Descartes simultaneously elaborating this crude schema, whilst progressively becoming aware of new or differentiated types of heuristic advice required by the schema.

[35] AT, X, p. 387 l.16— p. 388 l.9 (This would also advance the function of 'reviewing for heuristic insight' (cf. above), but here Descartes is centrally concerned with this 'problem' about memory and certainty.)

[36] AT, X, p. 389 l.8—15.

[37] AT, X, p. 388 l.18— p. 389 l.7 should be carefully examined in this regard.

[38] Subdivision occurs, for example, at AT, X, p. 390 l.6—20; cf. rule 11, *ibid.* p. 407 l.8— p. 409 l.10. Alquié, *op. cit.* (Note 24), p. 110 n. 1 and p. 111 n. 1 comments pertinently on the relations between these senses of 'enumeration'.

[39] One should perhaps speak of 'material in rules 8 to 11' rather than the rules themselves as far as the period 1619—20 is concerned. It is quite clear that there is some material in rule 8 which could only postdate Descartes' discovery of the law of refraction of light, an achievement most likely dating from 1626—27. See Schuster,

op. cit. (Note 17), pp. 51—5, 58—9; also Weber, *op. cit.* (Note 16), pp. 80—108, 205, although Weber joins several other scholars in dating the discovery of the law of refraction from the period 1621—23, *ibid.*, p. 94. There are also some remarks about rule 9 in the newer (that is, post-law of refraction) portion of rule 8 which suggest that the material in rules 8 to 11 was not distributed into its present form until after 1626. See Schuster, *op. cit.* (Note 17), pp. 58—9 and n. 85. However, the clear connection of the heuristic advice in these rules with the thrust of the main heuristic rules (5 to 7) indicates an early provenance for the material; and, even if it could be proven that all material in rules 8 to 11 dates from after 1626, their intentional and conceptual continuity with rules 5 to 7 would still deserve comment here.

[40] See AT, X, p. 392, l.1—7.

[41] E.g. Gilson, *op. cit.* (Note 5), pp. 158—9, 227—8; Sirven, *op. cit.* (Note 15), pp. 129—33; Gouhier, *op. cit.* (Note 15), pp. 60—1; G. Rodis-Lewis, *L'oeuvre de Descartes* (Paris, 1971), pp. 47, 50; Weber, *op. cit.* (Note 16), Chapter 1.

[42] In the *Olympica* are described (and interpreted) the famous three dreams which Descartes had on the night of 10—11 November 1619, during a period of work on the 'foundations of a marvelous science'. (AT, X, pp. 179—88). The 'marvelous science' is not identified, but Descartes' presentation/interpretation of the dreams can be read as bespeaking a recent concern with the fundamental premisses of the method (J.-P. Weber, *op. cit.* (Note 16), pp. 16, 41—7). On the status of the dreams and their interpretation, see Gouhier, *op. cit.* (Note 15), pp. 37ff.; Rodis-Lewis, *op. cit.*(Note 41), pp. 46ff. and A. Browne, 'Descartes' Dreams', *Journal of the Warburg and Courtauld Institutes* **XL**, 1977, pp. 256—73.

[43] J.-P. Weber, *op. cit.* (Note 16). I have developed and criticized Weber's chronological approach to the *Regulae* in detail in Schuster, *op. cit.* (Note 17), esp. pp. 41—2 and Note 4 thereto, and pp. 58—9.

[44] AT, X, p. 371 l.25 – p. 372 l.4.

[45] This and the other textual points in the following three paragraphs derive from Weber, *op. cit.* (Note 16) as employed in Schuster, *op. cit.* (Note 17).

[46] Rule 4A (on method) covers AT, X, p. 371 l.1 to p. 374 l.15. Rule 4B (universal mathematics) covers AT, X, p. 374 l.16 to the end of the rule. These parts of rule 4 were first so designated by Weber, *op. cit.* (Note 16) who maintains that the two portions are divided by a clear boundary of of conception, intention and chronology. J.-L. Marion maintains the essential identity of universal mathematics and method and does not grant the Weberian hiatus in rule 4. See Marion, *op. cit.* (Note 26), pp. 55ff. and the notes to his French translation of the *Regulae, René Descartes, Regles utiles et claires pour la direction de l'esprit en la recherche de la verité* (La Haye, 1977), pp. 155—7. To signify the claimed unity of universal mathematics and method, Marion insists that the former be denoted by Descartes' Latin term *mathesis universalis*. Throughout this discussion I employ the term 'universal mathematics', thereby signaling my acceptance, and attempted reinforcement of Weber's position. For further argument on this point see Schuster, *op. cit.* (Note 17), p. 43 and n. 12 thereto, and *passim*.

[47] Weber, *op. cit.* (Note 16) pp. 5—7. This does not mean that the two projects are not related, they are, as is shown below. This is a textual point; the two projects are introduced and discussed independently in rule 4.

[48] Weber, *op. cit.* (Note 16), pp. 5—6. Schuster, *op. cit.* (Note 17) argues that the later rules (roughly 12 to 21), written between 1626 and 1628 do attempt an articulation of the universal mathematics of rule 4B.

[49] AT, X, pp. 374, 377–8: HR, I, pp. 11, 13. Weber, *op. cit.* (Note 16), pp. 7–8, 40, 43.

[50] AT, X, p. 374 l.8–10.

[51] AT, X, p. 374 l.10–13.

[52] Weber, *op. cit.* (Note 16), pp. 1–4; cf. Schuster, *op. cit.* (Note 17) p. 46. These points also are textual (cf. above Note 47); they do not purport to show that there was no genetic relation between universal mathematics and the method.

[53] Weber, *op. cit.* (Note 16), pp. 8, 9, 15–7; Schuster, *op. cit* (Note 17), pp. 46–7.

[54] The 'treatise' is mentioned in rule 4B (AT, X, p. 379; HR, I, p. 14).

[55] AT, X, pp. 377–8.

[56] G. Crapulli, *Mathesis Universalis, Genesi di un'Idea nel XVI Secolo* (Rome, 1969); The relation of rule 4B to this tradition is discussed in Schuster, *op. cit* (Note 17), pp. 44–5 and Notes 14 and 15.

[57] On Beeckman see, *Journal tenu par Isaac Beeckman de 1604 à 1634*, C. de Waard (ed.), IV vols. (The Hague, 1939–1953); C. de Waard, *L'experience barometrique, ses antécédents de ses explications* (Thouars, 1936); R. Hooykaas, 'Science and Religion in the Seventeenth Century – Isaas Beeckman 1588–1637', *Free University Quarterly* **I**, 1950–52, pp. 169–83; and K. van Berkel, *Isaac Beckman (1588–1637) en de Mechanisering van het Wereldbeeld* (Amsterdam, 1983).

[58] 'Physico-mathematici Paucissimi' AT, X, p. 52. Descartes and Beeckman complimented each other on being the only true physico-mathematicians in Europe. Beeckman was to write in 1628 that 'Crediderim enim Verulamium [Bacon] in mathesi cum physica conjugenda non satis exercitatus fuisse; Simon Stevin vero meo judico nimis addictus fuit mathematicae ac rarius physicam ei adjunxit'. (*Journal* **III**, pp. 51–2). Van Berkel, *op. cit.* (Note 57), while confirming a good deal of the interpretation of the Descartes-Beeckman relationship discussed in Schuster, *op. cit.* (Note 8), takes exception to the idea that for Beeckman 'physico-mathematics' entailed the necessity of always adducing a corpuscularian explanation (pp. 293–4). See also, Van Berkel, 'Beeckman, Descartes et "La philosophie physico-mathématique"', *Archives de Philosophie* **XLVI**, 1983, pp. 620–6.

[59] See the *Physico-mathematica*, AT, X, pp. 67–78 and some of Descartes' cognate notes in his *Cogitationes privatae*, AT, X, pp. 217–28.

[60] These remarks are not intended to slight or whiggishly misrepresent the efforts of Beeckman and Descartes, singly or jointly. If one accepts, for Beeckman at least, Van Berkel's contention that physico-mathematics need not necessarily have applied to the corpuscular realm (cf. Note 58), then Beeckman's work (with Descartes) on the law of fall qualifies as first rate physico-mathematics, as does Beeckman's research on the laws of collision (thought of as applying to corpuscles, but not really capable of being deployed in the solution of further puzzles concerning the phenomena of that realm). The claim that physico-mathematics includes the requirement of corpuscular-mechanical explanation has not been questioned in the case of Descartes, so he has little physico-mathematics to show, with one powerful exception: The mechanical explanation of the law of refraction presented in the *Dioptrique* (1637) through the vehicle of an explicitly macroscopic model of the motion of 'tennis' balls. This explanation can be shown to rest on an implicit, and earlier, attempted explanation at the level of those corpuscular 'tendencies to motion' which for Descartes actually constituted light. Schuster, *op. cit.* (Note 8), pp. 299–352, argues that Descartes discovered the law of

refraction in the period 1625—7 by deploying the resources and principles of traditional geometrical optics. Then, acting in the style of physico-mathematics, he tried to elaborate an account of the mechanics of corpuscular tendencies to motion which could explain the already derived law. Whilst this arguably was physico-mathematics in the grand style, it should be noted that the core of the achievement rested with the techniques and resources of geometrical optics. The supposed physico-mathematical explanation being fitted, *post-facto*, to suit the frame of the solution as embodied in the ray diagrams through which Descartes came to recognize the law. There is no sense in crediting Descartes' physico-mathematics, in 1619 or 1927, with an independent problem solving efficacy: it did not solve the problem of refraction; rather, it suggested a style of *post facto* glossing. It further follows that this greatest of 'achievements' of physico-mathematics in no way followed from the structures of some general method of discovery — except in the ironically revealing sense that since methods are mainly tools for public glossing of claims, physico-mathematics was functioning here in just the way grand method doctrines often do! [I thank Dr. Alan Gabbey for demanding some clarification of the claims made in the text.]

[61] *Aquae Comprimentis*, AT, X, pp. 67—74, which dates from late 1618 or very early 1619. For full analysis and evidence on dating see Schuster, *op. cit.* (Note 8), pp. 93—111.

[62] S. Stevin, *De Beghinselen des Waterwichts*, reprinted and translated in *The Principal Works of Simon Stevin*, E. J. Dijksterhuis *et al.* (eds.), vol. I, (Amsterdam, 1955), pp. 415—7.

[63] *Physico-mathematica*, AT, X, p. 68.

[64] *Ibid.*, pp. 67—8; cf. Descartes to Beeckman, 26 March 1619, AT, X, p. 159; Descartes to Beeckman, 23 April 1619, AT, X, p. 162.

[65] *Physico-mathematica*, AT, X, pp. 69—73.

[66] *Cogitationes privatae*, AT, X, pp. 234—41. Material in the next four paragraphs is discussed in detail in Schuster, *op. cit.* (Note 8), pp. 111—27.

[67] The compass is described in Book Two and Three of the *Géométrie* in terms corresponding to the more crude figures and implied mode of use in the *Cogitationes privatae* of 1619/20 (AT, X, pp. 234—41). The compass consists of two main branches, AX and AZ, pivoted at A. Set inside the branches are a series of rulers, of which BC, DE and FG are set at right angles to AX, while CD, EF and GH are set at right angles to AZ. BC is fixed to AX at B, but the bases of the rest of the rulers can slide along the inner side of the branch to which they are set. As the compass is opened BC pushes CD along AZ, and CD in turn pushes the base of DE along AX, and so on. The compass is 'a machine for generating series of magnitudes [line lengths] in continued geometrical proportion' (J. Vuillemin, *Mathématiques et métaphysique chez Descartes* (Paris 1960), p. 112), for by similar right triangles it is the case that:

$$\frac{AB}{AC} = \frac{AC}{AD} = \frac{AD}{AE} = \frac{AE}{AF} = \frac{AF}{AG} = \frac{AG}{AH}.$$

[68] *Géométrie*, Bk III, AT, VI, pp. 442—4.

[69] Descartes to Beeckman, 26 March 1619, AT, X, pp. 154—60.

[70] *Ibid.*, pp. 154—6.

[71] *Cogitationes privatae*, AT, X, pp. 234—9.

[72] *Ibid.*, p. 234.

73 It also may arise from Descartes' still considerable naïveté concerning the principles of (cossist) algebra.

74 AT, X, pp. 234—5. This is obvious from the geometry of the compass: Let $AC = x$, $AD = x^2$ and $AE = x^3$, because

$$\frac{1}{AC} = \frac{AC}{AD} = \frac{AD}{AE} \quad \text{or} \quad \frac{1}{x} = \frac{x}{x^2} = \frac{x^2}{x^3}.$$

Then, since $AE = AC + CE$, or $x^3 = x + CE$, when CE is opened to two units the root, x, can be read off the compass.

75 This is clear from the figure. AE or x^3 can only be constructed as the sum of AC ($= x$) and CE ($= b$).

76 Consider a further entry in the *Cogitationes* (AT, X, pp. 244—5 and Note *a* to p. 245): Descartes reduced cubics of the form $x^3 = ax^2 + bx + c$ to the form $x^3 = b_1x + c$, which he believed to be suitable for solution on the compass. The root, he asserts, is to be extracted '*ex invento nostro*', most likely referring to the compass and to the manipulations discussed above. Once again, therefore, Descartes pursues general modes of solution, to be accomplished on the compass, to the exclusion of otherwise obvious difficulties: his method of reducing the original equation was of limited value and was subject to the same algebraic errors discussed above. In addition he still failed to see that the compass could only be used on 'species' of the equation with positive terms only.

77 The arguments in this section are developed in greater detail in Schuster, *op. cit.* (Note 17), pp. 51—4.

78 Milhaud, *op. cit.* (Note 1), p. 43; Schuster, *op. cit.* (Note 17), p. 51; but cf. the interpretation of L. Gäbe, *Descartes' Selbstkritik, Untersuchungen zur Philosophie des Jungen Descartes* (Hamburg, 1972), Anhang 1, pp. 113—20.

79 See above p. 50; cf. Schuster, *op. cit.* (Note 17), pp. 45, 53.

80 Of course, any gloss of a discipline is a discursive construct; that is not quite the point. Descartes was glossing a discipline, universal mathematics, which had no social and technical/practical density in itself, and which owed its entire 'existence' to the literary devices of extrapolating and generalizing from (misleading) descriptions of 'physico-mathematics' and of 'the real import of the compass'.

81 AT, X, pp. 384—7; HR, I, pp. 17—9. Given that we are dealing with a basic ratio of 2:1, applied first to 3, Descartes points out that there are three orders of difficulty of problems:

(1) To find 12, given 3 and 6. Or, to find 24, given 3, 6, and 12.
(2) To find 6, given 3 and 12.
(3) To find 6 and 12, given 3 and 24.

But, to find 6 and 12, given 3 and 48 is really easier than it might look, for it is of order (2). The route to solution is: First find 12, the mean proportional between 3 and 48, and then 6 and 24 the mean proportionals between 3 and 12 and 48 respectively. *Vive la méthode*!

82 I am not suggesting that all the heuristic rules arise this way; see Note 83 and pp. 45—6 above.

83 Strictly speaking it is only necessary that the general notion of 'heuristic rules' arise in this connection along with sketches of some few of the detailed rules. Close study of

rules 6 to 11 would show a second rule-generating phenomenon in which emerging problems and tensions in Descartes' method discourse invited and conditioned the formulation or reformulation of rules. We have seen something of this process in the analysis of rule 7 above.

[84] This, of course, is the conventional view, deeply entrenched in Cartesian studies since the advent of their modern phase at the turn of the century. Cf. for example, A. Boyce Gibson, 'La géométrie de Descartes au point de vue de sa méthode', *Révue de métaphysique et de morale* **IV**, 1896, pp. 386—98; L. Liard, *op. cit.* (Note 15).

[85] This point can be reinforced by the following considerations: Descartes' insights of November 1619 — the doctrine of intuition/deduction, the vision of the latticework and the perceived need for heuristic rules — mutually imply each other as interrelated elements in the overriding programme of developing the method. But, as with any system of concepts, the obvious structural relations between the elements do not fully account for their respective contents. They did not assume shape solely in relation to one another. Pre-given materials, concepts, resources, goals were moulded to give the desired interrelation. For example, although the latticework is correlative with 'intuition', the latter concept does not dictate much about the precise explication of the former, beyond the assertion of valid deduction links between terms. It cannot account for the specific and problematical discourse on 'absolutes', 'relatives', 'series' and 'degree of relation' which emerges. Similarly, the correlated notions of intuition and deductive latticework might or might not suggest the notion of heuristic aides; they certainly do not narrow the field of specific candidates for the title 'heuristic rule'. The point is that the latticework and the idea of heuristic rules were explicated by Descartes with precise exemplars in mind, exemplars, I suggest, which were drawn from reflection upon universal mathematics, or to be precise, from reflection upon the structure of purported examples of that discipline.

[86] Weber, *op. cit.* (Note 16), pp. 15—7, 40—7.

[87] This eliminates claims for the efficacy of a method which rest on an arguable misconstrual of its own proclaimed resources and goals, for example, claims that Descartes' method facilitated the discovery of articles of *faith*, or that Popper's method was of use in the *discovery* of a fact or law.

[88] The sociological literature supports this point to the extent that it establishes the socially negotiated and revisable character of 'facts', 'phenomena', 'adequate' or 'suitably replicated experiments', and the 'adequate implementation of standards' or 'meeting of goals'. For an overview of the literature to 1979 see M. Mulkay, *Science and the Sociology of Knowledge* (London, 1979), esp. pp. 49—95. In addition, B. Latour and S. Woolgar, *Laboratory Life, The Social Construction of Scientific Facts* (London, 1979), esp. pp. 151—86; K. Knorr-Cetina, *The Manufacture of Knowledge: An Essay on the Constructivist and Contextual Nature of Science* (Oxford, 1981), esp. Chapters 2 and 5; B. Barnes, 'On the Conventional Character of Knowledge and Cognition', *Philosophy of the Social Sciences* **XI**, 1980, pp. 303—33; A. Brannigan, *The Social Basis of Scientific Discoveries* (Cambridge, 1981); H. M. Collins, 'The Seven Sexes: A Study in the Sociology of a Phenomeon, or the Replication of Experiments in Physics', *Sociology* **IX**, 1975, pp. 205—24; A. Pickering, 'The Hunting of the Quark', *Isis* **LXXII**, 1981, pp. 216—36;

[89] One reservation must be registered to this claim. It is perfectly true that bits and pieces of 'methodological discourse', including putative glosses of the field in question, can be deployed in practice as resources in debate, negotiation and adjudication of the

content and acceptability of knowledge claims. Scientists can appeal to methodological principles to attempt to substantiate or undermine such claims. However, such deployments of method discourses within scientific debates are merely small portions of the total structure of action and belief through which knowledge is made and unmade. Such deployments do not represent clear and accurate meta-level versions of the specific practice of that field. Indeed, the deployment of method claims in scientific debate in no way whatsoever constitutes even *prima facie* evidence of the efficacy of that method. The issue must be turned on its head – how are such claims, as discursive phenomena, shaped by the resources of method discourses and what literary effects of genuine efficacy are thereby created? Recent sociological studies of methodological discourse in scientific debate and negotiation include: N. Gilbert and M. Mulkay, 'Contexts of Scientific Discourse: Social Accounting in Experimental Papers', in *The Social Process of Scientific Investigation. Sociology of the Sciences* **IV**, K. Knorr, R. Krohn, and R. Whitley (eds.) (Dordrecht, 1980), pp. 269–94; M. Mulkay and G. N. Gilbert, 'Accounting for Error: How Scientists Construct Their Social World When They Account for Correct and Incorrect Belief', *Sociology* **XVI**, 1982, pp. 165–83; G. N. Gilbert and M. Mulkay, 'Warranting Scientific Belief', *Social Studies of Science* **XII**, 1982, pp. 383–408; and M. Mulkay and G. N. Gilbert, 'Putting Philosophy to Work: Karl Popper's Influence on Scientific Practice', *Philosophy of the Social Sciences* **XI**, 1981, pp. 389–407.

[90] One presumes this comes as no surprise to readers of T. S. Kuhn, *The Structure of Scientific Revolutions*, 2nd ed. (Chicago, 1970), especially the 'Postscript'; T. S. Kuhn, *The Essential Tension: Selected Studies in Scientific Tradition and Change* (Chicago, 1977), especially chapter 13; J. R. Ravetz, *Scientific Knowledge and its Social Problems* (Oxford, 1971), pp. 71–240; G. Bachelard, *op. cit.* (Note 13); G. Bachelard, *Le nouvel esprit scientifique* (Paris, 1975); and B. Barnes, *T. S. Kuhn and Social Science* (London, 1982).

[91] Rule 12, AT, X, p. 427; Rule 13, AT, X, pp. 430–1.

[92] I leave aside the problem, obvious to anyone familiar with the literature cited in Notes 88 and 89, of the criteria by which Descartes selects as 'adequate' and 'reliable' Gilbert's own selection of a set of experiments, their performance and their glossing in his book.

[93] On the complexity of interpreting Descartes' remarks see G. Buchdahl, *Metaphysics and the Philosophy of Science* (Cambridge, Mass., 1969), pp. 85–8, 126–47; Schuster, *op. cit.* (Note 17), pp. 74–5, pp. 74–5 and notes 150, 151.

[94] Buchdahl, *op. cit.* (Note 93), pp. 97, 118–26; A. I. Sabra, *op. cit.* (Note 3), pp. 21–45; D. Clarke, 'Descartes' Use of "Demonstration" and "Deduction"', *The Modern Schoolman* **LIV**, 1977, pp. 333–44.

[95] D. Clarke, *op. cit.* (Note 94).

[95] The principles and laws of Descartes' dynamics are included here, not because there is scholarly agreement that Descartes intended all of them to be deducible from his metaphysics, but rather because they are foundational for all his detailed model building and particular explanations, *and* because he often gives strong indications that they were meant to be deducible from first principles.

[97] Cf. literature cited in Note 88, also H. Collins and T. Pinch, *Frames of Meaning, The Social Construction of Extraordinary Science* (London, 1982); on Descartes' handling of 'factual' evidence see Sakellariadis, 'Descartes' Use of Empirical Data to Test

Hypotheses' *Isis* **LXXIII**, 1982, pp. 68–76, and S. Sakellariadis, 'The Role of Evidence in Descartes' Scientific Method' (Ph.D diss., University of Pittsburgh, 1980).

98 How, for example, can the behaviour of Descartes' 'first element' comport with the metaphysical principle that extension is the essence of matter; or, how does Descartes' account of the internal sensations and passions of the soul comport with his ontological dualism? (Descartes' seems not to have attended to the former but devoted much effort to the latter.)

99 Descartes' judgments about these similarity/difference relations would at least in part depend upon already established patterns of interpretation. For example, Descartes thoroughly accepts the broad difference Gilbert had sought to draw between magnetic and electrical phenomena, although he does not accept this as an ontological distinction (between a spiritual and a corporeal cause respectively), and yet à la Gilbert he still promotes magnetism as a phenomenon of 'cosmic' scope and importance, contrasting it to the rather trivial role of the known electrical phenomena.

100 For example, in Cartesian mechanical philosophy aspects of the theory of magnetism and of the theory of light have crucial bearing on the structure of the theory of celestial motions (cf. Note 99); but, in a different mechanical world picture this particular linkage might not have been valued or sought.

101 Cf. Harold Garfinkel, *Studies in Ethnomethodology* (Englewood Cliffs, 1967); B. Barnes and J. Law, 'Whatever Should Be Done with Indexical Expressions?', *Theory and Society* **III**, 1976, pp. 192–222.

102 Feyerabend, *op. cit.* (Note 12), effectively first identified this phenomenon and isolated its mythopoeic character through what amounted to a structural delineation of Newton's methodological discourse. See below, Section 6.2.

103 If this move were accompanied by the claim that the results of research are justified only if such a *post facto* gloss is possible, then we would have the typical modern methodologist's tactic of appealing to the possibility of 'rational reconstruction'. On the view to be advanced here, 'rational reconstructions' guided by systematic methodologies are simply a species of methodological story or account, and share in their mythopoeic character. Existing studies which suggest this criticism of Popperian-Lakatosian conceptions of method include: P. K. Feyerabend, *op. cit.* (Note 11), 1978, pp. 201–2; Schuster, *op. cit.* (Note 9), esp. Section VIII; M. Mulkay, J. Potter, and S. Yearley, 'Why an Analysis of Scientific Discourse is Needed' in *Science Observed,* K. D. Knorr-Cetina and M. Mulkay (eds.), (London, 1983), pp. 171–203 (see esp. pp. 172–82 where the authors produce a brilliant ethnomethodological critique of a recent quasi-Lakatosian attempt to offer an historical 'rational reconstruction' of some developments in particle physics). Reasons for the mythic character of Lakatosian reconstructions are also discussed in J. A. Schuster, 'Kuhn and Lakatos Revisited', *British Journal for the History of Science* **XII**, 1979, pp. 301–17. As for Descartes' own *post-facto* glossing of practice, there is a wonderful example in the *Regulae* (Rule 8, AT, X, pp. 393–5) where he discusses (in the carefully chosen subjunctive mood) how one *might have discovered* the law of refraction of light by following the method.

104 Recent perspectives in the sociology of scientific knowledge would, I think, correctly reject the sort of static representation offered above, insisting that the contents of the spaces in the matrices, the structures of the matrices, and even the existence of this as opposed to some other set of matrices, are the (continually revisable) outcome of the modes of negotiation, interpretation and coercion by which evidence, facts,

arguments and evaluations are constructed in these fields. But this complication, which really amounts to the contemporary challenge to theorize the social construction of scientific knowledge, cannot and need not be resolved here. Cf. the literature cited in Notes 88 and 89.

[105] Schuster *op. cit.* (Note 9), especially Section V.

[106] In this connection I particularly rely on the researches of Harvey Sacks (e.g. 'On the Analysability of Stories of Children' in *Directions in Sociolinguistics*, J. J. Gumperz and D. Hymes (eds.), (New York, 1972), pp. 325—45); read against the background of A. Schutz, *Reflections on the Problem of Relevance* (New Haven, Conn., 1970) and A. Schutz and T. Luckman, *The Structures of the Life World* (London, 1974).

[107] In some cases the second task is involved in the first. For example, in cases where actors were arguably indoctrinated into a method, part of the account of the indoctrination process has to do with the mythological mechanics of method discourses. Early modern figures, such as Descartes, present a more complex picture. They moved in an intellectual culture deeply committed to the in principle existence of efficacious methods of discovery and appraisal (proof) in natural science. The problem was to devise, select and enforce the 'correct' one (cf. above Note 30).

[108] AT, X, p. 381 l.25, p. 382 l.9.

[109] See above Sections 4.3, 4.4.

[110] *Géométrie*, Liv. III, AT, VI, pp. 444—61.

[111] M. S. Mahoney, *op. cit.* (Note 30); M. S. Mahoney, *The Mathematical Career of Pierre de Fermat* (Princeton, 1973).

[112] Consider a parallel case, Descartes' discovery of a general constructive solution for all cubic and quartic equations, presented in the third book of the *Géométrie* as the solution of all the 'solid' problems of the ancients. I have reconstructed Descartes' path to this construction (Schuster, *op. cit.* Note 8, pp. 131—49), showing how this achievement grew out of Descartes' early compass researches, recounted in part above in Section 4.3, and how the analytical skills involved elude pat generalizations in terms of the simple heuristic rules of the method.

[113] The only way out of this problem would be to maintain with Cassirer (and some of his latter day acolytes) that Descartes' (discursive) corpuscularianism was some sort of mistake or illusion, and that Descartes did (or should) have entertained an 'ontology' for physics of coordinates and functions of physical quantities thereof. This, however, would hardly establish the efficacy of Descartes' method. If true it would only show that mathematics provided genuine tools for the pursuit of Descartes' physics. In any case the historical Descartes resolutely stuck to grinding out qualitative (and remarkably ineffective) corpuscular-mathematical discourse. See. E. Cassirer, *Leibniz System in seinem wissenschaftlichen Grundlagen* (Marburg, 1902), Einleitung, 'Descartes' Kritik der mathematischen und naturwissenschaftlichen Erkenntnis'.

[114] This may well help explain Gerd Buchdahl's acute observation (*op. cit.* Note 93, pp. 126—7) that Descartes seems to use terms such as 'analysis' and 'absolutes' in several contrasting ways; for example, the latter may be (a) elements to which explicanda are to be reduced; (b) elements into which explicanda are decomposed; (c) elements to which explicanda are referred; (d) elements from which they can be physically generated, or (e) logically derived. The cause of this proliferation is not so much a philosophical profundity struggling with inadequate terminology, or underdeveloped conceptual distinctions. Rather, it is the structural product of the way the method discourse

devours and denatures domain-specific discourses in the interests of (a pseudo) generality.

[115] 'Serious' denotes here the accounts given by 'properly trained' historians and sociologists of science, of course!

[116] Limitation of space precludes and discussion of the development of Descartes' method doctrine and its relation to his conception of universal mathematics after 1620. See Schuster, *op. cit.* (Note 8), pp. 235—53, 510—30; and Schuster, *op. cit.* (Note 17).

[117] See above Note 106.

[118] For example, in the culture of Popperian method discourse you cannot acceptably say the following: Popper's method could not really have been used as a Level II story claims it was used, because Popper's method cannot be straightforwardly applied [since a putative falsifying instance never speaks for itself and always requires interpretation of its status and relevance]. On the other hand, you *can* say that Popper's method could not really have been used as a Level II story claims it was used, because there is independent evidence that the actors in the story really are 'Marxist dogmatists' who do not understand what rational procedure is [as defined by Popper's Level I].

[119] Systematic methodologists, being reasonable people, do notice gaps between stories and counter-accounts, when these are brought to their attention. But this does not induce them to become mythologists of method. Consider Descartes, writing to Père Mersenne (27 February, 1967): '. . .je n'ai su bien entendre ce que vous objectez touchant le titre; car je ne mets pas *Traité de la Méthode*, mais *Discours de la Méthode*, ce qui est le même que *Préface* ou *Avis touchant le Méthode*, pour montrer que je n'ai pas dessein de l'enseigner, mais seulement d'en parler. Car, comme on peut voir de ce que j'en dis, elle consiste plus en pratique qu'en théorie . . . [in F. Alquié, *op. cit.* (Note 24), pp. 521—2].

[120] Feyerabend, *op. cit.* (Note 12), p. 150.

[121] *Ibid.*, p. 159 Note 7.

[122] Cf. above p. 77 and Note 118.

[123] Feyerabend, *op. cit.* (Note 12), pp. 155, 168—9.

[124] *Ibid.*, pp. 161—5.

[125] *Ibid.*, p. 163, Note 11.

[126] Cf. Feyerabend's remarks, *ibid.*, section VIII, p. 165.

[127] Schuster, *op. cit.* (Note 9).

[128] One of the best available studies of how a contemporary methodology is (and must be) deployed by scientists in fluid, context- and interpretation-dependent ways is to be found in Mulkay and Gilbert, *op. cit.* (Note 89), 1981. They conclude that: 'Because scientists' technical reasoning depends on highly personal judgements and because the meaning of the Popperian rules depends on such judgements, each individual scientist continually disagrees with his colleagues' Popperian interpretations. In situations of cognitive uncertainty Popperian rules have no single meaning. They cannot, therefore, operate as useful prescriptions, but only as evaluate retrodictions.' (p. 404)

Mulkay and Gilbert's line of analysis complements that advanced in this chapter. By focusing on empirical cases of the Level III deployment of method discourse, they have exposed examples of the slippage between Level III practice and the claims embodied in Level II stories, and they have explained the genesis of that slippage in the manner indicated in the passage cited. My analysis focuses on the generic character of method discourse which explains the possibility of these deployments and these slippages.

H. E. LE GRAND

STEADY AS A ROCK: METHODOLOGY AND MOVING CONTINENTS

1. INTRODUCTION

Alfred Wegener proposed in 1912 that the continents had once been united in a vast supercontinent which had begun breaking up in the late Cretaceous or early Tertiary. The pieces had then slowly moved to their present positions. Additionally, he advocated polar wandering in connection with paleoclimatic evidence. He elaborated his theory and supported it with additional arguments and evidence in his monograph, *Die Entstehung der Kontinente und Ozeane* of 1915 and in further German editions and translations including two English editions of 1924 and 1929. Geologists agreed that his theory of continental drift (hereafter, CD), were it accepted, would revolutionize the science. R. T. Chamberlin, a vigorous opponent, characterized the theory as follows: 'If we are to believe Wegener's hypothesis we must forget everything which has been learned in the last 70 years and start all over again'.[1] A firm proponent, Alexander Du Toit, concurred: '. . . the principles advocated by the supporters of continental drift form generally the antithesis of those currently held. The differences between the two doctrines are indeed fundamental and the acceptance of the one must largely exclude the other.'[2] First in the German and then in the English language scientific press debate over Wegener's theory raged in the 1920s and 1930s.

An occasional feature of these early debates is the use of 'method talk'; that is, the coupling of methodological remarks and discussions with arguments about the value and validity of CD. An examination of the literature suggests that 'method talk' filled a number of roles in these debates. First, prior methodological convictions could play a substantive role in the evaluation of rival theories. For example, geologists who regarded low-level inductive generalizations as the only acceptable form of theorizing would likely be quite antipathetic to any 'global' theories cast in hypothetico-deductive form or a theory which postulated as yet unobserved structures or forces. Second, geologists could employ 'method talk' in accounting for why scientific beliefs they

J. A. Schuster and R. R. Yeo (eds.), The Politics and Rhetoric of Scientific Method, 97–138.
© 1986 *by D. Reidel Publishing Company.*

thought to be erroneous might be held by other geologists. Third, 'method talk' was used opportunistically to defend favoured positions and to 'score points' against contrary views. Fourth, recourse could be had to 'method talk' to transcend or defend lines of demarcation among specialties within a discipline, among disciplines, or between a discipline and a wider social setting. Fifth, geologists on occasion altered the presentation of their theories in order to launch or rebut methodological criticisms. 'Method talk' could in this way play a constitutive role in theory development. Finally, 'method talk' could cloak arguments which might not normally be acceptable in formal scientific discourse: private criticisms based on race, nationality, or scientific credentials could be transformed into public criticisms phrased in methodological terms.

Mulkay and Gilbert have recently argued that there is an asymmetry in the accounts offered by scientists for correct and incorrect belief.[3] Their examination especially of interviews, 'informal texts', indicates that there is a commitment to the 'standard view' of scientific rationality. 'Correct' belief is unproblematic and is the product of proper scientific procedures. 'Incorrect' belief is problematic and is explained by scientists as arising from the influence of non-scientific factors. One effect of accounting for error in this manner is to reinforce the impression that the 'standard view' is justified and that the scientific procedures which it embodies are efficacious. Mulkay and Gilbert's particular techniques and selection of material are confined to contemporary examples. This analysis of the controversy surrounding Alfred Wegener's theory of CD suggests how a historical case using publications, 'formal texts', can give a finer texture to their somewhat schematic approach even though the more forthright ways of social accounting for error expressed in informal texts tend in formal texts to give way to more oblique ones. In particular, the use of methodological claims and counterclaims by opponents and proponents of CD is developed as one type of 'error accounting'.[4]

Further, this focus on 'method talk' in the debates of the 1920s and early 1930s supplements recent studies of the history of CD theory as an instance of theory change.[5] These studies, particularly in so far as they partake of the standard view of scientific rationality and methodology, provide no satisfactory account of methodological rhetoric or methodological issues. Indeed, the methodological utterances and concerns of the scientists involved are usually passed over as either unproblematic or irrelevant to their decision to adopt or reject Wegener's theory. This is symptomatic of the ritualistic responses of many historians of science, even philosophically-informed ones, who either accept at face value or conjure away 'method talk' when it is encountered in their sources.[6] 'Method talk' was certainly a prominent feature of the moving continents controversy — as it often is of scientific controversies.

Attention to methodological rhetoric and, ultimately, to broader social and political forms of analysis should not be taken as implying that Wegener's specific theory was not rejected for 'good scientific reasons' and, moreover, reasons which were often rooted in methodological judgments and practices. Such reasons included discrepancies between theory and data, the apparent failure of the much-touted geodetic measurements, and conflicts with physics and seemingly well-established parts of geophysics. The acceptance of CD would have resolved some geological problems but only at the cost of generating empirical problems and internal and external conceptual problems.[7] The mostly negative response — particularly in the Anglophonic geological community — was *over*determined. My emphasis on methodological rhetoric rather than on other theoretical-empirical issues raised by the participants is reflected in the inclusion of a large number of direct quotations. These are neither ornaments nor evidence *per se*, but objects of study.

2. PREVAILING VIEWS

In the first decade of the twentieth century there were several 'big theories' in geology. 'Permanentism', particularly strong in North America, regarded the existing continents and ocean basins as crustal features which had remained essentially permanent since the early stages of earth history. The continental masses were composed of 'sial' and floated in isostatic equilibrium in the denser ocean floor material and underlying substrate composed of 'sima'. This fundamental tenet could be harmonized with either a contracting or a steady state earth. The theory focused attention on structural geology, petrology, geomorphology, and, broadly speaking, regional and descriptive geology. One difficulty was the explanation of the distribution of similar fossil and living flora and fauna across extant ocean basins.

European geologists in particular were 'interchangers'; that is, they placed somewhat greater emphasis upon the vertical movements of the crust and argued for the appearance and disappearance of large land-bridges or even continents in order to account for the past and current distribution of flora and fauna. This view bears a similarity to Lyell's nineteenth-century views concerning the ceaseless elevation of sea-floors to dryland and the reverse processes of subsidence and denudation, but stress was placed on the 'progressive' development of the earth rather than cyclical stasis. Attention was also often given to lateral displacements, for example in the influential work of the Viennese geologist, Eduard Suess.[8] Generally, this view was teamed with a contracting earth although a steady-state earth was not excluded. One problem of this theory was its reconciliation with isostasy; that is, how could the lighter continents or landbridges sink into the denser sima?

Neither of these schools was monolithic. They could best be thought of as theoretical confederations. Within their broad confines were a number of variants. In addition to these two major groupings, there were a number of more or less radical alternatives put forward in the early twentieth century as the grand Suessian synthesis of the late nineteenth century gradually crumbled. M. T. Greene, reflecting upon the proliferation of theories immediately prior to Wegener's version of CD, concludes that Wegener's theory in 1912 was '. . . one of many different hypotheses created from the same materials.'[9]

Geologists often espoused the uniformity of geological agents as a fundamental assumption and methodological principle; that is, the forces in action today are the same in kind and degree as those of the past and it is these forces which shaped in the past, and are shaping today, the geological features of the earth. What was meant by 'the principle of uniformity' or 'uniformitarianism' was no more uniform either in detail or in particular applications in the early twentieth century than it had been in the hands of Lyell and Whewell.[10] One frequently-expressed reservation was that there might be some variance in the degree of these geological agencies, for example, the permo-carboniferous ice-age versus the limited glaciation of today. Proponents of extreme uniformitarianism could argue that variations in degree were cyclical or fell within a limited range. However, many continental geologists, following on from the work of Elie de Beaumont, Henry De la Beche and especially Suess, rather than Lyell, adopted a far less rigid attitude. In brief, they stressed the progressive development of the earth from an incandescent mass to the present, connected specific surface features to the gradual contraction of the earth, and occasionally invoked violent geological changes to account for such features as mountain structures.[11] These views had some support in Britain and elsewhere as well. Within the general geological community and even within the British sub-community there was a degree of tension. Archibald Geikie warned against a too-ready acceptance of the assumptions that forces in the past acted on the same scale or in the same manner as those we observe today or that in the past there were no forces other than those we have so far discerned.[12] G. A. J. Cole in his 1915 address as president of Section C (Geology) of the British Association suggested that pure uniformitarianism was overly simplistic and did not take into account the geological evolution of the earth.[13] William W. Watts two decades later in his 1936 BAAS Presidential Address echoed this note. He criticised strict uniformitarianism of a Lyellian stripe, accepted a variation of forces and phenomena 'in place, intensity, phase, and time' and argued for an evolutionary view of the earth rather than a static one.[14] Catastrophic events on a global scale or the action of large, unknown, 'one-off' agents were frowned upon but the ambiguity, vagueness — or better yet, flexibility — inherent in the definition of 'uniformity' suggests that its meaning and relevance in particular instances could vary according to the predilections of its users. As a methodological principle, it contained dissonances.

There was also a measure of disagreement on the relative importance of theorizing and fact-gathering in geology. Although some held that speculative theories could be valuable in connecting apparently disparate facts or suggesting directions of research, most placed a greater premium on the collection of facts themselves: only a small portion of the uppermost surface of the earth is accessible and processes occuring within the earth or over long periods of geological time cannot be duplicated in the laboratory. Therefore, any speculation is untrustworthy; only the accumulation of facts will lead to better theories. This reflected not just an inductivist view of proper scientific method, but a quite narrow, 'naive inductivist' version which could play into an almost anti-theoretic stance. This attitude was rather well captured by J. W. Gregory who wrote in 1915 in reference to the foundation of the Geological Society that 'the rash pursuit of cosmogonies did not suit the British mind, which preferred facts that could be established by observation to the uncertain products of speculation'[15], and when he asserted in 1930 that the geological method consisted of 'the interpretation of the direct contemporary evidence of the rocks'.[16] Thomas C. Chamberlin, chairman of the geology department at Chicago and founder of the *Journal of Geology*, advocated a rather influential middle position known quite descriptively as the "method of multiple working hypotheses".[17]

Generally speaking, Anglophonic geologists seem to have been oblivious to the outcome of the mid-nineteenth-century methodological debates between inductivists and hypothetico-deductivists which, sparked by theories of electromagnetic phenomena, had resulted in the enthroning of the method of hypothesis in physics.[18] There were some dissenting voices. Reginald Aldworth Daly admitted the controlling character of facts but argued as follows:

> Inasmuch as cosmogony and geology are both young sciences, consensus of opinion about the earth's origin and history is still reserved for the future. Meantime these sciences are advancing through the erection and testing of competing hypotheses; in other words, through speculation. . . . Science progresses by systematic guessing in the good sense of the word.[19]

Daly, it may be noted, was one of few Anglophonic supporters of CD in the 1920s and this defence of hypothesis prefaced a work in which he put forward a case for CD. European geologists were, in contrast, relatively open to speculative, global theories even if, as was the case with Suess, these grand schemes were sometimes a product of library work than of laboratory or field-work.[20] As with the principle of uniformity, then, if dispute arose as to the 'proper mix' of speculation and fact in a given geological theory, the issue could not be settled by appeal to a precise standard: it would be a matter for opinion, argument, and negotiation.

3. WEGENER'S THEORY: ITS ORIGIN, EVIDENCE, AND PRESENTATION

Wegener had been preceded in his postulation of large-scale continental drifting by several authors. He arrived, he tells us, at his view by first noticing the obvious congruence of the eastern coast of the Americas with the western coast Europe-Africa, but that he had not pursued this until he came across a set of references to the evidence of presumed former land connections between Brazil and Africa. 'This induced me', he relates, 'to undertake a hasty analysis of the results of research in this direction in the spheres of geology and palaeontology, whereby such important confirmations were yielded, that I was convinced of the fundamental correctness of my idea'.[21] He made no reference in this account of discovery to the theoretical ferment then agitating some geologists. Whether or not historians place confidence in the accuracy of this account, this was the public version given by Wegener which could serve either as a resource for proponents (that is, Wegener was forced to his theory by the data), or as a focus for methodological criticism (that is, he disobeyed the canons of inductivism). In the presentation of his theory which followed, he explicitly addressed his comments to permanentists and interchangers, as though there were two dogmatic schools of geology, and virtually ignored the plethora of potential rivals. Occasional comments and notes make clear, however, that he had a good grasp of several alternatives and he explicitly dealt with the drift-type theories of William H. Pickering and Frank B. Taylor, but stated that these two had not been known to him when he had elaborated his own hypothesis of drift.[22]

In brief, Wegener agrees with the permanency of the continental blocks, agrees that there must have been extensive land connections between continents now separated by oceans, and argues that these can be reconciled by the assumption of large-scale lateral movement of the continents. The evidence which he adduces in support of his theory is usually not the product of his own fieldwork. It is culled from the literature of many branches of geology: stratigraphy, paleoclimatology, paleontology, biogeography, geophysics, and geodetics to name some of the more important. To cite one example, his argument for the previous joining of New World with the Old across the present Atlantic Ocean depended upon the matching of geological features on the corresponding coastal margins, the geological descriptions for which he relied upon other authors. Wegener read widely in geology but he was not a professionally-trained geologist. His thesis was in astronomy, his fieldwork was mostly connected with meteorology, and his specialization was in meteorology and paleoclimatology.[23] The perspective and the tools he had to hand in constructing his theory were not those of the average field or academic geologist. His first published presentation is straightforward: he proposes his theory as a working hypothesis and then draws upon the literature for support. Only in the later, monographic presentations do his convictions seem to harden. In these, he straightforwardly states his theory in the first chapter and in each of the successive chapters presents evidence supportive of his theory drawn from a particular geological specialization.

3.1. *Early Response of the German Geological Community*

The early German response to Wegener's theory was decidedly mixed. His first papers elicited brief but favourable notice from Johannes Klöcking who was primarily interested in the support he thought drift theory and pole wandering afforded Heinrich Simroth's oscillation theory.[24] Karl Andrée criticized Wegener on the postulated connection between drift and the formation of folded mountains on the continental margins but was otherwise moderate in his comments.[25] Edgar Dacqué very swiftly took up drift as providing solutions to problems confronting permanentism and in his textbook on paleogeography championed a Wegenerian version against the supporters of Suess.[26] Carl Diener published the first severe criticisms. He catalogued a number of detailed objections which were frequently cited by later authors. He apparently viewed an enumeration of 'negative' facts as a sufficient answer, for 'Wegener's hypothesis, as impressive as it may seem at first glance . . . is after all only a playing with possibilities. It lacks a foundation of positive evidence. . . .'[27]

Wegener's 1915 monographic presentation of drift was the occasion for others to enter the fray. Wolfgang Soergel took a meta-theoretical stance from which to judge the acceptability of the new theory: to be acceptable, it had not only to provide solutions to the problems facing the older ones, but also to provide satisfactory explanations of all that which the older ones had successfully explained.[28] Citing Diener with approval, Soergel then sought to demonstrate that drift theory was inadequately supported and indeed contradicted by a detailed comparison of the two sides of the Atlantic 'split'. Like Diener, he rested his case on this marshalling of data. He did not direct specific or general methodological barbs at drift theory or its author. Andrée countered these empirical attacks by an appeal to methodological matters in a second publication. He iterated Diener's criticisms and characterized them as deriving from an 'actualist' approach: that is, as entertaining for the past configurations of the earth's surface only minor changes from the present, a variant of strict uniformitarianism. This, he asserted illustrated the dependency of historical reconstructions on initial assumptions as well as on available evidence which can be interpreted in more than one way.[29] Theodor Arldt took a similar tack, adding the interesting comment that even if a present westward drift of Greenland be confirmed, 'it is not admissable to extrapolate to a similar movement hundreds of thousands or even millions of years into the past'.[30] Both Andrée[31] and Arldt[32] concluded that there is a 'valid kernel' in Wegener's theory.

Max Semper's scathing rejection of drift stands apart from the other early German formal texts in its vitriolic tone and in its substantial use of 'method talk'. He opens with the odd comment that Wegener's theory is 'wonderfully suited to explain the character of a working hypothesis — by counterexample', a comment which has bearing on the title of his paper 'What is a Working Hypothesis?', but which at first seems unrelated to what then follows.[33] He cites copiously from the literature to make the point that the citations and sources used by Wegener to buttress his case are problematic, unrepresentative, erroneous, misleading or otherwise inadequate as evidence or argument. There is here an asymmetry: Semper apparently assumes that any source mentioned by Wegener is open to suspicion since Wegener's theory is wrong whereas any source or authority invoked by himself must be self-evidently true in so far as it is used to oppose Wegener.[34] This common phenomenon in the drift debate may be generalized in the following form: *A* in building a case against *B* may cite the work of *C* but typically provides no argument for why he trusts *C* rather than *B* except (inferentially) that *C* agrees with *A*. Wegener's misrepresentations and distortions account for the erroneous character of his drift theory and are presumably attributable to his careless use of scientific methods and his ignorance in various fields of geology.[35]

The trap is then sprung. Semper rhetorically asks if drift theory should not be pursued by some geologists as a 'working hypothesis'? It would then be legitimate for them to set aside for the moment the theoretical difficulties and inconsistencies facing drift theory in order to concentrate on empirical investigations suggested by it — an approach, he notes, that is consonant with 'the English way of thought'[36] and with the analytic function of working hypotheses. Absolutely not! Wegener's hypothesis is a synthetic one and if it be used as though it were an analytic working hypothesis 'One truly sews a rotten bag with good thread'.[37] In Semper's view, then, drift theory not only fails the test for a synthetic theory of geology; as a synthetic theory it is ineligible for consideration as even an *Arbeitshypothese*.

A flurry of articles and reviews greeted the second (1920) edition of *The Origin*. Wilhelm Schweydar believed that drift was geophysically possible but cautioned that the claimed present movement of Greenland had yet to be confirmed and that, in any case, 'Geographical and geological findings must first decide if continental drift has actually occurred (in the past)'.[38] Similar views were expressed by Friedrich Burmeister who gave Wegener's cited Greenland data a thorough raking but added that this did not imply his opposition to the theory as a whole.[39] Bruno Schulz,[40] Leopold von Ubisch,[41] and Franz Kossmat[42] all gave qualified endorsement to Wegener's theory. Bruno Kubart, though critical of many aspects, suggested that a combination of ideas drawn from the older theories and from Wegener could form a suitable basis for further work.[43] For these geologists, *contra* Semper, drift was a promising hypothesis despite some reservations.

Staunch opposition came from Friedrich Nölke, Albrecht Penck, and Erich Jaworski. Nölke rebutted the geophysical arguments and criticized the forces employed by Wegener. He concluded his attack with a methodological sally directed at Wegener's claim that even if the physical foundations as advanced be unsound, the theory as a whole would not be jeopardized:

> As long as a theory which relies predominantly on empirical proofs but at the same time makes use of physico-mechanical concepts cannot be physically established beyond question, every other hypothesis which is internally consistent and without contradiction and perhaps lacks only certain empirical grounds, is to be preferred to it.[44]

Further research can eliminate empirical problems but faults in the use of physical theory are irremediable. This implied claim for the primacy of consistency with physics in evaluating geological theory did not play a large part in the German reception of Wegener's theory. It loomed much larger, as we shall see, in the response of the Anglophonic community. Penck cited the earlier papers of Diener and Soergel in his rejection of Wegener's evidential base, particularly pattern-matching and the Greenland data: the theory, though extremely 'seductive', 'is built up not on conclusive proofs but rather mostly on circumstantial evidence . . . [which leads] in many cases, to other conclusions'.[45] Jaworski too drew heavily upon previously-stated objections in his 'critical summary' which was directed especially at the pattern-matching issue. He admitted that the theory did have some advantages but claimed that these were more than offset by problems: 'It serves no purpose to exchange old notions for new ones which explain, it is true, what hitherto was not understood, but heap up in return fourscore new difficulties.'[46] His most savage comments were, however, directed not at Wegener but at Nicolaas Wing Easton[47] who had taken up drift theory and extended it to argue that the continental blocks themselves were mosaics of smaller blocks.[48]

The early German response to drift theory could be summarized as follows: a sizeable group endorsed it with few reservations; a few were openly hostile; the largest group saw considerable merit in it but acknowledged the various problems it raised and cautioned that much further research was needed. There was a general attitude that Wegener's 'hypothesis' yielded attractive explanations of certain gross features of the earth's surface but there was considerable disagreement over whether or not it did — or could — explain these features in detail. Most believed that 'facts' were — or would be — sufficient to decide the issue. German critics of CD, excepting Semper and a few others, did not deploy 'method weapons' and, in particular, again excepting Semper, there were no attempts explicitly to attribute the construction and defence of this (erroneous) hypothesis to improper methodology.

German and, more broadly, European geologists seem to have been rather more attuned to current methodological views than their Anglophonic counterparts, and to have been sympathetic toward or at least tolerant of the method of hypothesis. Wegener presented his theory in something of the style of a hypothetic-deductive argument and this alone did not send his European opponents away to sharpen their method weapons. Rather, his hypothesis seems to have been evaluated according to criteria associated with the method of hypothesis: the requirement of independent or collateral support, broad explanatory scope, a wide range of confirming instances, the progressive character of CD as compared with its rivals and predecessors, possible 'crucial experiments', and so forth. This attitude was manifested in the concern of both Wegener and his critics with the Greenland measurements as an '*experimentum crucis*', the disputes over pattern matching as constituting confirming or disconfirming instances, and the thrust of Semper's, Soergel's, and Nölke's 'method talk', directed as it was not against hypothesis *per se* but at the inconsistencies and failures — as they perceived them — of CD in following the method of hypothesis. It was otherwise with many Anglophonic critics of drift theory who usually argued from a naive inductivist position. It is noteworthy that neither the German nor the Anglophonic critics of CD, even when deploying method rhetoric and method weapons, made reference to the formal methodological literature then available. This suggests that methodological treatises as such had little direct influence either as resources or as sources for prescriptive and normative judgments in the CD debates.[49] When some basis for these criticisms was advanced, it was usually in terms of previous successful applications.

3.2. *Early Reactions of the Anglophonic Geological Community*

Wegener's theory was largely unknown in the Anglophonic world until the 1920s. Reviews of his publications and the appearance of a translation of his major work sparked off intense debates and discussions in England, South Africa, and the United States. Topics of dispute extended from the interpretation of a geological deposit near Boston, Mass., to the origin and physical structure of the earth. Underlying, and sometimes surfacing in, many of these debates were methodological claims with purported reference to issues and conflicts which ranged from the admissibility of specific procedures and classes of data to the nature of a proper methodology of geology and of science, and to the mutual relations of the various branches of geology and of geology to other sciences. These methodological disputes seem to have been more widespread and more heated among British and American geologists than among their European colleagues.

The earliest reactions to Wegener in the English-speaking world foreshadow much of the later, fuller discussion. Philip Lake in a brief paper in 1922 criticized Wegener's presentation:

> Wegener himself does not assist his reader to form an impartial judgment. Whatever his own attitude may have been originally, in his book he is not seeking truth; he is advocating a cause, and is blind to every fact and argument that tells against it.[50]

In short, he accuses Wegener of following an unscientific methodology, at least in the presentation of his case. Harry Fielding Reid reviewed the work of Wegener on CD and that of Wegener's father-in-law, Köppen, on polar wandering and reached a similar but more explicit conclusion:

There have been many attempts to deduce the characteristics of the earth from a hypothesis; but they have all failed. . . . This is another of the same type. Science has developed by the painstaking comparison of observations and, through close induction, by taking one short step backwards to their cause; not by first guessing at the cause and then deducing the phenomena.[51]

These comments of Lake and Reid have no exact counterpart in the corresponding German literature. Their charges that Wegener was deductive, speculative, and highly selective with regard to the total array of data were often repeated by Anglophonic opponents of CD and constitute a first distinctive feature of the early critical response. These charges were not altogether without foundation, given Wegener's own description of the invention and elaboration of his theory and his adoption of a hypothetico-deductive style. Rather than dismissing this form of criticism as a simple instance of the genetic fallacy it is more useful to regard it as an instance of the deployment of 'method talk' as a weapon to defend preferred theories and as a shorthand statement of belief that competitor theories are unacceptable on other grounds. Some support is given this latter view by the observation that neither Lake nor Reid rested his case solely on broad methodological principles: each accompanied his 'method talk' with a catalogue of detailed objections to specific data and generalizations employed or ignored or mis-stated by Wegener. This combination of 'method talk' and references to specific conceptual and factual flaws contributes not only to their 'demonstration' that CD is false, but also to the explanation of why Wegener and others could nonetheless hold to it. On this account, Wegener proposed and defended his (erroneous) theory, firstly because his 'facts' were wrong, and secondly because he failed and continued to fail to adhere to the canons of the proper scientific method. He had sinned methodologically and — irredeemably — he refused to repent.

The import of the remarks of Lake and Reid should not be underestimated. For many Anglophonic geologists the labelling of a theory as deductive, speculative and selective was tantamount to labelling it worthless and liable to lead to serious errors. Many of their specific methodological objections could be interpreted as standard gambits against any hypothetic-deductive theory.[52] However, the vehemence with which Lake, Reid and others put their case, the passionate language which they sometimes employed and the tenor of the claims they pressed, suggest that some opponents of CD believed more than a preferred theory to be at risk. This may have contributed to the bitterness of some of the discourse in the famous 1926 debate in New York.

4. THE 1926 AAPG DEBATE

General methodological criticisms aimed at Wegener's style of theorizing and narrower methodological ones targeting specific areas of his theory were a constant refrain in the ensuing fuller, more formal discussions in the later 1920s and the 1930s. The most famous episode was the 1926 symposium on CD held under the auspices of the American Association of Petroleum Geologists and chaired by W. A. J. M. van Waterschoot van der Gracht, a 'pro-drifter'. Wegener himself participated as did a number of the major figures in the Atlantic geological community. The pattern of criticism was as expected: Several participants accused Wegener of violating methodological norms. Rollin T. Chamberlin, professor of geology at the University of Chicago, caustically asked 'Can we call geology a science when there exists such difference of opinion on fundamental matters as to make it possible for such a theory as this (CD) to run wild?'[53] After a reference to 'Wegener's own dogmatism'[54] Chamberlin listed a number of objections to the theory, in particular that it did not seem to be part of a comprehensive geological philosophy nor a general theory of earth behavior but rather an *ad hoc* 'explanation' of unrelated facts. He concluded:

> Wegener's hypothesis in general is of the foot-loose type, in that it takes considerable liberty with our globe, and is less bound by restrictions or tied down by awkward, ugly facts than most of its rival theories. Its appeal seems to lie in the fact that it plays a game in which there are few restrictive rules and no sharply drawn code of conduct.[55]

Bailey Willis commented that Wegener's book 'leaves the impression that it has been written by an advocate rather than by an impartial investigator'.[56] Charles Schuchert, the doyen of American paleontologists, after disputing numerous items of data employed by Wegener, caustically referred to his methodology as follows:

> . . . the whole trouble in Wegener's hypothesis and in his method is, as we have said, that he generalizes too easily from other generalizations, and that he pays little or no attention to historical geology or to the time of the making of the structural and biologic phenomena discussed. It is not, as he says, that the detailed worker cannot see the

forest because of the many different trees, or that the paleontologists need a geophysicist to show them the road on which they should travel. Facts are facts, and it is from facts that we make our generalizations, from the little to the great, and it is wrong for a stranger to the facts he handles to generalize from them to other generalizations.[57]

Good inductions are based on an adequate set and a proper understanding of 'the facts'; bad inductions, on an inadequate or poorly understood set or upon an ignorance or rejection of the qualifications attached to the inductions of others. The correct use of the proper method yields valid results. Schuchert, in his summary, admits the possibility of limited lateral crustal movement, but continues to question the large-scale movement postulated by Wegener. Indeed, he implies that the acceptance of Wegener's theory would amount to a surrender of the precious principle of uniformity:

We are on safe ground only so long as we follow the teachings of the law of uniformity in the operation of nature's laws. The battle over the permanency of the earth's greater features introduced by James D. Dana has been fought and won by Americans long ago. In Europe, however, this battle is not yet fought to a conclusion . . . .[58]

Edward Berry, professor of paleontology at Johns Hopkins, made perhaps the most intemperate judgment of Wegener's method:

My principal objection to the Wegener hypothesis rests on the author's method. This, in my opinion, is not scientific, but takes the familiar course of an initial idea, a selective search through the literature for corroborative evidence, ignoring most of the facts that are opposed to the idea, and ending in a state of auto-intoxication in which the subjective idea comes to be considered as an objective fact. You will doubtless recognize this sequence as exemplified by several American geological philosophers.[59]

He concluded his very short and dogmatic presentation with the excuse that he felt 'that it is as futile to discuss the interior of the earth until we have more facts, as it is to attempt a 'scientific' proof of a future life, or the divine inspiration of the Pentateuch'.[60] In brief, Wegener was not following an acceptable scientific method nor was his theory deserving of the attention of real geologists. A similar line was taken by Joseph Singewald.[61]

These texts are sufficient to establish that heavy methodological artillery was fired at Wegener. In addition, almost every piece of evidence he used was called in question by one or more opponents, as was the validity of every technical method which he employed; for example, geological pattern-matching or paleoclimatic pole-mapping. Digests of these criticisms formed the staple for most of the discussions of CD through the late 1920s, 1930s, and later. Rather than dwelling further on these matters, I wish to examine some of the methodological issues raised with special reference to the efforts of Wegener and his supporters to rebut or redirect them.

5. SPECIALISM VS GENERALISM

One feature of the critical response to CD may be characterized as specialism versus generalism. It may be conjectured that geologists themselves regarded increasing specialization as a sign of the growing maturity of their science. Almost without exception, introductory textbooks reeled off lists of subdivisions within geology. It is not without a trace of pride that Gregory remarked on the difficulties of keeping up with the literature even in one's own corner of the discipline.[62] Specialism might be construed as a sign of professional maturity but it did not come without some problems, problems which surfaced in the CD debates. Henri Poincaré remarked apropos of physics at this time:

> The English scientist does not try to erect a unique, definitive, and well-arranged building; he seems to raise rather a large number of provisional and independent constructions between which communication is difficult and sometimes impossible.[63]

The description might well have been applied to geology as well.

Wegener stressed that one of CD's great virtues was that it brought into relation not only geological data but also generalizations and hypotheses hitherto regarded as unconnected. He tended to concentrate on broad, plausible correlations rather than detailed, precise fitting of theory with data and to accept – perhaps uncritically – the conclusions of others in their specialist areas as grist for his generalist mill. As Du Toit was to later put it, CD 'represents the holistic outlook in geology'.[64] In retrospect, Wegener's stance could be construed as an attempt to ward off the slings and arrows of outraged specialists, for though he makes copious reference to the various specialist literatures, he himself was a specialist in meteorology rather than geology and believed that his theory should be judged as a whole rather than as a collection of details. Even if this construal be correct, the tactic was generally unsuccessful both in the early stages of the debate[65] and later. Some few geologists did accept the point that the theory should not be rejected on the basis of matters of detail.[66] This position could not easily be squared with a naive inductivist methodological stance. For these few, commitment to a theory took precedence over methodological rigour or, perhaps, they were of the view that apparent empirical inconsistencies could be dissolved by further research.

Geologists, however, were more typically concerned with the nuts and bolts — or rather the sands and stones — of the science than with bold, overarching synthetic theories. Wegener's opponents might well substitute the term 'superficial' for 'holistic'.[67] Arthur R. Hinks in a symposium on the problems of the earth's crust referred to 'specious theories of continental drift and wandering of the poles' which had been 'built on slight foundations, largely of convenience'[68] and condemned 'geologists who do not scruple to play fast and loose with the principles of every other science than their own'.[69] He went on to imply that Wegener and his supporters had violated the quintessentially British virtue of playing by ' . . . the Rules of the Game', arguing that the cardinal rule was that no player 'may adopt a principle contrary to the best opinion in another man's subject'.[70] This Wegener had surely done. His specialist critics, each of whom by definition was in his own area more knowledgable than Wegener, had little difficulty in finding him guilty of greater or lesser crimes of commission or omission in his presentation. Each could then conclude that since CD was unsatisfactory in his own limited area for a variety of reasons, it should be rejected *in toto*.[71]

Methodological rhetoric may well have been deployed in this connection by some specialist critics. Confident each might be of the inadequacies or inaccuracies of Wegener's theory for his own specialty, but to go outside one's special cranny of competence to make a general judgement of the theory as a whole could present problems. A specialist in, say, paleontology might well use 'method talk' in conjunction with specific 'empirical' rebuttals of Wegener in that field. That is, he might assign the specific failure(s) to methodological flaws.[72] The method weapon thus forged could then be aimed at Wegener's approach and his theory as a whole.[73] By 'transference', then, specific failures might give rise to general methodological rhetoric. This provides a plausible account of some of the methodological discourse.

It should be noted that some 'specialist' opponents of CD displayed seeming inconsistencies between their methodological rhetoric and their practice, at least in formal texts. Schuchert, at the level of general method discourse, attributes both the errors in CD and Wegener's proposal of it to his penchant for generalizing from the generalizations of others. But at the level of 'empirical' rebuttal of the details of the theory, he freely concatenates arguments drawn from the writings of specialists in other areas and cites their disapproval of CD as support for his own negative judgment; for example, Diener, H. D. Washington, Lake, A. P. Coleman, Reid, Berry, and others. Not only is this a divergence between principle and practice, it is a further illustration of the asymmetrical assumption that sources criticizing or conflicting with Wegener are reliable whereas those cited by or in agreement with him are open to question.

5.1. *Importation of Physics-Style Arguments*

An important methodological issue arising in the course of debate over CD was the perceived relation of geology to the other sciences, especially physics. Prior to 1900 physics, with the notable exception of Kelvin's determination of the age of the earth, had had limited intercourse with geology. Geology was certainly not considered an exact or mathematical science. At the turn of the century, however, the discoveries that radioactive decay released heat and that there were substantial deposits of radioactive material in the crust of the earth brought into question a number of the assumptions underpinning the simpler versions of contractionism.[74] T. H. Holland in his "Presidential Address" to Section C (Geology) of the 1914 British Association meeting colourfully described the impact of these findings upon a multiplicity of geological theories as follows:

> Thus, we see, whole fleets of hypotheses have been launched on this sea of controversy: some of the craft . . . have foundered after bombardment by the heavy missiles classically reserved for use by geologists; others, though built in the dockyard of physicists, have suffered from the spontaneous combustion set up by an inadvertent shipment of radium.[75]

Other geologists initially took a more conservative view of how these developments might impinge upon geological theory. Gregory, for example, remarked in 1915 that ' . . . this new branch of science is still so speculative that geologists may be excused from any immediate modification of their fundamental principles'.[76] Wegener, however, did not wait for a final, unanimous, authoritative verdict from physics.

Taking his cue from Joly and others, he regarded contractionism as decisively refuted by geophysics. Although he borrowed the conclusions of physics in this regard, he certainly did not borrow the style or form of physics arguments in propounding CD.

A major blow to CD came from this quarter in the person of Harold Jeffreys, later Plumian Professor of Astronomy at Cambridge, who imported not only the conclusions of physics but also its form and rhetoric into the CD debate. In his highly influential 1924 monograph, *The Earth*, Jeffreys admitted the difficulties presented by a lack of empirical data but argued that on the basis of well-grounded physical laws he could extrapolate to a model of the structure and processes of the interior of the earth although it was necessary to introduce 'some *ad hoc* assumptions'.[77] His stance was that physics should lay down the boundaries for any acceptable geological theory, and that primacy should be placed upon quantitative agreement 'of theory with fact'.[78] It is the task of the geophysicist to provide a general theory; that of the geologist to fill in the details.[79] He admits implicitly that this results in a new style of discourse about the earth, remarking that the book may be too mathematical for geologists but expressing the hope that 'if the geologist cannot follow a part of the book, . . . he will omit it and go on to the next non-mathematical passage, trusting that someone else will point out any intervening mistakes'.[80] There follows page after page of mathematical physics, with only occasional forays into geological data, leading inexorably — so it appears — to the conclusions that a somewhat more sophisticated contractionist model is in accord with all of his computations and that CD from the viewpoint of physics is an 'impossible hypothesis' unless 'forces enormously greater than any yet suggested are shown to be available'.[81]

It is clear that Jeffreys regarded his approach as the only methodologically sound one to follow and that his conclusions could not be challenged. The tone adopted throughout the argument is that of an impartial physicist merely applying well-established mathematical tools in an effort to remove the undergrowth of wild speculation in a juvenile science not yet schooled in proper scientific procedures. One cannot argue with numbers: they are exact, true, and impartial. Physics, the senior science, has spoken even if not in a language understandable to geologists. Had this approach been followed earlier, ' . . . many wild theories might have remained unpublished or, if published, have received only their fair share of attention . . . '.[82] The adoption of his stance means that 'Geophysics is no longer a field for uncontrolled speculation . . .'.[83]

Jeffreys' staking of a hegemonic claim for physics in the CD debate was generally well-received despite a number of potentially countervailing circumstances. It was clear to some of the proponents of CD that his methodologically-packaged claims were disingenuous or, more kindly, self-delusive. The mathematical clothing of his presentation disguised his practice of selecting from a range of available data and assumptions those which produced the desired result. Moreover, geologists both before and even after Jeffreys' entry into the fray were very hesitant about drawing firm conclusions about geological processes and structures within the earth from the acknowledged scanty evidence available to them from the study of surface features, bore holes, mines, seismological measurements, gravity irregularities, and similar data.[84] This hesitation perhaps was in accord with the naive inductivism espoused by many geologists, but it contrasted rather sharply with an acceptance of Jeffreys' free use of '*ad hoc* assumptions' and mathematical deduction in an argument which in form was analogous to a demonstration in geometry. Finally, it was ironic — though the irony was surely lost on geologists — that Jeffreys pushed forward the methods he used as the methods of physics and therefore as the preferred methods of all the sciences at a time when developments in relativity theory, and especially in quantum mechanics, could be seen as undermining this methodology. Jeffreys' pronouncements were nonetheless welcomed and accepted by many opponents of CD. We may well suspect that this co-optation of Jeffreys into the community of geologists owed less to the force of his arguments or to his methodology than to the fact that the conclusions of those arguments were inimical to CD and favourable to the older theories.

This uncharitable interpretation of the use made of Jeffreys is given some credence by the remarks of the opponents of CD, particularly those of Hinks at the 1931 meeting of Section E (Geography) of the British Association. Hinks, an astronomer and geographer, after promulgating his version of the methodological 'Rules of the Game', claimed that the best opinion in geophysics — which must not be contravened by those working in other fields — was Jeffreys', whose second edition of *The Earth* 'is an impressive canon of the laws of geophysics as they are formulated by the mathematicians'.[85] Having set up physics and, more particularly, Jeffreys as the jury for any theory of geology, Hinks then cites Jeffrey's conclusions as *prima facie* refutations of Joly,[86] Wegener,[87] and – inferentially – Arthur Holmes.[88] Hinks's tactics did not go unchallenged. In the discussion which followed, S. W. Woolridge and Holmes, geophysicists, and G. C. Simpson, a meteorologist, all of whom were pro-CD, rejected Hinks's stand completely. Indeed, Simpson went so far as to say that until mathematicians could provide better answers instead of merely criticizing those that had been offered by Wegener, he proposed to continue 'not playing the game'.[89] In short, both sides of the debate seemed disposed to accede to the hegemony of physics if it accorded with their preferences in geological theory and, conversely, to reject it if it seemed to conflict with them. This is, however, not to say that Jeffreys had no influence on the course of the debate nor that the defenders of CD closed their minds to his case.

Jeffreys' presentation did have a powerful impact upon the CD debates. It was readily acknowledged even by the supporters of CD that the mechanisms for drift that Wegener had put forward were quantitatively inadequate. The counter-strategies utilized included the following: the postulation of alternative mechanisms, a path pursued by Daly, Du Toit, and Holmes; the methodological claim that the lack of a known mechanism should not be cause for rejection of CD provided there were ample evidence that CD had in fact occurred, a claim made by several including van der Gracht[90] and Daly;[91] attempts such as those of Holmes and Rastall to criticize or rebut Jeffreys' argument; or some combination of these.

It would be fair to say that few of Wegener's supporters were equipped with sufficient physics to carry the fight to Jeffreys. Holmes, who had begun as a contractionist but who moved gradually to Wegener's camp, took issue with Jeffreys in 1925. Jeffreys had calculated that contraction of the earth was almost exactly enough to account, for example, for estimated compressions of 118 kms for the Alps, 100 kms for the Himalayas, and 70 kms each for the Rockies and the Appalachians. These estimates were quite selective in that they represent the lower end of a series of estimates which range in the case of the Alps up to 1200 kms. Holmes criticized Jeffreys on this point,[92] one crucial to Jeffreys' claims of quantitative agreement between theory and fact. The 'facts' were rather elastic. He also disputed Jeffreys' assumptions concerning the heat gradient and the distribution of radioactive material, key elements of his computations.[93] His counterattack, though well aimed, did little to defuse the geophysical objections to Wegenerian CD. On the one hand, most of Holmes's objections were couched in terms of objections to application of theory, data fit, and 'fudging'; that is, objections perhaps more typical of discourse in physics than in geology. On the other hand, Holmes's writings usually did not have the full-dress mathematical physics style of Jeffreys. Even a superficial comparison of page after page of Jeffreys' equations and calculations with Holmes's articles, in which geological discussions are buttressed but not dominated by complex mathematics, suggests that Holmes wrote geophysics so that it would be understandable to geologists but that Jeffreys wrote in the style of mathematical physics and indeed wrote deliberately 'over the heads' of most of those involved in the CD debate. Finally, Holmes may have made a tactical error in freely admitting that the work of Jeffreys, for all its faults, had clearly established that Wegener's twin forces of tidal drag and 'pole flight' were manifestly inadequate to power the continental blocks through the sima.[94]

For Holmes, this did not entail an acceptance of Jeffreys' counter-position. He flatly rejected it, arguing that it failed to account for several important phenomena, that Jeffreys' reasoning on other matters was unsound, and that Jeffreys' version was quantitatively inadequate with reference to the magnitude of known crustal folding. Finally — and revealingly — Holmes pointed to 'the fact that under the hypothesis (as presented by Jeffreys) continental drift is manifestly impossible'.[95] The inference is that even if his specific objections had been met, any form of contractionism which made no provision for large-scale lateral motion of the continents would be unacceptable. Holmes devoted himself to the development of the theory of convection currents in the mantle as the mechanism for drift.[96] He was well aware of the difficulties he faced:

> Unfortunately, once the straightforward conception of a continuously cooling earth is abandoned, the possibilities become hard to visualize, unwieldy in their complex interrelations, and difficult to check except by the most complete details of geological history, and of the physical properties of materials under familiar conditions.[97]

Those who were not already well-disposed toward continental drift might well regard such an attempt as not worth the effort. Jeffreys' book became the standard advanced text on geophysics, going through some six editions. His physical arguments against continental drift were frequently cited, both in the research and the textbook literature. Holmes' criticisms, though expanded in later publications[98] were picked up only by a few CD proponents.[99]

Robert H. Rastall mounted a more general rebuttal of physics-based objections. He pointed out three instances: Kelvin's determination of the age of the earth in which mathematical theory 'led to completely wrong results'; ice ages, for which it 'failed to produce an explanation'; and the process of nappe formation during orogenesis with respect to which it 'has not yet tackled one of the fundamental points of tectonic geology'.[100] Opposition based on physics was therefore methodologically ill-founded as it mistook the inability to discover a mechanism for drifting for the impossibility of drift itself. The latter had to be judged on the basis of geology, not physics. He argued for a suspension of 'mathematical judgment' on the problem of continental drift until 'geology and other sciences of observation have shown whether such drift did or did not take place'.[101]

5.2. *Wegener's Counter-Moves*

Wegener himself attempted to parry Jeffreys' attack by downgrading the importance of geophysics to his case. In his early publications he had argued that geophysics should occupy an exalted position in the evaluation of geological theories[102], and as late as 1927 he had affirmed that 'Were geophysics to come to the conclusion that the drift theory is wrong, the theory would have to be abandoned by the systematic earth sciences as well, in spite of all corroboration . . . '.[103] In 1929, however, he explicitly retracted these words and put forward the very different argument that no one specialty of the earth sciences had primacy, that evidence from all specialties had to be combined, and that ' . . . the geophysicist easily finds himself on the wrong track if he does not keep constantly in mind the results provided by these branches of science [paleontology, zoogeography, phytogeography] in order to check his own'. In sum, the apparent failure of his theory in one area should not result in its rejection.[104] This principle could generate serious methodological problems for highly synthetic theories: it could be driven to the questionable conclusion that to be acceptable a general theory need have only some measure of plausibility in a scattering of branches of a discipline even though seemingly decisively refuted in several others. Whatever be the merits or demerits of this principle, it is clear that Wegener's reformulation of this article of his *public* methodological creed came in response to the empirical and methodological brickbats he had received.

As a second line of defence Wegener tried to occupy the tactical high ground of inductivism. He argued in later editions of *The Origins* that, inductively, one could only thus far establish that drift had taken place. To demand a mechanism or — inferentially — to focus attention on the inadequacies of the mechanisms he had tentatively put forward was misguided. There was not yet sufficient information and, more particularly, the deductive demonstration whereby drift and associated consequences could be derived from a geophysical formula was not yet possible. As he put it, 'The Newton of drift theory has not yet appeared. His absence need cause no anxiety; the theory is still young . . . .'[105] This remonstrance and the odour of inductivist sanctity which it emanated smell a bit fishy. Later in this same edition, in discussing Jeffreys' work, he admits that the 'pole-flight' force was inadequate to explain both drift and orogenesis. In line with the method talk he had deployed earlier he claims that this should not tell against his theory. Rather than resting on the high ground, however, he proceeds to argue in a most non-inductive manner that the existence of fold-ranges 'is evidence that, in the course of the earth's history, at least at times, displacement forces have arisen which were much larger than the pole-flight force',[106] a suggestion which doubtless would have sent shivers down rigid uniformitarian spines. Wegener no less than his opponents seemed to have used the notion of the proper scientific method — in public statements — in a very flexible fashion.

Wegener regarded as damaging to his case the criticisms that his method was deductive, selective, and therefore unscientific. His rebuttal further suggests the flexibility and presumptive importance of 'method talk' as a resource in inter-theoretic debates. He added the following bald assertion to later versions of his presentation:

> The determination and proof of relative continental displacements have proceeded purely empirically, that is, by means of the totality of . . . data without making any assumptions about the origin of these processes. This is the inductive method. . . . The formulation of the laws of falling bodies and of the planetary orbits were first determined purely inductively, by observation; only then did Newton appear and show how to derive these laws deductively from the one formula of universal gravitation. This is the normal scientific procedure, repeated time and again.[107]

This statement is obviously intended as a refutation of the charge that he had followed an unscientific or, at least, an improper method in arriving at his theory and in presenting it. But it stands in seeming contradiction with his earlier-quoted account in this same edition of his path of discovery and development. It is tempting to treat this passage as a series of rhetorical flourishes aimed at defending his theory rather than as a seriously-intended methodological statement — unless we conclude that he was insensitive to the apparent conflict with his autobiographical comments. There is either an element of irony or of an underlying methodological tension evidenced in this injection of naive inductivism into an argument which had a hypothetico-deductive structure and flavour. It may be suggested that Wegener inserted this passage in order to disarm or placate his inductivist critics but no evidence is at hand to support this inference. Even had this been the case, the tactic was unsuccessful. Alternatively, we may read the autobiographical comments as an account of how he 'actually' arrived at his theory and the later, inductivist version as his own methodological reconstruction — his 'method like' version of this same process of discovery. The two accounts would no longer be in conflict because they are accounts at different levels of discourse.

The attempted refutation was, in any case, unsuccessful. Even authors who presented an impartial or moderately favourable treatment of drift theory in the 1930s continued to comment disparagingly on Wegener's approach. Charles Nevin's widely-used text on structural geology describes it as 'a type of special pleading whereby he glosses over or completely ignores many of the difficulties and at the same time presents his viewpoint in the best possible light'.[108] Similarly, in his relatively judicious discussion of recent geological theories Steers commented that Wegener 'make some really brilliant suggestions, but frequently spoils them by [errors] . . . , or by the omission of evidence of considerable import which tells against his own views'.[109] Even Holmes acknowledged that 'There is a general agreement that Wegener's methods in advocating his particular group of hypotheses are to be condemned. His plausible selection of data, frequently erroneous age determinations, faulty analyses of causes and devious reasoning, have undoubtedly had the effect of weakening his case'. This admission made, he argued for the careful examination and assessment of the general theory of drift as opposed to Wegener's particular version of presentation.[110] He stressed that 'proving Wegener to be wrong is by no means equivalent to disposing of continental drift'.[111] Such a policy — had it been adopted — would not only have given breathing room for his own or another's version of CD, it would also have put the onus on the critics of CD virtually to construct a general version to demolish.

6. BEHIND 'METHOD TALK'

It is clear that methodological claims and counter-claims were a prominent feature of the early CD debate. What functions did such claims serve? What clues do these methodological disputes provide for an analysis of the episode beyond the surface inter-theoretic and theoretic-empirical conflicts?

Sufficient exegesis has been given to one function of method claims: they serve as resources for defending preferred positions and attacking opposing ones in intertheoretic or intratheoretic conflicts. Let us consider this function with regard to the multi-level analysis which we might give of such conflicts. At a simple level – indeed, a naive one – we could accept the arguments of Wegener's critics at face value and conclude as did they that his theory was incorrect (as it was in many details, judged by either current or contemporary standards) and that it was incorrect because he did not follow the canons of *the* scientific method. We may today dismiss such a view as absurdly untenable in the light of modern history and philosophy of science, but we should take care not to dismiss as absurd this response by Wegener's critics in our historical analysis. At least some early twentieth-century geologists, it appears, believed that there was one efficacious scientific method, that its employment had enabled the construction of modern geology, and that Wegener had reached different and erroneous conclusions because he had employed a different (and erroneous) method. For this group, 'method talk' provided a satisfactory explanation for Wegener's erroneous beliefs. I shall not enquire further into why they may have held such beliefs; that is, why they were captured by the illusion of methodological efficacy. Rather, I shall simply note this as one level on which methodological claims which might or might not bear a relationship to the methodology actually employed by the protagonists could function as a register of error-accounting. For others, it may have been no more than a handy weapon.

At a different level, the remarks and actions of the protagonists in the early CD debates raise a general question about the relation between theories and what are regarded as appropriate methods of generating and evaluating those theories. There is some correlation between opposition to CD and professions of inductivism and between support for, or at least tolerance of CD as a legitimate piece of science, and the method of hypothesis.[112] Larry Laudan has argued that methodologies are often parasitic upon scientific practice and, more to the point here, 'That the reception of methodological doctrines . . . has been determined more by the capacity of those doctrine to legitimate a preferred scientific theory than by their strictly philosophical merits. Correlatively, when methodologies are rejected it is often because of their inability to rationalize what is regarded intuitively as exemplary scientific practice'.[113] If this view be correct, and this analysis of the CD debates certainly does not contradict it, *then methodological issues can have some force in the evaluation of a new theory provided that there be doubt that the theory meets some criteria of exemplary scientific practice.* On this view, had the evidence for CD been thought to be overwhelming, not only would methodological criticism have been muffled but it is conceivable that Anglophonic geologists would have scrapped naive inductivism in favour of the method of hypothesis. As we have seen, this was not the case. This view casts into somewhat different relief the arguments over the merits and demerits of CD with respect to the available geological and geophysical evidence.

The passion and direction of some of the methodological rhetoric hurled at Wegener suggests that the some of the protagonists saw more to be at stake than the question of which theory was better. Some clues as to these issues are provided by passages already cited; others, by methodological side-issues not yet raised. Again at a simple level, there is a recognition by some of Wegener's critics that if his views be accepted, much of geological theory would become outmoded and the 'empirical base' itself would have to be reconstituted. For example, the Squantum 'tillite' near Boston simply cannot be 'tillite'.[114] Again, I leave this aside as an obvious reason for anxiety and, possibly, for the stridency of some of the rhetoric. A deeper source of anxiety may have been a putative connection between CD and geological catastrophism.

Wegener was believed by some — though certainly a minority — to pose a threat to the broad foundational principle of uniformitarianism, whatever that might have meant to different geologists at the time.[115] The history of the uniformitarian-catastrophist struggle and the relatively recent victory of uniformitarianism in American ensured that even had this principle been a purely internal, scientific one any perceived threat to it would be actively resisted. Moreover, the question had not been and was not an 'internal' matter and was therefore all the more likely to rekindle strong expressions of views.[116] Whether Wegener was or was not a uniformitarian is not at issue; the point is that several members of the geological community claimed that he was tinged with catastrophism. Given the broad range of the contemporary definitions of 'uniformitarianism', it is certainly possible that some of Wegener's opponents may have sincerely believed him to have transgressed their limits of an acceptable geological theory. It is also possible that other, more opportunistic opponents saw advantage in claiming such transgression of an ill-defined boundary. In retrospect it is not surprising that the taint of catastrophism could be sniffed in CD. Earlier proponents for the large-scale horizontal motion of continents include Richard Owen in 1857, Antonio Snider in 1859, W. H. Pickering in 1907, and F. B. Taylor in 1910. All had postulated some form of catastrophe as a mechanism; usually, the formation or capture of the moon. Arthur Keith of the the U.S. Geological Survey reviewed Wegener's theory in 1923 and concluded as follows:

> The hypothesis apparently must, therefore, fall back on a cause which is of the order of the *special convulsion of nature appealed to in the early stages of geologic work*. Such a convulsion, which affected a whole hemisphere, could have no probable cause except one of an astronomic nature. A cause of similar nature is mentioned by him in the form of a shifting of the poles.[117]

This attitude is also implicit in Schuchert's remark apropos of Wegener that 'We are on safe ground only so long as we follow the teachings of the law of uniformity in the operation of nature's laws'.[118]

Wegener was at some pains to distance himself from such associations and to avoid overtones of catastrophism. At the end of the first formal presentation of his theory he suggested that if the continents are moving at the same rate as they had since the breakup of Pangaea, this should be detectable by accurate measurements of longitude over time. In the case of Greenland, the displacement would amount to 14 to 28 meters per year. He claimed that sets of measurements which had been performed were in rough agreement with this but admitted that most of these were not particularly exact. Indeed, one commentator at a meeting of the Geological Section of the British Association, Professor H. H. Turner, expressed the view that most of these longitude measurements were 'so much waste paper',[119] and Jaworski, Penck, Burmeister and others dealt severly with Wegener's use of this data. In this connection it is notable that none of the Anglophonic opponents of CD picked up Arldt's argument[120] that even if Greenland were found to be moving *now*, this constituted no proof of its drift in the geologically recent past.

Arldt's contention ran counter to the strict uniformitarian view that the processes and agencies we observe now — and only those — are paradigms for past geological changes. Less strict uniformitarians opposed to drift might have regarded Arldt's concession of the possibility of current drift as far too generous.

Wegener was confident that future, more exact measurements would confirm continuing drift.[120] It is clear from his discussion that he did not entertain a catastrophist view of drift: once Pangaea had broken up, the continents continued to move uniformly along time's arrow. Similar discussions were included near the end of the first few editions of his monograph. He attempted to turn the presumed continuance of drift to his advantage in methodological arguments as well. In the third edition he advanced the claim that 'The displacement theory possesses the advantage over all other theories, with similarly far-reaching ideas, that it can be proved by accurate astronomical determinations'.[122] The methodological and evaluative importance of such a 'crucial experiment' was conceded by his critics.[123] In his face-to-face confrontation with Anglophonic critics at the AAPG symposium, the major positive argument he advanced concerned the Greenland drift.[124] He gave geodetic arguments a much more prominent place in the final edition of his work, moving this section to the head of his chapters of evidence:

> We begin the demonstration of our theory with the detection of present-day drift of the continents by repeated astronomical position-finding. . . . The majority of scientists will probably consider this to be the most precise and reliable test of the theory.[125]

This increased emphasis upon the measurement of continuing drift doubtless owed much to longitude determinations conducted in 1927 which gave an apparent drift of Greenland of 36 meters per year,[126] but its possible value in warding off methodological criticisms should not be underestimated. 'Method talk' played in this sense a *constitutive* role in his articulation of the theory. The *post hoc* analytical dividing lines between theory, method, and presentation — never sharply drawn — are erased by Wegener's alterations in the presentation of his case and in his implied research programme in an attempt to turn aside method attacks. His methodological point was a strong one, but it was severely undercut by 1936 observations which indicated that the figure of 36 m/year was widely inaccurate (as it was, by several orders of magnitude).[127]

The question of uniformity was more often raised indirectly as a methodological side issue. David White of the National Research Council devoted much of his paper at the 1926 AAPG symposium to asking why the continents broke up when they did — during a relatively peaceful and geologically recent epoch — and not earlier?[128] Others pursued this same point.[129] One implication was that either the continents had not broken up in the epoch stated by Wegener or that some cataclysmic event which had left no other traces had precipitated the breakup. A second implication was that drift had only operated within relatively recent geological time. Wegener himself did not directly address this point, but it is clear that he deliberately confined his remarks to the later stages of earth history. His supporters hoisted the flag of inductivism and espoused the view that the evidential base for earlier epochs was simply too fragmentary to allow defensible conclusions to be drawn; 'There is . . . nothing contained in those writings to suggest that such kind of drift would have been restricted to any particular period and would not have operated from an early stage in the earth's history, only such remote episodes are too obscure to be elucidated just yet.'[130]

Parallel and related to the charge of catastrophism was the association made between Wegener and, as Berry had sarcastically dubbed them, 'Geological philosophers'. Consider the following passage which bears some similarities to Wegener's style and claims:

> If I succeed in demonstrating the truth of the somewhat startling propositions with which I commenced this chapter, it will only be by bringing to bear upon the question . . . a thousand converging lines of light from a multitude of researches made by scholars in different fields of modern thought.[131]

There follows a lengthy catalogue of geological data, faunal and floral distribution patterns, archaeological data, and myths and legends purporting to establish a connection between Africa and South America. The author was Ignatius Donnelly (Edmund Boisgilbert); the book, *Atlantis: The Antediluvian World*, arguably one of the 'blockbusters' in the popular science literature of the late nineteenth century. No one in the sources consulted explicitly tarred Wegener with the Donnelly brush but it is unlikely that British and American geologists of the early twentieth century would have been unaware of Donnelly's mixture of fact and fancy or oblivious to the parallels with Wegener's presentation.

The contribution of a variety of social factors to the methodological exchanges occasioned by Wegener's theory is moot. Wegener's nationality may have played some part in the reception of his views by the non-German geological community, but it probably had only tenuous connection with the methodological criticisms advanced by members of that community. During the 'Great War' German science and scientists were harshly criticized for their presumed ambition to dominate the scientific world and their role in the German attempt to conquer the political world. This attitude continued after the war, finding official expression in the creation of the International Research Council, which had as one of its aims the exclusion of German scientists from the international scientific community, and unofficial expression in the opposition to the use of the German language at conferences and in publications. Support for such measures came not only from European spokesmen but also from Anglophonics on both sides of the Atlantic.[132] Hostility toward German scientists may have been instrumental in the delay until 1922 of discussions of Wegener's views in English-language journals.[133] Nutting probably spoke for many when he editorialized: 'In penance we now burn our German books and studiously avoid reading anything in that language. We are surprised to find how well we can get along without anything in that language and for how little we are really indebted to that nation.'[134] It is less clear, however, if this hostile attitude played into the methodological criticisms by non-Germans of CD from 1922.

In the war-time and post-war attacks on 'German science' there were suggestive barbs cast at methodological targets. Rignano wrote of 'the excessively analytical spirit of the Germans'.[135] Nutting complained of some of the methods of German pure and applied science as 'entirely indefensible by any standard'[136] and doubted 'whether a third of even the more eminent German scientists were free from this taint [of piracy and plagiarism]'.[137] William Ramsay, editor of *Nature*, commented on the general assent to the proposition 'that German science has not been remarkable for originality'.[138] The arsenal of Allied scientists became well-stocked with anti-German rhetorical ammunition in which methodology, epistemology, and morality were mixed together.[139] There is, however, no case yet examined in which a methodological attack on Wegener is explicitly coupled with broader, anti-German sentiments. Nonetheless, this atmosphere may indirectly have borne in several ways on the evaluation of his views by non-German members of the geological community in the 1920s and 1930s.

One possible move would be to call into question Wegener's originality. Opponents could make this move in order to denigrate or dismiss as nothing novel his proposal of continental drift; proponents, in order to separate the idea of large-scale horizontal continental displacement from any taint of 'teutonism'. Either move would be in accord with propaganda directed against the 'creativity' or 'originality' of German scientists. There is, however, no evidence that those who championed Frank B. Taylor or Osmond Fisher[140] consciously did so for such reasons. Moreover, questions of priority did not seem to have had overt relation to methodological criticisms of Wegener.

The perceived role of German science and scientists in German society may have contributed in some small way to opposition to Wegener. German scientists before and during the war were thought to be influential in German political affairs.[141] This notion was strengthened in the immediate post-war years by German academic and political leaders who regarded science as a major resource left to the prostrate nation and as an important vehicle for regaining national prestige.[142] Any theory put forward by a German scientist might therefore be seen by some German and non-German scientists as a potential weapon in the struggle for national re-assertion. As in the case of priority questions, however, this line of argument bears on the general opposition to Wegener rather than on the specific methodological criticisms, the latter being merely one aspect of the former.

The background of hostility to German science may have had a further but more diffuse, more subtle, but more significant influence upon the methodological criticisms of Wegener. Let us assume that geologists of the period believed that there was a unique, efficacious methodology for the discipline which, if carefully followed, would yield objective, correct, morally neutral knowledge. If German science and scientists were regarded as having been twisted and distorted, this must have occurred through distortion or abandonment of the canons of this presumed proper methodology. The argument could then run as follows: Wegener has not only arrived at an incorrect conclusion, he has also failed to put a proper case for it; therefore, he has not followed the method in that he has not been objective and has not considered all the relevant data and so forth; therefore, he is unscientific; and this is precisely what might be expected from a German scientist!

Each of the three possibilities outlined may have been realized, particularly in view of the stridency of anti-German propaganda in the years preceding the early discussions of Wegener's theory of continental drift. No direct evidence has yet come to hand to indicate that such matters did in fact impinge upon the methodological discussions. There was a degree of national differentiation in the reponse to Wegener and this was noted by some of the participants.[143] Moreover, the methodological criticisms which were made were couched in or supported by reference to specific geological (as opposed to national) matters. The most that could be said is that whatever might have been the actual nationalistic contribution to methodological criticisms, such criticisms were 'overdetermined' by the way in which Wegener presented his case and by the prevailing views of the non-German geological community.

7. CONCLUSION

This examination of the early debates over Wegener's CD theory has shown that 'method talk' could function in a variety of ways. At a simple level methodology was a flexible resource which could be used to fortify theoretical and empirical positions, to undermine opposing camps, and to repulse method sallies by enemies. In accord with the work of Mulkay, Gilbert, and others, 'method talk' served as one register for scientists' accounting for (erroneous) beliefs held by others. Moreover, as in the case of the wave theory, scientists of differing methodological persuasions might evaluate quite differently the same theory according to its apparent exemplification of or conflict with the canons of their preferred methods. At a somewhat more complex level, method discourse in the instance of 'uniformitarianism' and the Greenland data may have played a constitutive role in the elaboration and presentation of the theory itself. Finally, method rhetoric was an acceptable mode of couching arguments and counter-arguments touching on the relationships among specialties in the same discipline, among different disciplines, and between science and the wider social order. Whether or not scientific methodology be efficacious in the generation and evaluation of knowledge claims, methodological claims and counter-claims certainly operated in and contributed to this episode of conflict over knowledge claims.

ACKNOWLEDGEMENT

I wish to acknowledge the very useful comments and criticism received from Rachel Laudan and the editors of this volume. The research was partially funded by an Arts Faculty Research Development Grant from the University of Melbourne.

NOTES

1 Rollin T. Chamberlin, 'Some of the Objections to Wegener's Theory', in: *Theory of Continental Drift: A Symposium on the Origin and Movement of Land Masses both Inter-Continental and Intra-Continental, As Proposed by Alfred Wegener*, W. A. J. M. van Waterschoot van der Gracht (ed.), (Tulsa, Okla., 1928), pp. 83—87, at p. 87.

2 A. L. Du Toit, *Our Wandering Continents* (Edinburgh, 1937), p. 3.

3 Michael Mulkay and G. Nigel Gilbert, 'Accounting for Error: How Scientists Construct their Social World When they Account for Correct and Incorrect Belief', *Sociology* **XVI**, 1982a, pp. 165—183; Mulkay and Gilbert, 'Warranting Scientific Belief', *Social Studies of Science* **XII**, 1982b, pp. 383—408.

4 Discourse analysis may be applied to a plethora of materials. 'Informal' texts might be secret tape-recordings of laboratory conversations or carefully-structured interviews with elaborate protocols. 'Formal' texts may include not only journal articles and monographs, but also published minutes of meetings, accounts of discussions of papers, and similar documents representing considerable variation in immediacy, 'polish', and editorial control. The division between 'informal' and 'formal' is both hazy and problematic. Such distinctions as 'spoken' vs 'written', 'unpublished' vs 'published', 'causal' vs 'deliberate' and 'spontaneous' vs 'reflective' may be adequate for the sociologist or the ethnomethodologist but may be unsatisfactory for the historian, particularly one whose materials consist solely of printed sources. This study, which *is* based on printed sources, employs the distinction of 'unpublished' vs 'published' and consequently is a study of 'formal' texts only. The implied difficulties are mitigated at least in this instance by the rich character of the publications examined. These extend from almost verbatim blow-by-blow accounts of discussions to specialist monographs. Within this range of formal texts appear many of the registers of error accounting alluded to in Mulkay and Gilbert's analysis of less formal texts. See *op. cit.* (Note 3, 1982a).

It is plausible that some types of social accounting found at informal levels might not be openly expressed at formal levels and further, that even if the same or similar register be employed at different levels, the actual wording might vary significantly. In the formal texts examined in this study 'method talk' seems to be the register most frequently used for error accounting. On reflection this is not surprising. The list of interpretative resources employed by scientists in informal texts according to Mulkay and Gilbert (*Ibid.*, p. 176) is quite extensive though not exhaustive and includes prejudice, extreme naivety, rushing in too quickly, subjective bias, irrationality, unfamiliarity with the literature, stupidity, concern for reputation, and many others. Most of them would be regarded as 'improper' or 'unscientific' or even libellous if deployed in formal texts but most could be framed more acceptably in 'method talk'. It is open to historians using even the most formal texts to essay a deconstruction along these lines.

A stronger argument for the use of such techniques on formal texts is that formal texts play essential roles in science. Within a research network informal communication is an occasion of paramount importance. This is less likely to be the case when broader issues transcending such networks arise; e.g., a dispute over a theory fundamental to a whole discipline. In the case to hand, only a small proportion of the geological community took part in face-to-face encounters between Wegener or his supporters and his critics. The attitudes of most earth scientists toward CD were informed by their reading of published materials or attending formal lectures. Formal texts — which contained implicit or explicit 'method talk' — constituted in the case of the early CD debate a major resource for the presentation and resolution of conflicting knowledge claims, a point certainly appreciated by their authors.

With respect to the uses of 'method talk' generally, there is a spectrum in which it may be deployed. This extends from the place of an individual's own 'method talk' in the construction of his science (J. Schuster, this volume) to a detailed critique of one scientist's work by another (Steven Yearley, 'Textual Persuasion: The role of Social Accounting in the Construction of Scientific Arguments', *Philosophy of the Social Sciences* **XI**, 1981, pp. 409—435, who gives a micro-analysis of a historical case of social accounting for error using a formal text in early nineteenth-century geology which is complementary to the approach followed in my investigation) to the uses of methodological rhetoric in the contexts of disputes within a subdiscipline, between subdisciplines, and between disciplines (this chapter) and to the place of science and scientists in the larger social setting (Richard Yeo, 'Scientific Method and the Image of Science: 1831—1891', in: *The Parliament of Science: The British Association for the Advancement of Science 1831—1981*, Roy Macleod and P. Collins (eds.), (Northwood, Middlesex, 1981), pp. 65—85). It is likely that there are significant differences in the form, substance, and aims of 'method talk' at these different levels. Nonetheless, method discourse may serve some of the same purposes at several of these levels and to this extent the conclusions of this study may be generalizable not only to other instances of inter-theoretic conflict but also to other aspects, both narrower and broader, of the history and social context of science.

[5] Anthony Hallam, *A Revolution in the Earth Sciences* (Oxford, 1973); David B. Kitts, 'Continental Drift and Scientific Revolution', *Bulletin of the American Association of Petroleum Geologists* **LVIII**, 1974, pp. 2490—2496; Rachel Laudan, 'The Recent Revolution in Geology and Kuhn's Theory of Scientific Change', in: *Paradigms and Revolutions*, Gary Gutting (ed.), (Notre Dame, Ind., 1980), pp. 284—296; and especially Henry Frankel, 'Arthur Holmes and Continental Drift', *British Journal for the History of Science* **XI**, 1978, pp. 130—149; 'Alfred Wegener and the Specialists', *Centaurus* **XX**, 1976, pp. 305—324; 'The Career of Continental Drift Theory: An Application of Imre Lakatos' Analysis of Scientific Growth to the Rise of Drift Theory', *Studies in History and Philosophy of Science* **X**, 1979a, pp. 21—66; 'The Reception and Acceptance of Continental Drift Theory as a Rational Episode in the History of Science', in: *The Reception of Unconventional Science*, Seymour H. Mauskopf (ed.), (Boulder, Colo., 1979b), pp. 51—90; 'Why Drift Theory was Accepted with the Confirmation of Harry Hess' Concept of Sea Floor Spreading', in: *Two Hundred Years*

of Geology in America, Cecil J. Schneer (ed.), (Hanover, N. H., 1979c), pp. 337—353; 'Hess's Development of his Seafloor Spreading Hypothesis', in: *Scientific Discovery: Case Studies*, Thomas Nickles (ed.), (Dordrecht, 1980), pp. 345—366; 'The Paleobiogeographical Debate over the Problem of Disjunctively Distributed Life Forms', *Studies in History and Philosophy of Science* **XII**, 1981, pp. 211—259.

[6] There are, of course, exceptions. Geoffrey Cantor in 'The Reception of the Wave Theory of Light in Britain: A Case Study Illustrating the Role of Methodology in Scientific Debate', *Historical Studies in the Physical Sciences* **VI**, 1975, pp. 109—132, argues persuasively that exponents of inductivism opposed the wave theory and advanced method-based criticism of it whereas hypothetico-deductivists thought the theory well illustrated the power of their approach. In this case methodological issues appear to have played a substantive role in theory choice, but it is not clear if Cantor has investigated the possibility that much of the method rhetoric may have been more of a *machine de guerre* than a substantive element. See also G. Cantor, *Optics after Newton: Theories of Light in Britain and Ireland, 1740—1840* (Manchester, 1983), pp. 177—186.

[7] Larry Laudan, *Progress and its Problems: Toward a Theory of Scientific Growth* (Berkeley, 1977), pp. 14—18 and 45—64; for examples of contemporary comments in this same vein see Chester R. Longwell, 'Some Physical Tests of the Displacement Hypothesis', in: Van der Gracht *op. cit.* (Note 1), pp. 145—157, at p. 146; and Erich Jaworski, 'Die A. Wegenersche Hypothese der Kontinentalverschiebungen: Eine kritische Zusammenstellung', *Geologische Rundschau* **XIII**, 1922, pp. 273—296, at p. 293.

[8] E. Suess, *Das Antlitz der Erde*, 3 vols (Prague-Vienna-Leipzig, 1883—1909); French ed., 4 vols. (Paris, 1897—1918); English ed., 5 vols (Oxford, 1904—1924).

[9] M. T. Greene, *Geology in the Nineteenth Century: Changing Views of a Changing World* (Ithaca, 1982), p. 289.

[10] Ibid., pp. 74—75.

[11] Greene *ibid.*, provides a clear account of these matters and indicates the extent to which Lyell's importance has been exaggerated by Anglophonic historians of geology.

[12] A. Geikie, *Textbook of Geology*, 4th ed., 2 vols (Macmillan, London, 1903), vol. 1, p. 3.

[13] Grenville Arthur James Cole, 'Presidential Address: Geology: Crustal Movements', *Report of the British Association for the Advacement of Science* (Manchester, 1915), pp. 403—419, at p. 404.

[14] William W. Watts, 'Presidental Address: Form, Drift, and Rhythm of the Continents', *Report of the B.A.A.S.* (Norwich, 1935), pp. 1—21.

[15] John Walter Gregory, *Geology of Today; a Popular Introduction in Simple Language* (London, 1915), p. ix.

[16] J. W. Gregory, 'The Machinery of the Earth', *Nature* **CXXVI**, 1930, pp. 959—963, at p. 959; see J. A. Steers, *The Unstable Earth* (London, 1932), esp. pp. 201—204 for a similar expression of such sentiments in light of the CD debates.

[17] See Stephen J. Pyne, 'Methodologies for Geology: G. K. Gilbert and T. C. Chamberlin', *Isis* **LXIX**, 1978, pp. 413—424.

[18] Larry Laudan, *Science and Hypothesis* (Dordrecht, 1981) provides a perceptive series of studies on this development.

[19] R. A. Daly, *Our Mobile Earth* (New York, 1926), p. xx.

[20] Greene *op. cit.* (Note 9), pp. 190—191.
[21] Alfred Lothar Wegener, *Die Entstehung der Kontinente und Ozeane* (Brunswick, 1915), pp. 8—9; quoted from Wegener, *The Origin of Continents and Oceans*, tr. J. G. A. Skerl of 3rd German ed., (London, 1924), p. 5.
[22] Wegener *ibid.*, (1915), p. 8; Wegener, 'Die Entstehung der Kontinente' [an abridgement of 'Die Herausbildung der Grossformen der Erdrinde (Kontinente und Ozeane), auf geophysikalischer Grundlage' read 6 Jan. 1912], *Geologische Rundschau* **III**, 1912, pp. 276—292, reference to p. 277.
[23] It is apparent from Wegener's publications that he had a good grasp of the geological literature insofar as it supported his theory; his opponents typically were to criticize him not for being an 'outsider' but for failing either to be familiar with, or, more strongly, for refusing to acknowledge data which they thought told against CD.
[24] Johannes Klöcking, 'Simroths biologische Entwicklungsgesetze im Lichte der A. Wegenerschen Hypothese von der Horizontalverschiebung der Kontinentalschollen', *Petermanns geographische Mitteilungen* **LIX**, 1913, pp. 121—122.
[25] Karl Andrée, *Über die Bedingungen der Gebirgsbildung* (Berlin, 1914).
[26] Edgar Dacqué, *Grundlagen und Methoden der Paläogeographie* (Jena, 1915).
[27] Carl Diener, 'Die Grossformen der Erdoberfläche', *Mitteilungen der K. K. geographische Gesellschaft in Wien* **LVIII**, 1915, pp. 329—349, at p. 342.
[28] Wolfgang Soergel, 'Die atlantische 'Spalte': Kritische Bemerkungen zu A. Wegeners Theorie von der Kontinentalverschiebung', *Zeitschrift der deutschen geologischen Gesellschaft* **LXVIII**, 1916, pp. 200—239, at p. 201.
[29] K. Andrée, 'Alfred Wegeners Hypothese von der Horizontalverschiebung der Kontinentalschollen und das Permanenzproblem im Lichte der Paläogeographie und dynamischen Geologie', *Petermanns geogr. Mitt.* **LXIII**, 1917, pp. 50—53 and 77—81, at p. 79. These points have bearing on David Kitts's arguments concerning retrodictive uncertainty in geology: 'Retrodiction in Geology', in: *Proceeding of the 1978 Biennial Meeting of the Philosophy of Science Association*, ed., Peter D. Asquith and Ian Hacking, 2 vols. (Michigan, 1981), vol. 2, pp. 213—226, esp. p. 223.
[30] Theodor K. H. Arldt, 'Die Frage der Permanenz der Kontinente und Ozeane', *Geographischer Anzeiger* **XIX**, 1918, pp. 2—12, at p. 7.
[31] Andrée, *op. cit.* (Note 29), p. 80.
[32] Arldt, *op. cit.* (Note 30), p. 10.
[32] Max Semper, 'Was ist eine Arbeitshypothese?', *Zentralblatt für Mineralogie, Geologie, und Paläontologie*, 1917, pp. 146—163, at p. 146.
[34] Yearly, *op. cit.* (Note 4) makes a similar point *vis-à-vis* Richard Kirwan's criticisms of Sir James Hall.
[35] Semper, *op. cit.* (Note 33), p. 157.
[36] *Ibid.*, pp. 160—161.
[37] *Ibid.*, p. 161.
[38] Wilhelm Schweydar, 'Bemerkungen zu Wegeners Hypothese der Verschiebung der Kontinente', *Zeitschrift der Gesellschaft für Erdkunde zu Berlin* nr. 3/4, 1921, pp. 120—125, at p. 122.
[39] Friedrich Burmeister, 'Die Verschiebung Grönlands nach den astronomischen Längenbestimmungen', *Petermanns geogr. Mitt.* **LXVII**, 1921, pp. 225—227.
[40] Bruno Schulz, 'Die Alfred Wegenersche Theorie der Entstehung der Kontinente und Ozeane', *Die Naturwissenschaften*, Heft **XV**,1921, pp. 241—250.

[41] Leopold von Ubisch, 'Wegeners Kontinentalverschiebungstheorie und die Tiergeographie', *Verhandlungen der physikalischmedizinischen Gesellschaft zu Würzburg* **XLVI**, 1921, pp. 57—69.
[42] Franz Kossmat, 'Erörterungen zu A. Wegeners Theorie der Kontinentalverschiebungen', *Zeitschrift der Gesellschaft für Erdkunde zu Berlin*, nr. 3/4, 1921, pp. 103—110.
[43] Bruno Kubart, *Bemerkungen zu Alfred Wegeners Verschiebungstheorie* (Arbeiten des phytopaläeontologischen Laboratoriums der Universität Graz II), 1926, 28 pp.
[44] Friedrich Nölke, 'Physikalische Bedenken gegen A. Wegeners Hypothese der Entstehung der Kontinente und Ozeane', *Petermanns geogr. Mitt.* **LXVIII**, 1922, pp. 114—115, at p. 115.
[45] A. Penck, 'Wegeners Hypothese der kontinentalen Verschiebungen', *Zeitschrift der deutschen geologischen Gesellschaft* **LXXII**, 1921, pp. 110—120, at p. 119.
[46] Jaworski, *op. cit.* (Note 7), p. 293.
[47] *Ibid.*, pp. 293—294.
[48] N. Wing Easton, 'On Some Extensions of Wegener's Hypothesis and their Bearing upon the Meaning of the Terms Geosynclines and Isostasy', *Verhandelingen Geologisch Mijnbouwkundig Genootschap voor Nederland en Kolonien* **V**, (Geol. ser.), 1921, pp. 113—133. It is noteworthy that other Dutch geologists with extensive field experience were favourably disposed toward the main points of Wegener's theory although frequently critical of details: Gustaaf Adolf Frederik Molengraaff, 'The Coral Reef Problem and Isostasy', *Nederlandse Akademie van Wetenschappen* **XIX**, 1916, pp. 610—627; G. L. Smit Sibinga, 'Wegener's Theorie en het Ontstaan van den Oostelijken O. I. Archipel', *Tijdschrift van het Koninklijk Nederlandsch Aardrijkskundig Genootschap* **XLIV**, (ser. 2), 1927, pp. 581—598.
[49] This concurs with the conclusions are generally and forcefully put by L. Laudan, *op. cit.* (Note 18), esp. pp. 8—9.
[50] P. Lake, 'Wegener's Displacement Theory', *Geological Magazine* **LIX**, 1922, pp. 338—346, reference to p. 338.
[51] H. F. Reid, 'Drift of the Earth's Crust and Displacement of the Pole', *The Geographical Review* **XII**, 1922, pp. 672—674, at p. 674.
[52] See for example the discussion of Thomas Reid's 1785 *Essays on the Intellectual Powers of Man*, in L. Laudan *op. cit.* (Note 18), pp. 90—94.
[53] Chamberlin, *op. cit.* (Note 1), p. 83.
[54] *Ibid.*
[55] *Ibid.*, p. 87.
[56] Bailey Willis, 'Continental Drift' in: van der Gracht *op. cit.* (Note 1), pp. 76—82, at p. 82.
[57] C. Schuchert, 'The Hypothesis of Continental Displacement' in: van der Gracht *op. cit.* (Note 1), pp. 104—144, at p. 139.
[58] *Ibid.*, p. 140.
[59] E. Berry, 'Comments on the Wegener Hypothesis' in: van der Gracht *op. cit.* (Note 1), pp. 194—196, at p. 194.
[60] Ibid., p. 196.
[61] J. Singewald, 'Discussion of Wegener Theory' in: van der Gracht *op. cit.* (Note 1), pp. 189—193, reference to pp. 189—190.

62 Gregory, *op. cit.* (Note 15), p. vii.
63 H. Poincaré, *Science and Hypothesis* (New York, 1952; 1st ed. 1905), p. 215.
64 Du Toit, *op. cit.* (Note 2), p. 4.
65 See for example Lake, *op. cit.* (Note 50); P. Lake, 'Wegener's Hypothesis of Continental Drift', *Nature* **CXI**, 1923a, pp. 226—228; P. Lake, 'Wegener's Hypothesis of Continental Drift', *The Geographical Journal* **LXI**, 1923, pp. 179—194; Ried *op. cit.* (Note 51); W. B. Wright *et al.*, 'The Wegener Hypothesis: Discussion at the British Association, Hull', *Nature* **CXI**, 1923, pp. 30—31; Harold Jeffreys, *The Earth: Its Origin, History and Physical Constitution* (Cambridge, 1924).
66 For example, van der Gracht, 'The Problem of Continental Drift' (1928a) in: van der Gracht *op. cit.* (Note 1), pp. 1—75, reference to p. 67; S. W. Woolridge and R. S. Morgan, *The Physical Basis of Geography: An Outline of Geomorphology* (London, 1937), at p. 48.
67 Chamberlin, *op. cit.* (Note 1), p. 86.
68 Arthur Robert Hinks, G. C. Simpson, J. W. Gregory *et al.*, 'Problems of the Earth's Crust: A discussion in Section E (Geography) of the British Association . . . '. *Geographical Journal* **LXXVIII**, 1931, pp. 433—455 and 536—544, reference to p. 435.
69 *Ibid.*, p. 436.
70 *Ibid.*
71 See Frankel, *op. cit.* (Note 5: 1976).
72 For example, Schuchert, *op. cit.* (Note 57), p. 134.
73 *Ibid.*, p. 139.
74 Joe D. Burchfield, *Lord Kelvin and the Age of the Earth* (New York, 1975).
75 T. H. Holland, 'Presidential Address', *B.A.A.S. Transactions of Section C* (1914), at p. 346.
76 Gregory, *op. cit.* (Note 15), p. ix.
77 Jeffreys, *op. cit.* (Note 65), p. 1.
78 *Ibid.*, p. vii.
79 H. Jeffreys, *Earthquakes and Mountains* (London, 1935), p. 168.
80 Jeffreys, *op. cit.* (Note 65), p. viii.
81 *Ibid.*, pp. 260—261.
82 *Ibid.*, p. 4.
83 *Ibid.*
84 For example, Richad Dixon Oldham, 'The Constitution of the Interior of the Earth as Revealed by Earthquakes', *Quarterly Journal of the Geological Society* **LXII**, 1906, pp. 456—475; Alfred J. Bull, 'Some Aspects of the Mountain Building Problem', *Proceedings of the Geological Association* **XXXVIII**, 1927, pp. 144—156; van der Gracht, *op. cit.* (Note 66), p. 5; Gregory *op. cit.* (Note 16); Arthur Holmes, 'The Thermal History of the Earth', *Journal of the Washington Academy of Science* **XXIII**, 1933, pp. 169—195, at p. 169.
85 Hinks, *op. cit.* (Note 68), p. 434.
86 *Ibid.*, p. 435.
87 *Ibid.*, pp. 437—438.
88 *Ibid.*, pp. 437—438, 536.
89 G. Simpson in Hinks, *op. cit.* (Note 68), p. 442.

[90] Van der Gracht, *op. cit.* (Note 66), p. 40.
[91] Reginald A. Daly, 'The Earth's Crust and its Stability', *American Journal of Science* **V** (series 5), 1923, pp. 349—377, at p. 371.
[92] A. Holmes, 'Radioactivity and the Earth's Thermal History. Part IV: A Criticism of Parts I, II and III', *Geological Magazine* **LXII**, 1925, pp. 504—515, at pp. 505—506; see also Holmes, *op. cit.* (Note 84), p. 181.
[93] *Ibid.*, pp. 506—515.
[94] A. Holmes, 'Radioactivity and Earth Movements', *Transactions of the Geological Society of Glasgow* **XVIII**, part iii 1928—29 [1930], pp. 559—606, at pp. 560—61.
[95] *Ibid.*, pp. 563—64.
[96] *Ibid.*, pp. 575—600; see Frankel, *op. cit.* (Note 5: 1978).
[97] Holmes *op. cit.* (Note 92), p. 515.
[98] For example, Holmes in Hinks, *op. cit.* (Note 68); Holmes *op. cit.* (Note 84).
[99] For example, Bull, *op. cit.* (Note 84), p. 150; Woolridge and Morgan, *op. cit.* (Note 66), pp. 21, 53, 121.
[100] Robert H. Rastall, 'On Continental Drift and Cognate Subjects', *Geological Magazine* **LXVI**, 1929, pp. 447—456, at p. 447.
[101] *Ibid.*, p. 450.
[102] For example, Wegener, *op. cit.* (Note 22: 1912).
[103] A. L. Wegener, 'Die geophysikalischen Grundlagen der Theorie der Kontinentenverschiebung', *Scientia* **LXI**, 1927, pp. 103—116, reference to p. 103; quoted from A. L. Wegener, *The Origin of Continents and Oceans*, tr. John Biram of 4th German ed., (New York, 1966; orig. ed., 1929), p. vii.
[104] Wegener *ibid.*, (1929), pp. vii—viii, 57, 97.
[105] *Ibid.*, p. 167.
[106] *Ibid.*, p. 174.
[107] *Ibid.*, p. 167.
[108] Charles M. Nevin, *Principles of Structural Geology*, 2nd ed., (New York, 1936), at p. 306.
[109] Steers, *op. cit.* (Note 16), pp.161—162.
[110] A. Holmes, review of van der Gracht, *op. cit.* (Note 1) in: *Nature* **CXXII**, 1928, pp. 431—433, at p. 431.
[111] Holmes in Hinks, *op. cit.* (Note 68), p. 449.
[112] This is analogous to Cantor's conclusions relative to the wave theory. Cantor, *op. cit.* (Note 6).
[113] L. Laudan, *op. cit.* (Note 18), p. 9.
[114] A. L. Wegener, 'Two Notes Concerning My Theory of Continental Drift' in: Van der Gracht, *op. cit.* (Note 1), pp. 97—103, at pp. 97—98; Wegener *op. cit.* (Note 103: 1929), pp. 134—136.
[115] Ursula B. Marvin, *Continental Drift: Evolution of a Concept* (Washington, 1973), p. 5 notes the possible importance of this issue.
[116] Incidentally, the so-called 'Velikovsky Affair' provides several parallels. See, for example, Alfred de Grazia (ed.), *The Velikovsky Affair* (London, 1966).
[117] A. Keith, 'Outlines of Appalachian Structure', *Bulletin of the Geological Society of America* **XXXIV**, 1923, pp. 309—380, at p. 361 (emphasis added).
[118] Schuchert, *op. cit.* (Note 57), p. 140.

[119] H. H. Turner in Wright, *op. cit.* (Note 65), p. 30.

[120] Arldt, *op. cit.* (Note 30).

[121] Wegener, *op. cit.* (Note 22: 1912), pp. 291—292.

[122] Wegener, *op. cit.* (Note 21: 1924), p. 112.

[123] For example, J. W. Gregory, review of Wegener *ibid.*, in *Nature* **CXV**, 1925, pp. 255—257, reference to p. 256. The argument also had appeal to 'neutrals'; for example, it probably played a role in John W. Evans's efforts to make available an English translation of Wegener: J. W. Evans, 'Introduction', in A. Wegener, *ibid.*, pp. vii—xii, esp. pp. ix—x.

[124] Wegener, *op. cit.* (Note 114), pp. 100—103.

[125] Wegener, *op. cit.* (Note 103: 1929), p. 23.

[126] *Ibid.*, pp. 29—30.

[127] N. E. Nörland, 'Astronomical Longitude and Azimuth Determinations', *Royal Astronomical Society Monthly Notices* **XCVII**, 1937, pp. 489—506; Chester R. Longwell, 'The Mobility of Greenland', *American Journal of Science* **CCXLII**, 1944, p. 624.

[128] D. White, 'Discussion of Floating Continents' in: Van der Gracht, *op. cit.*, (Note 1), pp. 187—188.

[129] Chamberlain, *op. cit.*, (Note 1), p. 83; Nevin, *op. cit.*, (Note 108), p. 307.

[130] Du Toit, *op. cit.* (Note 2), p. 1; see also van der Gracht, 'Remarks Regarding the Papers Offered by the Other Contributors to the Symposium', in van der Gracht, *op. cit.* (Note 1), pp. 197—226, at p. 203; Steers, *op. cit.* (Note 16), pp. 160—61; Woolridge and Morgan, *op. cit.* (Note 66), p. 47.

[131] Ignatius Donnelly (Edmund Boisgilbert), *Atlantis: The Antediluvian World* (London, 1884), p. 4.

[132] For example, P. G. Nutting, 'National Prestige in Scientific Achievement', *Science* **XLVIII**, 1918, pp. 605—608; Lord Walsingham, 'German Naturalists and Nomenclature', *Nature* **CII**, 1918, p. 4; Eugenio Rignano, 'A Plea for a Scientific Quadruple Entente', *Nature* **XCVIII**, 1917, pp. 408—409. *Cf.* the excellent study by Brigitte Schroeder-Gudehus, *Les Scientifiques et la paix: La communauté scientifique internationale au cours des années 20* (Montreal, 1978).

[133] The case of Einstein could be used to argue either for or against this inference. Einstein's general theory of relativity was disseminated in England through the effects of A. S. Eddington even before the end of the war and there was an enormous upsurge of interest manifested in the popular and scientific press following the 1919 British eclipse expeditions. However, one must weigh such factors as Einstein's highly successful visit to England against his snub by the French Academy; Einstein's outspoken pacifism and his circulation of a petition opposing the infamous manifesto to German scientists against Wegener's service of the Western Front; the apparent success of Einstein's 'crucial experiment' against the apparent ambiguity or failure of Wegener's; and similar particulars. Cf. John Earman and Clark Glymour, 'Relativity and Eclipses: The British Eclipse Expeditions of 1919 and their Predecessors', *Historical Studies in the Physical Sciences* **XI**, 1980, pp. 49—85, esp. 81—84 and, for a general coverage, R. W. Clark, *Einstein: The Life and Times* (London, 1973), esp. chaps. 9—10.

[134] Nutting, *op. cit.* (Note 132), p. 605.

[135] Rignano, *op. cit.* (Note 132), p. 408.

[136] Nutting, *op. cit.* (Note 132), p. 605.

[137] *Ibid.*, p. 607.

[138] William Ramsay, 'The War', *Nature* **XCIV**, 1914b, pp. 527—28, at p. 527; see also W. Ramsay, 'Germany's Aims and Ambitions', *Nature* **XCIV**, 1914a, pp. 137—39; and, for remarks specifically directed at German originality in geology, Joseph Barrell, 'The Growth of Knowledge of Earth Structure', *American Journal of Science* **XLVI** (series 4), 1918, pp. 133—170, at p. 170.

[139] Of various calibres are the essays edited by Gabriel Petit and Maurice Leudet, *Les Allemands et la science* (Paris, 1916). The contributors included distinguished French intellectuals. Particularly relevant are Stanislas Meunier, 'La Géologie à la Prussienne', pp. 263—273 and Pierre Duhem, 'Science Allemande et vertues Allemandes', pp. 137—152.

[140] For example, Taylor: Daly, *op. cit.* (Note 91), pp. 365—66; van der Gracht, *op. cit.* (Note 66: 1928), p. 34; Fisher: R. D. Oldham in Lake, *op. cit.* (Note 65: 1923), p. 189; Thomas Crook, 'The Wegener Hypothesis and the Origin of the Oceans', *Nature* **CXI**, 1923, pp. 255—256.

[141] Schroeder-Gudehus, *op. cit.* (Note 132), pp. 80—85; see also Fritz K. Ringer, *The Decline of the German Mandarins: The German Academic Community 1890—1933* (Cambridge, MA., 1969).

[142] Schroeder-Gudehus, *op. cit.* (Note 132), pp. 233—247; *cf.* Paul Forman, 'Scientific Internationalism and the Weimar Physicists: The Ideology and its Manipulation in Germany after World War I', *Isis* **LXIV**, 1973, pp. 151—180.

[143] For example, Daly, *op. cit.* (Note 91), p. 371.

T. D. STOKES

METHODOLOGY AS A NORMATIVE CONCEPTUAL PROBLEM: THE CASE OF THE INDIAN 'WARPED ZIPPER' MODEL OF DNA

1. INTRODUCTION

As Larry Laudan observes, it is 'widely maintained that the methodology to which a scientist subscribes is really little more than perfunctory window dressing, which is honored more in the breach than in the observance'.[1] This is a view shared by T. S. Kuhn, and common in much post-Kuhnian writing. Kuhn makes it clear why he lacks interest in scientists' commitments to methodological canons in the following comparatively rare reference to them:

> The men who transformed scientific theory during the seventeenth century sometimes talked like Baconians, but it has yet to be shown that the *ideology* which a number of them embraced had a major effect, substantive or methodological, on their central contributions to science.[2]

Notwithstanding these examples of methodological inefficacy, Laudan insists that his own historical research,[3] that of Cantor,[4] and many others,[5] provides 'overwhelming evidence that the methodological beliefs of scientists often do profoundly affect their research and their appraisals of the merits of scientific theories.'[6] Laudan points out that philosophers of science have focused their attention on the role of empirical problems in science, especially stressing that of the empirical anomaly, or failed prediction. Against this, Laudan urges that what he calls '*conceptual problems*' have been '*at least as important* in the development of science as empirical problem solving.'[7]

Discussion of the role of scientists' methodological beliefs commonly conceives method in grand, abstracted terms — inductivism versus deductivism, for example. Scientists do sometimes conduct their controversies on this elevated plane, but not very often. For them, debate is likely to be couched much more concretely, with general principles instantiated as particular ways of dealing with specific problems. In these circumstances, *technical* matters arise at the same time as methodological issues. Indeed, it is frequently difficult and unhelpful to try to consider them separately. Argument over what is technically desir-

J. A. Schuster and R. R. Yeo (eds.), The Politics and Rhetoric of Scientific Method, 139–165.
© 1986 *by D. Reidel Publishing Company.*

able, and over the signification of results is the context in which scientists conduct their everyday method talk. Thus it is in close examination of techno-methodological thought and controversy that we can best examine the effect of methodological commitments on scientific practice.

James Watson and Francis Crick's Nobel Prize winning double-helical model of the structure of DNA (deoxyribonucleic acid) is very widely considered to have been one of the great scientific discoveries, seminal to the development of an understanding of the molecular basis of genetics. In 1976, however, a group in New Zealand, led by G. A. Rodley and R. H. T. Bates, and another in India, headed by V. Sasisekharan, published independently devised but essentially identical radical alternatives to the double helix.[8] This Side-By-Side (SBS) conformation is non-helical. Its two exoskeletal polynucleotide strands have been likened to a 'warped zipper'.[9] Unlike Watson and Crick's helices, but like a zip fastener, they are topologically independent. As a result, the two strands may separate laterally during replication. This avoids the longitudinal 'unwinding' required of the intertwined double helix, a process frequently seen as entailing difficulties.

I examined elsewhere the genesis of the SBS structure in New Zealand and India, utilizing recorded and transcribed recollections of the inventors.[10] I argued that the origin of the New Zealanders' version of the alternative structure (which has slight priority) was a response to an empirical anomaly which seemed irresolvable on any model of DNA entailing, as Watson and Crick's does, the need to unwind during replication.[11] According to the testimony of the scientists involved, however, the SBS structure devised in India was inspired by adoption of methodological norms at variance with those practiced by the specialist community.[12] I have shown that explaining the Indians' motivation in these terms is consistent with Kuhn's views, which offer a theoretical perspective in many respects fruitful in understanding the hostile reception accorded the novel proposal.[13] On such a view the Indians are seen as mavericks, having taken seriously an explict methodology which served essentially an ideological function for most scientists, contrasting sharply with the methodological norms implicit in their practice. Kuhn is concerned to account for group consensus among scientists and, provided that condition is present, can afford to allow a few dissidents. Dissension, however, must be untypical, uninfluential and thus extrinsic to a Kuhnian understanding of the dynamics of any particular 'disciplinary matrix'.

In contrast, Laudan's notion of a conceptual problem gives a central place to tension between explicit methodological norms and scientific theories. There are two genera in his taxonomy – internal conceptual problems, generated by inconsistency or unclarity within a theory; and external conceptual problems, which arise when a theory '*T* is in conflict with another theory or doctrine, T^1, which proponents of *T* believe to be rationally well founded . . .'[14] There are three differently generated species of external conceptual problems. However, we need only consider one, begotten by what Laudan calls 'normative difficulties' – conflict between a scientific theory and a methodological theory (metatheory).[15] In Laudan's view:

If a scientist has good grounds for accepting some methodology and if some scientific theory violates that methodology, then it is entirely rational for him to have grave reservations about the theory.[16]

Nonetheless, he considers that resolution of the conflict may lie in modification of either to accommodate to the other. Here, I propose to explore Laudan's views on the role of method with respect to work on the structure of DNA and, especially, the origins of the version of the SBS model of DNA that was developed at the Indian Institute of Science by V. Sasisekharan and his co-workers N. Pattabiraman and Gautam Gupta.

The analysis will reveal that the problem which impelled Sasisekharan on the course which led to the Indian variants of the SBS model of DNA *was* a conceptual problem sparked by a clash between a theory and a metatheory. Specifically, there was conflict between the way in which the Watson-Crick structure for DNA had been devised, refined and tested, and a methodological perspective which required a different, and more stringent approach. Although Laudan's notion of a conceptual problem helps elucidate this episode, his proposed criteria for deciding the significance a particular clash, and whether the scientific theory or the methodology should be altered, prove inadequate. An attempt is made to improve upon them by taking into account the interests of the appraiser.

I will resist a sharp distinction between social and cognitive senses of 'interest' and, in so doing, seek to eliminate the contradiction between internal and external explanations of the role of method in science. Thus, for example, my reconstruction relies significantly upon the account of events rendered sometimes afterward by the participants. But it is not accepted uncritically. Rather I argue that their background and circumstances lend credence to the claim that, by and large, their story is reasonable, and *not* a rationalization.

2. A PARADOXICAL BEGINNING

The development of the 'warped zipper' structure for DNA in India had a paradoxical beginning, with Sasisekharan defending the Watson-Crick model against an earlier heresy. In 1970 Sasisekharan was on sabbatical leave (from the Centre of Advanced Study in Biophysics, University of Madras) as Visiting Professor in the Department of Biochemical Sciences at Princeton. While he was there, a group which included two scientists from Princeton, Y. Mitsui (the lead co-author) and and R. Langridge (the senior co-author), published a paper suggesting that one kind of DNA might have a different handedness from that of the Watson-Crick structure. Watson and Crick's double helix is right-handed, the two strands twist clockwise. Mitsui *et al.* suggested that their example had an anti-clockwise, *left*-handed sense.

When fibres of semi-crystalline hydrated DNA salts are exposed to X-rays in cross-section, they produce a diffraction pattern which is recorded photographically. The pattern is characteristically cross-shaped, but its detail varies quite widely as changes are made to the humidity, saturation and cations present in the sample of the DNA. These variant crystalline forms were labeled, as they were discovered, A-DNA, B-DNA, C-DNA and D-DNA. B-DNA is thought to most closely resemble the non-crystalline *in vivo* structure of the molecule. A sample can be interconverted from A-DNA to B-DNA, and *vice versa*. Another indicator of variations in crystal structure is circular dichroism (CD) spectra. This technique measures the differental absorption of left and right-handed circularly polarised light. A review published in 1970 noted that theoretical calculations of the optical activity of B-DNA as measured by the CD were 'in good agreement with experiment',[17] and went on to claim that, accordingly, 'optical activity is a considerable asset in identifying nucleotide samples and a sensitive probe of the conformation of nucleic acids'.[18]

Watson and Crick defended the right-handedness of their model by arguing from the outset that left-handed double-helical structures are stereochemically unviable because they entail unacceptably close adjacent, unbonded atoms.[19] This view was widely accepted for many years. The situation is, however, more complex. In 1965, Fuller *et al.* admitted that a left-handed double-helical structure for B-DNA is stereochemically viable.[20] But, they maintained, the A-form of the molecule resists such a configuration. And so they reasoned that

Since the conformation of DNA may be changed reversably from A to B by change of humidity, it is most unlikely that the transition involves such a radical change in structure as alteration of helical sense. We may therefore conclude that DNA in the B form (which is found *in vivo*) consists, like the A form, of right-handed helices.[21]

Mitsui *et al.* examined a synthetic DNA consisting of an alternating sequence of deoxyriboinosinic and deoxyribocytidylic acid, poly d(I-C). poly d(I-C).[22] Inosine (I) and cytosine (C) were the two bases present, the former not normally occurring in natural DNA. There were difficulties for a conventional Watson-Crick view of the structure because 'right-handed models either have unacceptably short nonbonded interatomic contacts or do not fit the X-ray intensity data'.[23] A left-handed model, however, though it had an unusual sugar conformation, had no short contacts and fitted the diffraction better.

Even more strikingly, this was 'the first known example of a double-stranded DNA with an intense negative long wavelength CD band'.[24] Indeed, it was precisely the inverse of the long wavelenght CD band of DNAs with the naturally occuring bases (adenine (A), thymine (T), guanine (G), cytosine (C) and uracil (U)) all of which had a strong positive long wavelength CD band. But, Mitsui *et al.* pointed out, the three left-handed model systems reported in the literature each had inverted CD spectra.[25] Poly d(I-C).d(I-C) is a D-DNA, but Mitsui and co-workers could not convert it into an A or B-DNA. Recall that Fuller *et al.* had argued that inter-convertability of A and B-DNA told against the possibility that they had different helical senses. Mitsui and colleagues concluded that their inability to interconvert a D-DNA to an A or a B-DNA, taken in conjunction with a CD spectra for the D-DNA opposite to that of an A or B-DNA, strongly suggested that poly d(I-C).poly d(I-C) was a left-handed double-helix.

Sasisekharan recalls

Now, this created ... a stir in the minds of people who were working on polynucleotides; particularly Struther Arnott, who said that this was a particularly bizzare looking structure. He even said that in print; and he said that he could build a right-handed structure that ... agreed with the X-ray data, so that there was no need to invoke a left-handed structure.[26]

A diffraction crystallograph of poly d(I-C).poly d(I-C) had appeared on the cover of *Nature*.[27] It wasn't a particularly good photograph, and Sasisekharan successfully set about improving upon it, obtaining a much clearer pattern. What was more, when Sasisekharan raised the humidity of the specimen, that pattern changed from a D to a B-DNA. One again, an argument of the sort used by Fuller *et al.* seemed applicable. As Sasisekharan explained it,

> A change of handedness cannot occur just by raising the humidity from 75% to 95%, or so. [This is] because all that you do, you keep the [DNA] fibre on a glass capillary, and at the end of the capillary you have some saturated salt solution to keep the humidity at the desired level. So, in [such] a solid state, it cannot change handedness.

Notwithstanding interconvertability, however, the CD spectra of the D-DNA remained a marked contrast with that of the B-form. And, as was noted earlier, optical activity was considered a reliable guide to conformation. So the difference needed to be explained away.

Sasisekharan approached Langridge and the others involved in the work on poly d(I-C).poly d(I-C) at Princeton with his results. As Sasisekharan saw it, the evidence was now against the left-handed conformation but, nonetheless, the CD spectra remained an outstanding puzzle. He found the response to this viewpoint distinctly odd. Langridge and his colleagues were evasive, and appeared to have lost interest in the problem. Perhaps they were somewhat stung by the hostile reception accorded their proposal from all quarters. Certainly Sasisekharan had proven a more technically proficient experimenter. In any event, Sasisekharan himself concluded:

> Maybe they didn't do a thorough job. And then I realized that it was very likely that they had not systematically explored the possibilities. Model building was not proceeding in a systematic way. A model was built just to calculate the theoretical Fourier transforms. And if it agreed with the [diffraction] data, one was happy and said 'OK I have the structure.' *But that does not mean that there cannot be another structure consistent with the data* (emphasis added).

This characterization was accurate enough and, as Sasisekharan realised, true not only of the work of Langridge and his colleagues, for their approach was that normally adopted *in practice.* Yet as we shall soon see, in what they *said* about method, the specialist community set its face against the Duhem-Quine thesis, according to which the search for a unique account of data is in principle doomed, and claimed to have sought, and achieved precisely that.

3. A NEW METHOD AND ITS CRITICS

Sasisekharan was by no means the first to question the methodology underlying the refinement and testing of models of DNA. Watson and Crick's views on the methodology appropriate to the study of biological macromolecules like DNA developed from their frank admiration of the approach used by Linus Pauling to solve the structure of the protein keratin.[28] Unsurprisingly, the methods used with such success by Watson and Crick were adopted in nearly all subsequent work. In essence, the idea was to begin from a structural hypothesis consistent with stereochemistry and see whether it fitted the X-ray diffraction crystallographs and biochemical assays of the molecule under examination. Neither Watson nor Crick were trained to produce this data. The diffraction evidence came from Maurice Wilkins and Rosalind Franklin working at Kings College, London (though Wilkins worked in loose collaboration with Watson and Crick in Cambridge). Determination of the base-ratios, crucial in deciding the base-pairing arrangements, was done by Chargaff in the U.S.A.

Both Franklin and Chargaff had reservations about Watson and Crick's work. Chargaff recalled his first meeting with them in typically acerbic terms:

> So far as I could make out, they wanted, unencumbered by any knowledge of the chemistry involved, to fit DNA into a helix. The main reason seemed to be Pauling's alpha-helix model of a protein.[29]

In lecture notes, Franklin remarked of Pauling's approach: 'The time has come to review evidence and assumptions — have we found *the* solution or *a* solution?'[30] Her poor relationship with Watson and Crick, and her opinion of their ideas about the structure of DNA, are explained by Olby in similar terms.

> Franklin was a professional structural crystallographer who distrusted intuitive guessing, and who wanted to solve the structure [of DNA] by direct methods, i.e., without introducing assumptions in the form of hypothetical structures. She was not against helices as such, but against assuming helices when the evidence, in her opinion, was inadequate.[31]

Watson and Crick's success secured their methodology against earlier approaches. It was further entrenched when successfully imitated. Thus, in 1968, Hamilton remarked:

The derivation of the A conformation of DNA greatly aided the elucidation of the RNA double helix which is very similar, and this conformation has recently been used . . . as the basis of a molecular model of the anticodon arm in the transfer RNA molecule.[32]

This illustrates the way in which Watson and Crick's solution for the structure of DNA, and the way they developed it, can usefully be regarded as an exemplar, in Kuhn's sense. For Kuhn, exemplars are the standard problem-solutions of a scientific community. And, according to him, such communities implicitly inculcate novice members with such exemplary practice. Kuhn's 'normal' science consists essentially in creative imitation of exemplars.[33] Indeed, Watson and Crick's approach was used by others to check and refine the details of their model — though without altering it in any essential element.[34] The ostensible motivation for this undertaking is described by Hamilton, who observes:

Proof of the structure was necessary because the preliminary rough agreement between the X-ray data and the model was not so good as to *exclude the possibility that a structure of some other kind might correlate better with the data*.[35]

Sasisekharan and before him (as we will shortly see) another scientist, Jerry Donohue, were to find this ambition laudable, but unrealized and unrealizable given the methodology deployed. First, though, let us hear from Hamilton how the task was undertaken:

The analysis was . . . conducted by atomic groups. Scale models of DNA were built with the stereochemical features of the nitrogen bases, sugar and phosphate groups determined from published structures of simpler compounds. No conformation was allowed in which non-bonded atoms approached one another at less than their van der Waals's distances. The intensity pattern produced by a crystal incorporating such a model was calculated, as well as the contributions made by the three component groups; all were compared with the observed diffraction. The groups were then adjusted in position to improve the agreement between the observed and calculated diffractions.[36]

By 1957, Wilkins took the view that 'the correctness of the structure is established with reasonable certainty'.[37] Nevertheless he felt that further work remained for, echoing Hamilton, he held that 'in view of the importance of the molecule, *it is desirable to eliminate all possible doubts and ambiguities concerning its structure*'.[38]

Attention now turned to the sub-variant crystal structures; first B-DNA and then A-DNA and C-DNA. In his 1968 review, Hamilton expressed confidence that, in each case, 'the diffraction data are satisfactorily accounted for by the basic double helical structure incorporating the same Watson and Crick base-pairing scheme'.[39] Further, he considered that this work had 'gone a long way to establishing that the double helical feature and base-pairing scheme is *unique*, and is a much more convincing demonstration of the correctness of the general structure hypothesis than had one conformation alone been studied'.[40] Indeed, Hamilton went so far as to say that 'X-ray diffraction and model building have elevated the Watson-Crick pairing hypothesis to the level of experimentally established fact'.[41]

Not quite everyone accepted that it was *fait accompli.* In 1956 the crystallographer Jerry Donohue (to whom Watson and Crick had acknowledged indebtedness in their original paper)[42] pointed out that that stereochemical contraints do not, by themselves, determine a Watson-Crick base-pairing scheme for DNA, going on to suggest an alternative mode.[43] Somewhat later, Hoogsteen proposed a third base-pairing scheme.[44] Hamilton, however, took the view that the new Fourier technique for calculating the theoretical diffraction pattern of a model structure had 'excluded any significant participation of base-pairs other than those of the Watson-Crick type from the DNA structure'.[45] In reply to this and similar claims, Donohue mounted a vigorous counter-attack, claiming that

> the published X-ray diffraction data from DNA are incapable of being analyzed by the Fourier method for the purpose of obtaining evidence as to what the structure of the substance is, not to mention refining a proposed model.[46]

In support of this contention, Donohue argued that Fourier analysis did not exclude a structure known — on grounds independent of the diffraction data — to be defective. He characterized Fourier analysis as 'pulling-yourself-up-by-your-bootstraps', citing the view of Pinnock *et al.* that 'Fourier synthesis . . . is not a good test of a proposed structure; [because] it always tends to support the hypothesis upon which it is based'.[47]

These claims provoked a flurry of responses occupying, together with a reply by Donohue, a substantial part of an issue of *Science*.[48] *Nature* joined in too, noting with evident approval that 'the establishment' had 'closed ranks to repel the assault'.[49] In their response, Wilkins *et al.* conceded to Donohue that 'it is undersirable to use Fourier syntheses to prove directly that a structure is correct'. However, they asserted Fourier methods 'can readily show that the alternative structures [of Donohue and Hoogsteen] are incorrect, *thereby providing proof by elimination*'.[50] Moreover, thought Wilkins *et al.*, Donohue and Hoogsteen base-pairing schemes could be eliminated on stereochemical grounds, without the need to resort to interpretation of diffraction evidence. In the most detailed, technical response, Struther Arnott said that, though he supported 'many of Donohue's general cautionary statements on the use of Fourier syntheses approach', nevertheless 'a decision can be made in favour of a DNA model of the Watson-Crick type', again by exclusion of specific alternatives using a refinement of Fourier analysis involving 'difference synthesis'.[51] Francis Crick, whilst admitting himself a 'biased witness', challenged Donohue to put up or shut up (or, as *Nature's* editorialist had it, Crick put 'an Olympian boot into Donohue's ribs').[52] Crick said:

> If Donohue thinks than an equally effective model for DNA could be produced with some alternative base-pairing, let him build such a model and publish the co-ordinates. The fit of this model with the observed data could be compared with that of the models described by Wilkins and his colleagues. We would then all see which model fits the data better, or whether there is nothing to choose between them.[53]

In his rejoinder Donohue contended that, contrary to Arnott's view, difference synthesis did not eliminate the bias in favour of the model under test.[54] Indeed Donohue insisted once again that difference synthesis would not eliminate a structure known on other grounds to be unacceptable. Leaving technical questions aside, Donohue observed that a 'basic difference in the philosophy of what constitutes proof of structure with X-ray data has been laid bare'.[55] Donohue attributed to Arnott and Wilkins *et al.* the view that, methodologically, 'it is sufficient to consider various models, and then choose as correct (after adjustment) one that gives satisfactory best agreement with experiment'.[56] Donohue felt, however, that in those circumstances 'one can never be certain that a model sufficiently close to the true structure has been constructed'.[57]

With Franklin, Donohue and Sasisekharan wanted to know whether they had '*the* solution or *a* solution?'[58] Similarly, as we have seen, Hamilton had urged the need to 'exclude the possibility that a structure of some other kind might correlate better with the data',[59] and Wilkins had maintained that 'it is desirable to eliminate all possible doubts and ambiguities . . . '[60] But the response of the specialists to Donohue's critique makes it plain that they were prepared to settle for something a good deal less unequivocal. *Nature's* editorialist put the matter clearly, if testily:

> It is presumably still true that even the helical nature of DNA has been demonstrated only inductively, and that it cannot be analytically disproved that some other arrangement of atoms might not exist which would give the observed X-ray diffraction patterns. Whereas the fundamentalist view that Donohue seems to represent may well carry technical weight, its application in molecular biology would not seem best likely to promote progress, for had it prevailed earlier, no models would have been built. . . .[61]

It is probably the views of Chargaff and/or Franklin which are alluded to, and with which Donohue is associated. It is difficult not to conclude, as the specialist community plainly did, that in the light of his alternative base-pairing scheme — an arrangement which everyone thought patently wrong — Donohue's high principled critique was tendentious. Yet the behaviour of the specialist community itself illustrates how little the official demand for a unique solution for the structure of DNA mattered, and how easily methodological orthodoxy could be altered to eliminate a tedious inconsistency between theory and practice.

With the whole episode safely in the past, Crick's felt able to say that, whilst Donohue had carried his criticisms too far, 'his doubts . . . at times had some justification'.[62] Indeed, still later, Crick and others were to admit:

> We consider it unwarranted to rely solely on the details of exact model building, our knowledge of stereochemistry, though now fairly good, may not be adequate to provide firm answers, nor is it advisable to put one's faith completely on the fine details of X-ray diffraction patterns. That of the B form has always been rather poor and may not yield a clear, unambiguous decision . . .[63]

Then, however, there was a new challenger, the SBS model of DNA. But Crick and his colleagues could afford to admit that there really were limitations in the techniques upon which they had relied for decades. In the same paper they announced the results of a 'crucial experiment', claiming that it eliminated the 'warped zipper' in favour of the double helix.[64]

4. THE GENESIS OF THE SBS MODEL IN INDIA

With Sasisekharan's methodological qualms set in their historical context, we may turn now to their consequences. As we have seen, despite having begun by buttressing orthodoxy against the left-handed heresy of Mitsui *et al.*, Sasisekharan was troubled by very much the same sort of methodological doubts as Donohue had raised — though he entertained them privately. Donohue's interest had begun when, in the mid 50s, he discovered that the stereochemistry of bases under-determined Watson-Crick base-pairing. Two decades later, before going to Princeton, Sasisekharan had done some work in Madras which had suggested that the bonding flexibility of the nucleoside (a base attached to a sugar) and nucleotide monomer (a sugar group bonded to both a base and a phosphate) was greater than had been generally assumed.

Bond angles between atoms may form at a fixed angle, or over a range of angles. Where a range of bond angles between two atoms is possible, the preferred angle is determined by considerations of overall energetic favourability. The calculation of energetic favourability for each bond angle of a complex molecular structure such as DNA is a formidable task which, to be done completely satisfactorily, requires simultaneous solution for all bonds. Before the powerful computers necessary if such an approach is to be contemplated were widely available, the custom was to start with an arbitrary bond angle (say, the mid-point of the permissible range) changing it only in order to ease close contacts that cropped up elsewhere. This had been done in the process of refining of the Watson-Crick model of DNA, described earlier.

In Madras, Sasisekharan had tried to see whether there was any flexibility available in the bonding of the constituent atoms of the nucleoside and nucleotide. It had been argued — but not demonstrated in Sasisekharan's view — that the bond angles in DNA were severely constrained, and that this rigidity compelled a right-hand sense for Watson and Crick's two helices. Initial indications were that some flexibility did exist. However, the work was restricted by computer equipment of inadequate size. In particular, Sasisekharan had been prevented from extending the work to include an examination of the ring flexibility of the deoxyribose sugar. Needless to say, Princeton boasted computing facilities far superior to those available in Madras. Sasisekharan determined to make use of them for, as he later said, he had concluded the rigidity or otherwise of the nucleotide monomer 'would play a very important role in model building of polynucleotides'.

Indeed, before beginning his sabbatical at Princeton, Sasisekharan concluded that,

> If the DNA structure is rigid, then right from the monomer, the building block itself, it should be rigid. If, on the other hand, if [the monomer] is flexible, then you won't be able to arrive at a unique structure.

After arriving at the conclusion that the rigour of conformational work on DNA left something to be desired, Sasisekharan consciously designed his conformational studies to be methodologically superior to the work begun by Wilkins at King's College, London, and pursued by his students and collaborators. Sasisekharan's view was that the most prominent contemporary product of this group, Struther Arnott, now effectively 'owned' the problem. Indeed Hamilton considered that Arnott's work had 'provided positional parameters for all DNA atoms correct within a few tenths of an ångstrom unit'.[65] Arnott's techniques were a benchmark. Even if his results supported the orthodox view, but especially if they did not, Sasisekharan had to be able to lay claim to have used an improved methodology. If the work were only to confirm Arnott's the claim would be needed to justify publication. If, however, Sasisekharan wanted to confound the results of Arnott and his predecessors, then a demonstrable methodological superiority would help him hurdle the referees.

Thus we can detect a confluence of factors which acted to generate and reinforce Sasisekharan's methodological views. His work in Madras led him to suppose that some bonding rigidities had been introduced to simplify model building that, when investigated, proved artificial and, possibility, misleading. The experience was repeated when Langridge's group's claims turned out to be founded on technical inadequacies. We have noted that there was a recurrent 'fundamentalist' reaction to the 'quick and dirty' methods of Watson and Crick, despite their productivity, so we should not be surprised when Sasisekharan, a professional molecular biophysicist, responded in the same way. Moreover, he had been given some reason by the work he had done in Madras to suppose rigour would prove productive, and not just pedantic. And, once launched upon a full scale reworking of the conformational studies on DNA, it made strategic sense to do the thing impeccably. In addition, Sasisekharan's perception of a need for a more rigourous methodology was bound up with the desire to take advantage of technological developments. He held that 'stereochemistry can never be satisfied by scale models alone, not in the late 1970s and 1980s. It might have been all right in the 1960s and 1950s'. Large computers now made it possible to determine the energetically favoured stereochemistry of a molecule with great certainty and accuracy. They allowed the totality of the intramolecular influences on the bond angles of a polymer structure to be examined, so that the details of the most likely (and not just a possible) conformation would be revealed. And it seemed to Sasisekharan that electronic data processing techniques permitted him to eliminate any doubt as to the possibility of good, alternative conformations for DNA radically differing from Watson and Crick's model. Contrary to the view of *Nature*, quoted earlier, Sasisekharan did think that it could be deductively established whether or not 'some other arrangement of atoms might exist which would give the observed X-ray diffraction patterns'.[66]

Accounting for the role of method in these terms does not explain away Sasisekharan's methodological commitments. Those views are made more, not less plausible by the historical and biographical circumstances surrounding them. In Donohue's case we may be inclined to look askance at loudly protested methodological purism. But this is because we can offer an *alternative* and, perhaps, *more plausible* motivation to the compulsion of reason. Donohue's peers suspected him of being disingenuous, accordingly they required him to put up — a better model — or shut up. The factors to which I have made reference are *not* an alternative explanation of Sasisekharan's actions. Rather, they help explain why he adopted and acted from a 'fundamentalist' methodological perspective.

Thus it was that, with a Japanese visitor to Langridge's laboratory, Sasisekharan turned the Princeton computing facilities to the question of the ring flexibility of the deoxyribose sugar. And, indeed, they found 'that the sugar linked with the phosphates can move like an arm by as much as 5 to 7 ångstroms'. This discovery led to an investigation of the effects of this flexibility on the preferred handedness of a bihelical structure. Preliminary results 'indicated that as far as the backbone conformational angles are concerned, both right-handed and left-handed double-helical structures are possible'. At this point, September 1972, Sasisekharan left Princeton to take the Chair of Molecular Biophysics at the Indian Institute of Science. Once there he resumed his study of the conformation of DNA — though not without experiencing considerable difficulty in persuading graduate students to collaborate with him on the project. Eventually however, N. Pattabiraman and, later, Gautam Gupta joined Sasisekharan.[67] Under Sasisekharan's direction, Pattabiraman focused his attention on the conformation of the sugar-phosphate exoskeleton, whilst Gupta gave his to the base-pairs.[68] Sasisekharan says that, in this research

> The criteria we followed were: Watson-Crick base pairing must be there because there is no doubt that it is the chemical basis of genetics, models must be stereochemically satisfactory, and they should agree with the [X-ray diffraction] data.

Having confirmed Sasisekharan's finding that the sugar-to-phosphate bond angles could vary over a wide range, Pattabiraman set out to discover the impact this would have on possible exoskeletal conformations. It was a formidable task; nine or ten parameters admitted of between 36^{10} and 36^{13} possible combinations. The necessity for a computer will be readily appreciated. Eventually, Pattabiraman recalls, 'I found good left-handed structures coming very smoothly'. Concentrating on these, he was able to develop a stereochemically viable left-handed double-helical structure, with A, B, C and D-DNA subvariants, and whose fit with the X-ray data was comparable with right-handed models.[69]

In the back of Sasisekharan's mind were two other considerations. Firstly, there were the recurrent obstacles to a satisfactory account of strand separation during replication which arose from the necessity for unwinding that was imposed by a bihelical model.[70] Secondly, there were the equally persistent problems in developing stereochemically viable Watson-Crick models of DNA molecules exhibiting superhelices (like overwound rope) and right-angled bends. Both these conformations were known in electron micrographs of DNA, and were thought to be normally present *in vivo*. The difficulties modelling them stemmed from what Sasisekharan characterized as the 'rod-like rigidity' of a double-helical structure (and regardless of whether it had left or right-handed helical sense).[71] Worries about unwinding had disturbed many throughout the nearly thirty years since Watson and Crick had advanced their model, provoking consideration of an alternative structure. Most, including Watson and Crick, gave thought to the incorporation of left-handed helical elements as a way around the problem. And most rejected the idea — or had their left-handed models rejected — on the basis of the arguments, first of Watson and Crick, and then of

Fuller *et al.*, that DNA could not be left-handed.[72] Watson and Crick's consideration and rejection of left-handed helices was commented upon in a review by Delbrück and Stent written in 1957.[73] A student stumbled upon this paper by chance, and brought it to Sasisekharan's attention. His response was

> inasmuch as we [can] have both right-handed and left-handed helical structures satisfying both stereochemical and X-ray data, why should we not combine the two — which would avoid tangling the two chains.[74]

Pattabiraman recalled that, in working out how best to do this, 'it occurred to us, in order to have a flexible model, you need both left and right [handed] helical sections' in each strand. They used the most regular mixture of semihelical sections, successive half-helices of opposite hand (that is, exhibiting a switch from right to left-handedness every five base pairs) Gupta's contribution lay in an examination of the base-sugarphosphate interactions. The flat, sheet-like bases are edge bonded to a sugar group at one end, and to the other base of the pair (A-T, or G-C). There are two distinct 'faces' to a base. Which face is oriented 'up' and which 'down' is specified by Watson and Crick base-pairing. Gupta found that this aspect of the stacking arrangements was not required by considerations of diffraction or local stereochemistry and so was, to that extent, arbitrary. It was possible to *invert* the bases from their customary orientation without difficulty. Moreover, if the bases were inverted at the bend region of an SBS structure — where the exoskeletal strand changes handedness — the overall conformation looked more energetically favourable.

Thus Sasisekharan, Pattabiraman and Gupta developed two versions of their alternative structure for DNA. Both utilized alternating left and right-handed half helices, but Type I had unaltered Watson-Crick base-pairing and Type II used inverted base stocking at the bend region.[75] Sasisekharan insisted on exhaustive double checking of all calculations. He said he was on his guard because it is 'so easy to be overwhelmed by enthusiasm', adding

> In fact, we hesitated to publish because I wanted to be sure that I hadn't made a mistake, and that none of my students had made a mistake. Ones does not want to rush into print when one's whole reputation is at stake. I've been in this field many years. I cannot afford to make a mistake.

Eventually, however, a paper was prepared and sent to the Indian journal *Current Science*. The referees' verdict came as shock for, in the interim, Rodley and his co-workers had advanced their SBS model — identical with the Indians' Type I structure. The referees drew the New Zealanders' model to Sasisekharan's attention, and insisted that he acknowledge their priority as a condition of publication.

5. SASISEKHARAN AND (METHODOLOGICAL) CONCEPTUAL CREATIVITY

Sasisekharan had perceived a clash between the strength of the evidence which supported the Watson-Crick double-helical structure for DNA and the strength of the evidence which he felt *ought* to buttress it. He began his inquiries in the belief that the techniques by which the fit between evidence and theory was appraised as adequate, indeed decisive, could and should be improved. Thus Sasisekharan had perceived a methodological conceptual problem of the sort described by Laudan — namely, a clash between a scientific theory and the norms under which it is supposed to be subsumed. The outcome suggests that Laudan may be correct when he argues that such problems are at least as important as empirical anomalies in stimulating research.

To begin with, however, Sasisekharan was not straightforwardly confronted by a choice between a theory and metatheory. Rejection of the Watson-Crick model was not immediately at issue. Sasisekharan certainly saw specific difficulties in replication theory, and the stereochemistry of larger scale molecular conformations. Nevertheless, these were niggling difficulties according to orthodox methodology, and Sasisekharan did not place any great emphasis upon them.[76] Rather, he wanted to improve the orthodox methodological standards, and then apply them to the Watson-Crick structure, to see what resulted. Sasisekharan did have some concrete evidence to think that the outcome might not be a mere vindication of orthodoxy; but this came nowhere near amounting even to a sketch of an alternative to the double helix. It was a direct result of the application of Sasisekharan's rigorous methodological precepts which led to an alternative model comparable to the double helix. The comparison, however, was done by reference to the novel methodological standards.

The improvements Sasisekharan sought to make were not merely technical — though they were enabled by advances in the technology available. As we have seen, the received view was that the Watson-Crick model of DNA was acceptable because it had been established as demonstably superior to those alternatives that happened to have been advanced against it. Of course, any model had to meet basic standards of stereochemical viability, reasonable fit with diffraction data, and consistency with Watson-Crick base-pairing rules. But the rigour of those standards was strongly influenced by the presence or absence of competition, as well as how stiff that competition was (i.e., how good an alternative to Watson and Crick). Thus *Nature* had observed, 'whether or not the reader finds the fit to the Watson-Crick model convincing in itself, the results for Hoogsteen and Donohue [base-]pairs seem to undermine the contention that [crystallographic] technique cannot eliminate alternatives'.[77] Contrary to this, Sasisekharan held that a structure for DNA could not be regarded as proven satisfactorily unless and until it had been established that it was superior, not merely to any model that *happened* to have been proposed, but superior to *any possible* structure.

It is a fine irony that the restrictions imposed by the limitations of the computing facilities at Bangalore meant that the grand aim of an exhaustive survey of the conformational possibilities for DNA was never completely achieved by the Indian group. Rather, the work had to be carved into pieces managable by the computer and linked together by 'old-fashioned' scale models. The Sasisekharan never quite managed the methodological certainty he began by looking for. On the other hand, he never claimed to have done so. When the SBS model was first advanced by the Indians, they made no great play of their techniques, arguing only that the evidence was insufficient to distinguish decisively between their new structure and of Watson and Crick, whilst the 'warped zipper' had greater stereo-chemical flexibility and told a simpler story about replication.[78]

Nevertheless, the very existence of such a state of affairs seemed to Sasisekharan to vindicate his suspicion of certitudinal complacency among the specialist community. Moreover, the SBS model was only one of several alternatives to various aspects of the conventional structure that turned up. For example, the Indian's claims to have established, in theory, the viability of left-handed double-helical structures and the influence of base pairs on the exact details of exoskeletal conformation was vindicated by the subsequent experimental discovery of the so-called Z-DNA.[79] In contrast with the regular Watson-Crick structure, this molecule exhibited a left-handed sugar-phosphate double helix puckered in 'Z' shape under the influence of the particular base-pairs present.

6. INTERESTS AND THE WEIGHTING OF CONCEPTUAL ANOMALIES

Both Sasisekharan and, before him, Donohue saw a conceptual problem in the methodological warrant for the Watson-Crick double helical model of DNA. But why did Sasisekharan and Donohue take such a different view of the adequacy of conformation methodology to that of their peers in the specialist community? Laudan's concept of a conceptual anomaly generated by a perceived clash between methodological norms and a substantive scientific theory admits of degrees of severity, and he provides three criteria by which to judge the 'relative weighting of conceptual problems'. To a certain extent, these standards assist us in understanding Sasisekharan's, Donohue's and the rest of the specialist community's judgments. However, they are based on an ill-founded assumption that there *ought* to have been consensus about the methodological warrant for models of the structure of DNA.

Laudan's first criterion for assessing the weight of conceptual problems is the difference between the *desired* and the *actual* logical relation between a theory and a methodological metatheory. He specifies five 'cognitive relationships' between any two such theories. In decreasing order, they are: entailment, reinforcement, compatibility (indifference), implausibility and inconsistency. The worst conceptual problem, according to Laudan, results where inconsistency obtains and entailment is desired. However, he says, difficulties are created by all possible relationships except entailment.[80]

Considered in these terms, Sasisekharan's problem was that the *reinforcement* provided by the evidence favouring the Watson-Crick double helix was less than should be required by acceptable methodological standards. According to London, a theory, *T*, is said to reinforce another theory, T_l, insofar as '*T* provides a 'rationale' for . . . T_l'[81] The consensus among specialists was that an adequate 'rationale' for accepting the Watson-Crick model existed. Sasisekharan, and before him Donohue, took the view that the 'rationale' was insufficient. Wilkins, Hamilton and Arnott, considered it enough that no alternative happened to have been advanced which was as good as or better than Watson and Crick's. Sasisekharan and Donohue wanted to be sure that none ever could be.

Donohue took matters no further than this – perhaps because his methodological critique had really been no more than a response to the earlier rejection of his alternative base-pairing scheme. Sasisekharan, with the research in Madras indicating that there might some profit in it, turned a methodological principle into a programme, and set out to find out whether any comparable models could be devised. Some one might suggest that Sasisekharan used a methodological principle to *legitimate* his programme. But the absence of a *polemical* (as distinct from *heuristic*) role for method in Sasisekharan's published work argues against such an interpretation. Eventually Sasisekharan's programme confirmed his suspicion that the evidence for the structure of DNA was capable of more than one plausible construction.

This first criterion describes the difference in judgement made by the Sasisekharan, Donohue, Hamilton, Arnott, etc., well enough. But who was correct? Was there an adequate 'rationale' for believing Watson and Crick or not?

Laudan explains his second measure of the strength of a conceptual problem in the following terms:

> When a conceptual problem arises as a result of a conflict between two theories, T_I and T_2, the seriousness of that problem for T_I depends on how confident we are about the acceptability of T_2. If T_2 has proven extremely effective at solving empirical problems and if its abandonment would leave us with many anomalies, then matters are very different for the proponents of T_I. If, on the other hand, T_2's record as a problem solver is very modest, then T_2's incompatibility with T_I will probably not count as a major conceptual problem for T_I.[82]

There is nothing modest about the problem solving record of the semi-conservative molecular theory of inheritance and replication which is based on the Watson-Crick model of DNA. Moreover, the techniques by which it was devised and refined, the very techniques whose underlying justification Sasisekharan was contesting, had achieved the status of exemplar with several successful creative imitations such as solving the structure of RNA. The crystallographer Hamilton argued explicitly that these successes counted strongly in favour of the Watson and Crick structure.[83] This second criterion, then, seems to help explain not the perception of a conceptual problem by Sasisekharan or Donohue, but the limited concern that Donohue's published critique of the methodology bolstering the Watson-Crick base-pairing engendered among his fellow specialists. Similarly, it explains why the specialist community, despite accepting some of Donohue's methodological qualms, nevertheless placed the onus upon him to produce a superior model of DNA. Donohue's doubts were discounted against the powerful – if indirect – evidence provided by the achievements based on the double helix. It was this undeniably considerable success which underwrote the specialist's confidence.

Are we, then, to conclude that Sasisekharan and Donohue were wrong in their assessment of the importance of the conceptual problem they perceived? I think not. Quite typically, both the scientist, Donohue, and the metascientist, Laudan, assume that any one who criticizes a scientific theory from a methodological point of view must – if they are correct – have reasons for doing so which will impress all or most members of a scientific community. This is a plausible enough view: Either a theory has good methodological warrant, or it does not. However, this is not so. The standards by which a scientist will, and should, assess the adequacy of a theory or hypothesis depend upon her or his purposes in doing so. By and large, most members of every scientific community are engaged in what Kuhn calls 'normal science' – refining, developing and applying already successful theories whose methodological well-foundedness is simply taken for granted. Neither will it become an issue for the community as a whole unless and until a competitor is proposed which permits normal 'puzzle-solving' and (perhaps above all else) practical application to continue.

Notwithstanding this, the organization of modern science provides for a few scientists who have the time, facilities and inclination to ponder on recalcitrant (though not particularly obstructive) problems – empirical or conceptual. Having a cognitive interest in such matters, generally speaking, requires that a scientist's social interests can be served thereby. A biochemist working for Roche, or a chemist at Du Pont, is not likely to be in such a position. But the fact that the great majority of scientists are engaged in such applied R&D, should not lead us to ignore the comparatively few in settings where the pursuit of intellectual curiously can have its material rewards because it serves Mammon more indirectly – institutions of 'pure' research and universities. Even those ostensibly engaged in work justified by its usefulness, such as medical research, can often contrive to persuade granting bodies of the necessity for some 'fundamental' or 'basic' investigations. Even though such science is comparatively rare, it is a mistake to deny its existence. It serves a variety of important training and public relations functions, and produces enough unintentionally profitable spin-offs – like genetic engineering – to provide a more tangible justification.

On such a view of things, the adequacy of the refinement and testing techniques used to appraise the Watson-Crick model of DNA depends on the interacting cognitive and social *interests* of the scientist judging them. Nowadays, especially with the explosive growth in recombinant DNA work, a very great number of scientists use Watson and Crick's model of the structure of DNA. But very few of them are competent to assess it, let alone interested in doing so. Indeed, only the small group who work on the conformation of biological macromolecules are really able to read critically the technical papers of the even smaller group who specialize on the structure of DNA. The rest must rely on review articles produced for a wider scientific audience by the specialists. Such an authority structure is typical of modern, ultra-specialized science.

In a review aimed at a wider audience, Hamilton made the specialists interests plain when he observed that whether the Watson-Crick model is

> the actual structure of DNA or still a structural hypothesis . . . [i]t is nevertheless relevant to current research; specifically, the derivation of the A conformation of DNA greatly aided the elucidation of the RNA double which is very similar, and this conformation has recently been used . . . as the basis of a molecular model of the anticodon arm in the transfer RNA molecule.[84]

We have seen that Donohue's methodological critique was seen by the specialist community as serving subterranean interests of his own which they did not share. And, since Donohue's cognitive interests were not directly methodological, he did not deploy an alternative methodology. To him, it seemed enough to cast doubt on the decisiveness of the techniques used to eliminate his base-pairing hypothesis.

Sasisekharan's professional interest in the ring flexibility of deoxyribose and the conversion of D-DNA to B-DNA led into more strictly methodological questions. He had some reason to suspect that, if a more rigorous method were applied, there might be room for an alternative structure. But he was as unsuccessful as Donohue had been in attempting to interest his peers. And the reason for this was the same too: The specialists' methodology was a means to an end that they seemed to have achieved — it underpined a spectacularly successful research programme. And, in those circumstances, scientists rarely see more than a few clouds on the horizon. But Sasisekharan knew what Donohue had been told — come up with the goods and people will sit up and take notice. To try to do so suited his cognitive, scientific interests — after all he worked in an institution with the function of training India's scientific elite and entrusted with keeping its capacity for first class science before the international scientific community. Anything which he did that produced Ph.Ds and papers in international journals would keep the salary checks coming in. Obviously, therefore, Sasisekharan was interested in the views which his fellow specialists held about him and his work. From the outset considerations of this kind acted to reinforce Sasisekharan's methodological purism. It also made him so cautious as to disasterously delay publication of the outcome of his conformational research.

Laudan's third criterion for assessing the importance of a conceptual problem is its 'age'. With others, for example, Kuhn and Imre Lakatos, he shares the belief that the past successes of a theory in eliminating problems warrant confidence in the eventual elimination of newly apparent difficulties.[85] Accordingly, for Laudan, in the case of conceptual problems, handicapping is 'weight for age' — the more persistent a problem the worse it is. It is hard to see how this criterion, unmodified, can get a grip on Donohue's and Sasisekharan's conceptual problem (or, for that matter, on the failure of the specialist community to accept its importance). But here, again, the interests of scientists must be taken into account. What is really relevant is the length of time a perceived conceptual problem has stood in the way of the perceiver's pursuit of her or his interests. Considerations of this kind emphasize the subjective character of conceptual problems. To relate them to a specialist scientific community, one must ask how any given individual's interests bear on those of the community as a whole, or identifiable interest groups within it. If really central concerns are thwarted, even a short-lived problem may seem intolerable. Conversely, if only peripheral interests are affected, earlier successes of the theory concerned will probably seem quite sufficient grounds for turning to other matters in the confident expectation that someone will figure out a solution eventually.

7. CONCLUSION

Finally then, Larry Laudan's notion of a normatively generated conceptual problem has proven helpful in understanding the role of method in the study of the structure of DNA. Nevertheless, the case study suggests that methodological qualms should not be seen in isolation from the context in which they arise and are received. And contrary to Laudan's view, the appraisal, indeed the perception of conceptual problems of this kind, seems to have a great deal to do with the individual and group interests of the scientists involved. Nevertheless, methodological commitments do appear as more than mere *post hoc* ideological legitimation — though that, too, has been seen to occur.

ACKNOWLEDGEMENT

The generous co-operation of V. Sasisekharan, N. Pattabiraman and Gautam Gupta, as well as the financial assistance of the Faculty of Arts and Department of History and Philosophy of Science, University of Melbourne, are gratefully acknowledged. I am indebted to the editors of this volume for their helpful advice.

Deakin University, Australia

NOTES

[1] L. Laudan, *Progress and its Problems* (London, 1977), p. 58.

[2] T. S. Kuhn, 'The History of Science', in his *The Essential Tension* (Chicago, 1977), pp. 116—117 (emphasis added).

[3] L. Laudan, 'Thomas Reid and the Newtonian Turn of British Methodological Thought', in *The Methodological Heritage of Newton*, ed. R. Butts and J. Davis (Toronto, 1970), pp. 103—131. L. Laudan, 'The Methodological Foundations of Mach's Opposition to Atomism', in *Space and Time, Matter and Motion*, ed. P. Machamer and R. Turnbull (Columbus, 1976), pp. 390ff.

[4] G. Cantor, 'Henry Brougham and the Scottish Methodological Tradition', *Studies in History and Philsoophy of Science* **II**, 1971, pp. 69—89; G. Cantor, 'The Reception of the Wave Theory of Light in Britain: A Case Study Illustrating the Role of Methodology in Scientific Debate', *Historical Studies of the Physical Sciences* **VI**, 1975, pp. 109—132.

[5] See Lauden *op. cit.* (Note 1), ch. 2 n. 13, pp. 230—231.

[6] *Ibid.*, pp. 58—59.

[7] *Ibid.*, p. 45.

[8] G. A. Rodley, R. S. Scobie, R. H. T. Bates and R. M. Lewitt, 'A Possible Conformation for Double-Stranded Polynucleotides', *Proceedings of the National Academy of Sciences* (USA) **LXXIII**, 1976, pp. 2959—2963; V. Sasisekharan and N. Pattabiraman, 'Double Stranded Polynucleotides: Two Typical Alternative Conformations for Nucleic Acids', *Current Science* **XLIV**, 1976, pp. 779—783.

[9] S. Arnott, 'Is DNA Really a Double Helix?', *Nature* **CCLXXVIII**, 1979, pp. 780—781.

[10] T. D. Stokes, *The Side-By-Side Model of DNA: Logic in a Scientific Invention* (Ph.D. dissertation, University of Melbourne, 1983). The interviews were recorded during

November and December of 1979 at, respectively, the University of Canterbury, Christchurch, New Zealand and the Indian Institute of Science, Bangalore.

[11] *Ibid.*, esp. Chapters I and X.

[12] *Ibid.*, esp. Chapters II and XI.

[13] T. D. Stokes, 'The Double Helix and the Warped Zipper — An Exemplary Tale', *Social Studies of Science* **XII**, 1982, pp. 207–240. The principal sources for Kuhn's views are his seminal *Structure of Scientific Revolutions*, 2nd. edition (Chicago, 1970), and collected essays, *op. cit.* (Note 2).

[14] L. Laudan, *op. cit.* (Note 1), p. 49.

[15] *Ibid.*, pp. 54–64.

[16] *Ibid.*, p. 61.

[17] I. Tinoco Jr. and C. R. Cantor, 'Application of Optical Rotatory Dispersion and Circular Dichroism to the Study of Biopolymers', *Methods of Biochemical Analysis* **XVIII**, 1970, pp. 81–200.

[18] *Ibid.*, p. 178.

[19] J. D. Watson and F. H. C. Crick, 'The Structure of DNA', *Cold Spring Habour Symposia on Quantitative Biology* **XVIII**, 1953, pp. 123–131.

[20] W. Fuller, M. H. F. Wilkins, H. R. Wilson and L. D. Hamilton, 'The Molecular Configuration of Deoxyribonucleic Acid', *Journal of Molecular Biology* **XII**, 1965, pp. 60–80.

[21] *Ibid.*, p. 74.

[22] Y. Mitsui, R. Langridge, B. E. Shortle, C. R. Cantor, R. C. Grant, M. Kodama and R. D. Wells, 'Physical and Enzymatic Studies on Poly d(I-C).Poly d(I-C), an Unusual Double-helical DNA', *Nature* **CCXXVIII**, 1970, pp. 1166–1169.

[23] *Ibid.*, p. 1168.

[24] *Ibid.*

[25] *Ibid.*, p. 1169.

[26] This and subsequent, unacknowledged comments of Sasisekharan's are from the transcripts of interviews recorded by the author at the Indian Institute of Science in December 1979.

[27] *Nature* **CCXXVIII**, December 19, 1970.

[28] See J. D. Watson, *The Double Helix*, Norton Critical Edition, ed. G. S. Stent (New York, 1980), originally published in 1968.

[29] E. Chargaff, *Heraclitean Fire* (New York, 1978), p. 101.

[30] Cited in Robert Olby, *The Path to the Double Helix* (London, 1974), p. 374, emphasis in the original.

[31] *Ibid.*

[32] L. D. Hamilton, 'DNA: Models and Reality', *Nature* **CCXVIII**, 1968, pp. 633–637.

[33] See T. S. Kuhn, 'Second Thoughts on Paradigms', in *op. cit.* (Note 2), pp. 293–319. Stokes, *op. cit.* (Note 13) develops this analysis at some length.

[34] The only major change made was to increase the diameter of the helix.

[35] L. D. Hamiton, *op. cit.* (Note 32), p. 634, emphasis added.

[36] *Ibid.*, 634–635.

[37] M. H. F. Wilkins, 'Molecular Structure of Deoxyribose Nucleic Acid and Nucleo-Protein and Possible Implications in Protein Synthesis', *Biochemical Society Symposia* **XIV**, 1957, p. 14.

[38] *Ibid.*, pp. 14–15, emphasis added.

[39] L. D. Hamilton, *op. cit.* (Note 32), pp. 634.

[40] *Ibid.*, p. 636, emphasis added.
[41] *Ibid.*
[42] J. D. Watson and F. H. C. Crick, 'Molecular Structure of Nucleic Acids: A Structure for Deoxyribose Nucleic Acid', *Natutre* **CLXXI**, 1953, p. 737.
[43] J. Donohue, 'Hydrogen-Bonded Helical Configuration of Polynucleotides', *Proceedings of the National Academy of Sciences* (USA) **XLII**, 1956, pp. 60—65.
[44] K. Hoogsteen, 'The Structure of Crystals Containing a Hydrogen-Bonded Complex of 1-Methylthymine and 9-Methyladenine', *Acta Crystallographica* **XII**, 1959, pp. 822—823.
[45] L. D. Hamilton *op. cit.* (Note 32), p. 636.
[46] J. Donohue, 'Fourier Analysis and the Structure of DNA', *Science* **CLXV**, 1969, p. 1095.
[47] P. R. Pinnock, C. A. Taylor and H. Lipson, 'A Re-Determination of the Structure of Triphenylene', *Acta Crystallographica* **IX**, 1956, pp. 173—178.
[48] *Science* **CLXVII**, 1970, pp. 1693—1702.
[49] *Nature*, Unsigned, 'The Fly in the Fourier', **CCXVI**, 1970, pp. 404—405.
[50] M. H. F. Wilkins, D. A. Marvin and L. D. Hamilton, 'Some Misconceptions on Fourier Analysis and Watson-Crick Base Pairing', *Science* **CLXVII**, 1970, p. 1693, emphasis added.
[51] S. Arnott, 'Crystallography of DNA: Difference Synthesis Supports Watson-Crick Base Pairing', *ibid.* pp. 1694—1700.
[52] *Nature, op. cit.* (Note 49), p. 404.
[53] F. H. C. Crick, 'DNA: Test of Structure?', *Science* **CLXVII**, 1970, p. 1694.
[54] J. Donohue, 'Fourier Series and Difference Maps as Lack of Structure Proof: DNA Is an Example', *ibid*, pp. 1700—1702.
[55] *Ibid.*, p. 1700.
[56] *Ibid.*
[57] *Ibid.*
[58] R. Olby, *op. cit.* (Note 30).
[59] L. D. Hamilton, *op. cit.* (Note 32), p. 634.
[60] M. H. F. Wilkins, *op. cit.* (Note 37), pp. 14—15.
[61] *Op. cit.* (Note 49), p. 404.
[62] F. H. C. Crick, 'The Double Helix: A Personal View (1974)', *Nature* **CCXLVIII**, 1974, pp. 766—771.
[63] F. H. C. Crick, J. C. Wang and W. R. Bauer, 'Is DNA Really a Double Helix?', *Journal of Molecular Biology* **CXXIX**, 1979, p. 451.
[64] Crick had been privately urging the need for a decisive test for some time. It should, he thought, be based on distinctly different predictions arising from the two competing models. The idea was to ascertain the number of complete helices, or 'linking number' of a circular DNA molecule — a very large number if Watson and Crick were correct, a much smaller number if the SBS structures were. Crick *et al.* claimed the result was in the order of magnitude expected on a double helical, and not a 'warped zipper' model. Sasisekharan comments: 'The linking number argument depends on the assummption that the molecule is 100% one handed, so that even if you have in a molecule 5% . . . which prefers lefthandedness . . . the linking number [argument] is out'.
[65] L. D. Hamilton, *op. cit.* (Note 32), p. 636. An ångstrom (Å) is one ten millionth of a millimetre ($1/10^{-10}$ metre).
[66] *Op. cit.* (Note 49), p. 404.

67 At the bottom of the difficulty lay the necessity for these students to work on a problem whose solution was sufficiently certain to promise a successful doctoral disseration. In the end Sasisekharan was obliged to offer a 'safer' topic if the research on the structure of DNA proved too troublesome.

68 Interviews with Pattabiraman and Gupta were recorded by the author at the Indian Institute of Science in December 1979. The transcripts are the source of all otherwise unacknowledged material, including direct quotation.

69 In essence, these models were the *same* as the refined Watson-Crick, except that they were *left*-handed.

70 Indeed, it was one such recalcitrant problem, explaining the formation of single-stranded regions in intact circular DNA, which initiated the work of Rodley and his co-workers in New Zealand. See Stokes, *op. cit.* (Note 10), esp. Chapters I and X.

71 No doubt the early work in Madras on ring flexibility helped make this consideration prominent in Sasisekharan's mind.

72 *Op. cit.* (Note 19 and 20).

73 M. Delbrück and G. S. Stent, 'On the Mechanism of DNA Replication', in: *A Symposium on the Chemical Basis of Heredity*, ed. W. D. McElroy and B. Glass (Boston, 1957), pp. 699—736.

74 From an undated draft reply to the author's initial letter to Sasisekharan (29/6/79). It was not sent because arrangements were made for the author's visit to the Indian Institute of Science, where he was given a copy of the corrected draft.

75 Eventually the Indians found that they had to use inverted base stacking at the bend regions because the normal orientation introduced close contacts which could not be eliminated. Consequently, they abandoned the Type I structure. See V. Sasisekharan, N. Pattabiraman and G. Gupta, 'Some Implications of an Alternative Structure for DNA', *Proceedings of the National Academy of Sciences* (U.S.A.) **LXXV**, 1978, pp. 4092—4096.

76 This provides a sharp contrast with the New Zealanders. Rodley and his co-workers regarded this fairly longstanding empirical anomaly as a *falsification*. Sasisekharan, together with the rest of the specialist community, acknowledged it as being worryingly recalcitrant, but they were certainly not prepared to abandon the Watson-Crick structure because of it. See Stokes, *op. cit.* (Note 10).

77 *Op. cit.* (Note 49), p. 404.

78 *Op. cit.* (Note 8 and 75).

79 A. H.-J. Wang, G. J. Quigley, F. J. Kolpak, J. L. Crawford, J. H. van Boom, G. van der Marel, and A. Rich, 'Molecular Structure of a Left-Handed Double Helical DNA Fragment at Atomic Resolution', *Nature* **CCLXXXII**, 1979, pp. 680 686.

80 L. Laudan, *op. cit.* (Note 1), p. 54. It is, of course, hard to see how a substantive scientific theory could be entailed by a methodological metatheory. However, Laudan devised his criteria to cover all types of conceptual problem, including clashes between two substantive theories where entailment is a more plausible relationship.

81 *Ibid.*

82 *Ibid.*, p. 65.

83 See text at Note 32 and ff.

84 L. D. Hamilton, *op. cit.* (Note 32).

85 L. Laudan, *op. cit.* (Note 1), pp. 65—66. It is not, however, commonly appreciated that this is, quintessentially, an *inductively* justified viewpoint.

JAN SAPP

INSIDE THE CELL: GENETIC METHODOLOGY AND THE CASE OF THE CYTOPLASM

I. INTRODUCTION

If a reliable set of procedures for the discovery and evaluation of knowledge existed — that is, if there were a single, efficacious scientific method — then the history of science would be complete. The growth of science would be the result of an unfolding, its future course predicted at the moment the method was discovered. But for anyone who understands that scientific thought and action are historically determined, for anyone who believes that scientific thought, like any other thought, is relative to time and social context, such a position seems wildly implausible.

One implication of such a perspective would be that only those deviant groups that have fallen by the wayside of the correct scientific method would actually have a history. Time and social context would be relevant only to those 'non-scientific' scientists whose non-scientific interests and institutional or social milieu created obstacles to the proper functioning of the 'correct scientific method'. In effect, within this framework, to attribute correct belief to correct scientific method amounts to nothing more than pure tautology. In other words, correct belief tends to be evaluated by 'scientific' criteria, incorrect belief by 'non-scientific' criteria. This asymmetry[1] in the way knowledge claims tend to be evaluated by historians is particularly acute in the history of the modern sciences where historians often begin with the explicit or implicit assumption that the dominant theories of our time are correct or at least represent a closer approximation to the truth than past theories. They come to believe in them so strongly that they often fail to recognize the social dimension of that very triumph.

The history of genetics has often been treated in this way. The Mendelian-chromosome theory is normally held to have been generally accepted by 1915, by which time it is judged to have been solidly grounded in observation and experimentation. Historians have subsequently followed the path of its development, from the convergence of diverse specialties on the 'evolutionary synthesis' of the 1930s and 1940s to the DNA doctrine of the 1950s and 1960s.[2]

J. A. Schuster and R. R. Yeo (eds.), The Politics and Rhetoric of Scientific Method, 167–202.
© 1986 *by D. Reidel Publishing Company.*

Only when Mendelian principles have been opposed have historians called upon 'non-scientific' criteria to account for such (incorrect) beliefs. William Bateson, for example, who was unable to follow the Mendelian track all the way to the chromosome theory, is judged an idealist and to represent 'conservative' thought in science.[3] Neo-Lamarckians tend to be seen as provincial dogmatists (in the case of French neo-Lamarckians),[4] as fools, or as frauds (like Kammerer). Lysenkoism is held to provide an example of how science becomes distorted and perverted when the political and the scientific become too intimate.[5] Those who focus their analysis on the deleterious effects of neo-Lamarckism and Lysenkoism can do this primarily because of their implicit assumption that science somehow follows its own immanent logic if left free of social and political interference. This assumption would be acceptable if one reliable method (or set of methods) existed that invariably produced objective knowledge. But the problematic character of such an assumption is recognized by many historians and philosophers, as well as by scientists themselves.

As long as the definition of what science is remains a central issue within science itself, then both social and intellectual criteria have to be taken into consideration when evaluating both rejected *and* accepted knowledge claims. As Pierre Bourdieu has argued, a scientific field is a place of struggle in which scientists compete for scientific authority, that is, for the 'socially recognized capacity to speak and act legitimately in scientific matters'.[6] What is at stake in this struggle is the delimitation of the phenomena, questions, answers and techniques that are considered to be interesting, important, acceptable and useful. In the attempts of scientists to impose a definition of science, each participant tends to uphold those scientific values that are most closely related to him or her personally or institutionally. However, although no value-free set of scientific methods exists, this does not deny that each field has an inherent social logic. The relative specificity of a scientific field results, as Bourdieu has argued, from a tendency among researchers to be able to gain the right to act and speak legitimately in scientific matters only from their fellow producers who, being their competitors, are the least inclined to grant recognition without discussion and scrutiny. It is largely through this dialectic that order is established out of the 'anarchic antagonism of particular interests'.

Within this context, formal methodological doctrines (inductivism, hypothetico-deductivism, falsificationism, etc.) are deployed in diverse and conflicting ways. These discursive resources are used as polemical tools to legitimate divergent objectives and techniques. They are summoned to help on both sides of scientific controversy, that is, in warding off and making room for alternative theoretical possibilities. They are also used in attempts to establish property rights in disciplinary disputes and in individual priority disputes. Thus, although method does not exist in the widely believed sense, formal method doctrines do play a discursive role in the social constitution of science.

But formal methodological doctrines are not the only resources available for these purposes. Scientists also use non-formal discursive resources to accomplish the functions just mentioned, including accounting for knowledge claims. The other chapters of this volume deal mainly with formal method doctrines. In this essay, I widen the category of 'methodological' accounting resources to include 'social' accounting, as discussed by Mulkay and Gilbert,[7] as well as a new category of resource involving ascriptions of 'technique-ladenness of observation'. All these types of resource, I argue, function in similar ways in the social construction of science, and hence all can be treated within a broadened category of 'scientific method'.

Scientists use interpretations of past celebrated scientific events and historical patterns as precedents in both understanding and attempting to legitimate their positions in scientific controversy. It will be shown in this account that scientists also point to the social pressures of the reward system in the field and the motivations of their competitors in an attempt to delegitimate a rival theory by showing that its dominance or prominence is due to the social power of its adherents rather than to its inherent cognitive superiority. One cannot exclude these types of social accounting resources from the complex corpus of scientific 'methods', once formal methods themselves are understood as accounting resources.

Scientists also (and perhaps more frequently in experimental sciences) call upon what I shall term the 'technique-ladenness of observations' in their attempts to advance alternative theories and to account for the 'false' beliefs of their competitors. By 'technique-ladenness of observations' I refer to the means by which scientific concepts become bound to phenomena studied by certain techniques. The term embraces the scientific equipment, procedures, materials, tools, and skills that are involved in the production of scientific results. In order to understand better how this problem may bear on scientific controversy, it is necessary to make a closer examination of the power relations in science.

Within a scientific field, with its diverse problems, theories, techniques and research strategies, and where several competing groups exist, there are invariably differences in their success as perceived by the competitors themselves. At any given moment there is a hierarchy of objects and techniques that strongly orientates practice. The dominant groups are those that lay down 'the rules of the game', and they dictate the realm of tools that are recognized as legal tender within the field. What constitutes a scientific 'proof' and 'correct scientific procedure' is itself socially determined by the immediate state of the field. When the outcome of scientific controversy is dictated by technical capacity and institutional power, the intellectual merit of a theory as judged by formal methodological criteria loses much of its significance. It is in these circumstances that scientists can call upon the technique-ladenness of observations to show that the dominant conceptions of their competitors are intellectually constructed objects created through specific modes of scientific production. The technique-ladenness of observations is also an element in the disciplinary distinctions drawn by scientists, based on divergent and often conflicting objectives, techniques, theories and doctrines.

Formal method doctrines, social accounting, and references to the technique-ladenness of observations all play essentially the same accounting role in the social making (and unmaking) of scientific knowledge. Contrary to the position expressed most prominently by Mulkay and Gilbert,[8] all these kinds of arguments, by which scientists account for the 'false' beliefs of their competitors, occur in the formal literature of science itself. Although these authors do note that many 'non-scientific' factors such as 'a defensive attitude', 'prejudice', 'dislike', 'failure to put enough effort', 'being trained in terms of a false theory' and 'accepting the views of an authorized figure' are frequently used by scientists in accounting for incorrect beliefs of their competitors, they attempt wholly to exclude these kinds of arguments from those found in the published scientific literature. Their claim that 'there is seldom, if ever, any *published* attempt to explain why false theories are accepted by one's colleagues' is based in part on the assumption that 'this would lead formal debate into the realm of "non-scientific" considerations'. Their conclusion is 'that scientific debates in the *formal* literature are radically misleading in that they omit all social theorizing or social accounting which takes place when members informally read, assess and discuss formal knowledge claims'.[9]

The above assertions, and the underlying claim that the best if not the only way to study scientific controversy is through the study of informal scientific discourse, are over-stated to say the least. It will be shown in this account that all the above-mentioned 'non-scientific' social factors *are* used by scientists in formal debates and, indeed, that they play instrumental roles in scientific controversy.

2. THE NUCLEAR/CYTOPLASMIC CONFLICT AND THE DISCIPLINARY STRATEGY OF GENETICS

To illustrate my argument I shall extract an example from a recent study of one of the central controversies in the history of genetics.[10] This centres on a debate about the relative importance in heredity and evolution of the nuclear chromosomal genes and the extra-nuclear or cytoplasmic constituents of the cell. Today, cytoplasmic genetics is an important domain of research, and since the advent of molecular biology in the 1960s various kinds of cytoplasmic inheritance have come to be generally recognized. Some of the principal and generally recognized cytoplasmic genetic factors are cytoplasmic DNA-based genes associated with the energy-generating organelles of the cell, the mitochondria and chloroplasts. A number of eukaryotic cells also exhibit cytoplasmic inheritance of characters due to the transmission of symbiotic micro-organisms. In addition to these mechanisms, there is another class of cytoplasmic inheritance based on self-perpetuating metabolic patterns and a remaining category, not readily associated with nucleic acids, that involves the inheritance of the supra-molecular structure of the cell.

Nevertheless, the significance and scope of cytoplasmic heredity has been the subject of intense controversy throughout the twentieth century. During the first half of the century investigations of cytoplasmic inheritance were carried out close to the margins of, and developed in constant tension and conflict with, the research programmes and doctrines of Mendelian genetics. They became allied with, or appropriated by, forces opposed to nucleo-centric genetics, namely, those advocating the Lamarckian notion of the inheritance of acquired characteristics and, during the late 1940s, those advocating Lysenkoism.

The significance of some of the more recent experimental results concerning the hereditary role of the supra-molecular structure of the cell remains a subject of controversy as directly conflicts with the nucleic-acid doctrine. Many biologists consider these results to represent a major challenge to neo-Darwinian doctrines.[11] In very recent years, even some leading neo-Darwinian evolutionists have admitted their concern about the significance of this class of cytoplasmic heredity. For example, at a recent symposium on evolution and development, John Maynard Smith stated that 'Neo Darwinians should not be allowed to forget these cases, because they constitute the only significant experimental threat to our view'.[12]

However, the development of cytoplasmic genetic research has been largely ignored by historians of genetics who have given one-sided attention to the development of Mendelism.[13] The significance of cytoplasmic heredity continues to be overshadowed by the writings of molecular biologists such as François Jacob who wrote an historical account of the 'logic of life' leading to the nucleic acid doctrine,[14] and of evolutionists such as Ernst Mayr, who insists that nothing new has happened in evolutionary thought since the neo-Darwinian synthesis of the 1940s and 1950s, and who attempts to disclaim or trivialize the evolutionary significance of various forms of cytoplasmic inheritance.[15]

Throughout most of the twentieth century the theoretical limits of this dispute were framed between two extreme positions. On the one hand, there was the view that the cytoplasm of the cell (transmitted largely through the egg) was primarily responsible for the 'fundamental' characteristics of the organism, that is, those characteristics common to the higher taxonomic groups (macro-evolution). According to this view, Mendelian genes on the chromosomes in the nucleus (transmitted biparentally) were largely concerned with the relatively trivial characteristics that distinguished individuals, varieties, or species (micro-evolution). These were traits such as eye colour, hair colour and wing shape, that 'topped off', as it were, the more 'fundamental' organismic features. This proposal was made by many embryologists and physiologists, including E. G. Conklin, J. W. Jenkinson, J. Loeb, A. Brachet, and M. Guyer, who were unable to reconcile embryological observations of early development and the behaviour of 'the organism as a whole' with the particulate confines of the chromosome theory. The belief that nuclear genes played only a minor role in heredity and evolution was also maintained by many naturalists who found it difficult to account for macro-evolution and the adaptive correlations of the different parts of the organism as a whole in terms of gene mutations.[16]

On the other hand, Mendelian geneticists in the United States, led by T. H. Morgan and his *Drosophila* school, upheld a predominant if not exclusive role for the Mendelian genes in heredity. By the 1920s most American geneticists had come to view genes as discrete physical units located on chromosomes. They claimed genes as the 'governing elements' of the cell, largely immune from the rest of the cell and dictating its activities. Morgan himself expressed this view succinctly in 1926: 'In a word the cytoplasm may be ignored genetically'.[17]

The controversy over the relative importance of the cytoplasm and the nucleus began to come to a head in the late 1940s and 1950s with the rise of physiological or biochemical genetics. Although relatively few examples of non-Mendelian, cytoplasmic inheritance had been reported by that time, apologists for cytoplasmic heredity united their work and formulated a coherent body of counter-arguments against the 'nuclear monopoly'. In order to understand better the problems that the cytoplasm posed for its investigators, a brief glance at the historical background to the dispute will be helpful. Before turning to a detailed analysis of the complex of methodological arguments that came to the fore in the 1950s, however, I shall first outline the discursive and social strategies employed by those who upheld the dominant role of the nuclear genes in heredity.

In the face of competing groups in the field, biologists supporting Mendelism adopted a disciplinary strategy, socially differentiating themselves from the larger field of the study of heredity. They established their own objects of inquiry, techniques, standards, doctrines, journals and societies. Early Mendelians used a series of polemical devices to legitimate their position in the field. They claimed that their views were based on experimentation and referred to non-experimental approaches based on statistics and natural history as being 'descriptive' or 'speculative' and based on 'folk' notions of heredity.[18] Although Mendelian geneticists could say nothing about the problem of ontogenetic development, early Mendelians advocated a pragmatic strategy and denounced the efficacy of 'experimental embryology' and larger objectives in the field. Geneticists, led by the Morgan school, maintained a controlled discipline, creating a variety of new problems that they could readily solve by Mendelian techniques: mapping genes on chromosomes, investigating gene changes, and associating them with a wide range of adult characteristics. Using rapid and effective techniques for analysing differences between individuals of an interbreeding group, and with the support of practical breeders, Mendelian genetics emerged along a progressive research front and grew rapidly in the United States. Indeed, Mendelian genetics may have had a limited technical capacity for solving existing scientific problems, but it had a great capacity for attracting young scientists interested in problems that could be successfully and rapidly solved by a set of effective techniques.

By the 1930s Mendelian geneticists had risen to an authoritative position in the field and had constructed a notion of heredity to suit their practice. 'Heredity' became largely understood in terms of the sexual transmission and exchange of genetic material from one generation to the next. As the American geneticist L. C. Dunn wrote in 1917,

> The working method of heredity is known, if heredity be properly only concerned with the way in which heredity factors are distributed in the germ-cells. For development, the mechanism is but grossly known, but we have learned enough to foster a suspicion that one day the governance of the chromosomes over development will be explained in physico-chemical terms.[19]

American geneticists used the Mendelian-chromosome theory to explain almost every case of the inheritance of characteristics. A large body of breeding data and cytological observations could be interpreted in terms of the grouping and distribution of genes. The lack of genetic evidence for cytoplasmic heredity was used to deny alternative hereditary mechanisms. As arguments with American geneticists continued, defensive attitudes and motivations became visibly intertwined in the formal debate. Dunn, for example, argued that the claim of those who postulated the existence of cytoplasmic heredity rested solely on their refusal to accept exclusive nuclear control, or on their refusal to admit it even though they recognized it:

> The whole case of the supporters of any theory which views the cytoplasm as determinative rests on either their refusal to go back and inquire the source of this cytoplasm, or on their refusal to give due emphasis to the source even though they recognize it.[20]

Most early American geneticists discursively objectified their procedures. They held them to be atheoretic and fully capable of detecting the existence of alternative hereditary mechanisms. The lack of evidence for cytoplasmic heredity would then be based on pure empiricism. Morgan himself used this tactic to separate his views from his motivations when he attempted to deny the distinction between fundamental and trivial characteristics so as to diminish the threat associated with the cytoplasm. His theoretical position, he claimed, was based on experience rather than on an attempt to magnify the importance of his work on chromosomal inheritance:

> Mendelian workers can find no distinction in heredity between characteristics that might be ordinal or specific, or fundamental, and those called 'individual'. This failure can scarcely be attributed to a desire to magnify the importance of Mendelian heredity, but rather to experience with heredity characters. That there may be substances in the cytoplasm that propagate themselves there and that are outside the influence of the nucleus, must, of course, be at once conceded as possible despite the fact that, aside from certain plastids, all Mendelian evidence fails to show that there are such characters. In a word, the distinction between generic versus specific characters or even 'specificity' seems at present to lack any support in fact.[21]

However, from the beginning of the controversy, advocates of cytoplasmic heredity questioned whether the experiments conducted by means of the Mendelian procedures should be expected to detect cytoplasmic inheritance when it occurred. The limited techniques used by Mendelians were held to be largely responsible for their narrow views. As one defender of the cytoplasm phrased the problem in 1916:

> . . . the cytoplasm often (perhaps usually) determines the type of cleavage, the early course of development, and in large measure the larval characters, while the adult characters are determined by the chromosomes. With the developmental stages the student of heredity using the usual breeding methods has little to do. He may be pardoned a bias in favor of the chromosomes because he rarely studies larval characters. To the physiologist and morphologist, on the other hand, the rigid conviction of the geneticist that the chromosomes contain all the tools of his trade has not unnaturally been viewed with skepticism.[22]

Although the views of the Morgan school predominated in the United States, this was not the case in France and Germany between the two world wars. The institutional structure of the university systems in the latter countries posed major obstacles to rapid specialization[23] and many biologists resisted what some later saw as Morgan's highly 'operational' approach to science.[24] Many Continental biologists maintained that the exclusive attention given to nuclear genes in heredity was largely due to the fact that cross-breeding analysis was so easy to do and chromosomes so easy to dye. The view of E. Guyénot was representative of many Continental biologists by the mid 1920s. Unlike Morgan who dogmatically asserted that the cytoplasm could be ignored genetically, Guyénot maintained:

> The results of embryonic mechanics show ... that it is necessary to enlarge the framework of heredity, and not to limit efforts to the *evidently more easy study of chromosomal factors*. Heredity is not a nuclear phenomenon, nor a cytoplasmic phenomenon; it remains, in its entirety, a cellular phenomenon.[25]

In Germany, between the two world wars, Carl Correns, Fritz von Wettstein, Otto Renner, F. Oehlkers, Peter Michaelis and many other geneticists attempted to challenge what they called the 'nuclear monopoly'.[26] They detected various cases of maternal inheritance of characteristics based on cross-breeding different species and genera of certain plants. The hereditary property of the cytoplasm was designated the *Plasmon* in contrast to the *Genom* located in the nucleus. However, investigations of cytoplasmic inheritance were slow and procedures were tedious compared to the highly efficient techniques for analysing chromosomal inheritance. Genetic investigations of the cytoplasm drew relatively few adherents and cytoplasmic heredity presented a minor discourse dissociated from the larger realm of Mendelian genetics that had become closely allied with natural selection in the 'evolutionary synthesis' of the 1930s and 1940s.

The evidence for the *Plasmon* was either attacked and then dismissed, or else ignored by most American geneticists. It was difficult to exclude genic mechanisms with the existing techniques. Those who upheld the predominant if not exclusive role of the nucleus in heredity could provide various alternative hypotheses which would confirm the results to the theory of chromosomal gene control. Although such interpretations were purely formal, there was, surprisingly, no genetic phenomenon that could not be explained, with due allowances, by Mendelian genes.

When the American geneticist E. M. East reviewed the evidence for cytoplasmic inheritance in 1934, his primary discursive tactic was to dismiss the question of the relative roles of the nucleus and the cytoplasm in heredity on formal methodological grounds. Unlike Morgan, however, East did not use empiricism simply to deny the distinction between 'fundamental' and 'trivial' hereditary traits. Instead, he attempted to push this outside the realm of science as an illegitimate question. East admitted that since Mendelian analysis was based largely on differences between individuals of a species, the possibility existed that the cytoplasm controlled the generalized characteristics of higher taxonomic groups while the nucleus controlled characteristics of the variety or of the species. However, in his opinion this hypothesis was 'not a very satisfying scientific proposal' since it could not be proved or disproved by existing techniques.

If true the fact could not be demonstrated, since parental *differences* only can be detected and followed. If untrue, evidence for its falsity ordinarily could be obtained only indirectly, and with difficulty by following the non-generalized differences between the greater taxonomic groups, and such crosses are rarely viable.[27]

East did not stop there. He also questioned the actual *motives* and *competency* of the investigators themselves. There had been a number of cases of individuals who had attempted to challenge genetic doctrines with examples of Lamarckian principles and whose work had been judged to be either sloppy or fraudulent. Suspecting that some German researchers may have been attempting to provide room for the possibility of the inheritance of acquired characteristics, he endeavoured to cast a shadow of notoriety over cytoplasmic inheritance research. There were several types of phenomena reported that were due to the direct transfer from cell to cell of 'alien matter' such as symbiotes and parasites capable of producing morphological changes in some organisms. East warned:

It is not to be supposed that modern biologists will cite such instances, when recognized, as examples of heredity. But since an earlier generation of students used them, before their cause was discovered, to support arguments on the inheritance of acquired characteristics, it is well to be cautious in citing similar, though less obvious, cases as being illustrations of non-Mendelian heredity.[28]

Cytoplasmic inheritance came to the fore of genetic research only after new techniques and objectives were introduced into the mainstream of genetic research.

3. 'POSITION EFFECT': NUCLEO-CYTOPLASMIC RELATIONS IN BIOCHEMICAL GENETICS

When, following World War II, micro-organisms were domesticated for genetic use, the face of genetics quickly changed. Many young geneticists were no longer content to analyse hereditary transmission of morphological secondary characters, such as wing shape or tail length, far removed from the intra-cellular level where the genes were located and where they exerted their primary effects. Instead, they turned to take advantage of micro-organisms in order to study genetic control of the physiological properties and chemistry of the organism. The pattern of research characteristic of biochemical genetics, beginning in the 1940s, relied on the cooperation of geneticists and biochemists. The geneticist isolated mutants that were found to be unable to grow or which grew poorly on a well-defined medium. The biochemist sought the reason for this inability. It was hoped that this double study would give a biochemical description of genetic control.

During this twin shift to the study of micro-organisms and of genic action and genic control, cytoplasmic heredity for the first time found a conspicuous place in genetic discourse. The rise of microbial genetics brought with it some unorthodox types of variations in cell lines inherited in a way not conforming to the standard rules of Mendelian inheritance. The work which brought cytoplasmic heredity to the fore of genetic discourse was led by T. M. Sonneborn and Boris Ephrussi, neither of whom were originally trained as members of the Morgan school.[29] Sonneborn, who maintained a strong interest in the inheritance of acquired characteristics, began to investigate the genetics of micro-organisms in the 1920s, many years before the mass migration of geneticists to exploit *Neurospora, E. coli* and other microbes during the 1940s and 1950s. In the 1920s and throughout most of the 1930s, cross-breeding analysis had not been effectively achieved in micro-organisms. In most cases they seemed to reproduce solely vegetatively and showed few signs of sexuality. Sexual reproduction was a *sine qua non* for genetic research, which was based on cross-breeding analysis. Micro-organisms therefore had been largely excluded.

After establishing control over the sexuality of the ciliate *Paramecium* in 1937 and constructing sex-like mating types — a task which took seven years of systematic investigation — Sonneborn provided the first demonstration in the United States that, like higher organisms, unicellular organisms possessed genes that behaved in the classical Mendelian way. He quickly rose to prominence in the American genetics community as a leading authority on the genetics of micro-organisms. At Indiana University, where he remained for the rest of his investigative career, Sonneborn wandered well off the main Mendelian track, investigating various novel modes of inheritance and developing a major school of *Paramecium* genetics. In the decade following the control of sexuality in *Paramecium*, he investigated three types of non-Mendelian inheritance involving different mechanisms. In all cases he reported an associated sensitivity of cytoplasmic traits to environmental influences.

Ephrussi was originally trained in embryology and is well known today for his earliest attempts to give a biochemical description of genic action in the mid-1930s by directly combining genetic and embryological methods in *Drosophila*. After World War II, Ephrussi was appointed to the first chair of genetics in France, and he chose a French strain of baker's yeast for genetic use in his Parisian laboratory. In the course of his investigations he stumbled upon the non-Mendelian inheritance of a deeply seated physiological change, respiratory deficiency. During the late 1940s and 1950s, working with other leading French-speaking biologists, including André Lwoff, Philippe L'Héritier and Jean Brachet, Ephrussi brought forth new and diverse evidence for cytoplasmic genetic properties and joined Sonneborn in supporting the theoretical significance of the cytoplasm in heredity. When an Institute of Genetics was formed in France following World War II, under Ephrussi's direction, the cytoplasm was pushed to the centre of genetic research as a sound strategy to compete with American and British geneticists predominantly preoccupied with the study of nuclear genes.

To Sonneborn and Ephrussi and their followers, the primary theoretical significance of the evidence for cytoplasmic heredity continued to lie in embryological problems of development, to which most geneticists paid little attention. To account for the regulative process of cellular differentiation, Sonneborn, Ephrussi and their associates postulated the existence of various sorts of cytoplasmic genes or 'plasmagenes', sorting out at cell division, multiplying at various rates, responding differentially in different environments, and interacting with each other, in various modes of competition and cooperation.

During the late 1940s, the evidence for cytoplasmic heredity and its intimate relations with the environment became entangled in the Lysensko affair.[30] However, it should be stressed that neither Sonneborn nor Ephrussi was Lysenkoist, and although the nuclear/cytoplasmic dispute cannot be considered in any way to have been value-free, it must be distinguished from the Lysenko affair. Ephrussi and Sonneborn did not argue on grounds that most Western geneticists would exclude as 'philosophical' or 'political'; they did not cite authority figures outside the field such as Marx or Engels; and they did not dismiss the Mendelian-chromosome theory entirely. Instead, they made their arguments in relation to the contemporary social and technical norms of the discipline, and they recognized the general occurrence of inheritance based on chromosomal genes.[31]

Although the experimental work and theoretical presentations of Sonneborn and Ephrussi brought the cytoplasm to the widespread attention of geneticists, genetic research on cytoplasmic heredity and attempts to understand the principles of somatic cell differentiation were still carried out only on a relatively minor level compared to the work on nuclear genes. By the mid-1950s, advocates of cytoplasmic heredity could only cite about twenty cases of cytoplasmic inheritance in the genetic literature.

During the 1940s and 1950s many former members of the Morgan school, including G. W. Beadle, A. H. Sturtevant, E. Altenburg, H. J. Muller and J. Schultz seized upon the evidence for cytoplasmic genetic particles in an attempt to dispel any threatening notion that plasmagenes possessed a degree of autonomy and frequency of occurrence comparable to that of nuclear genes. They attempted to explain some cases in Mendelian terms and they dismissed other cases as being due to the transmission of parasites or more complex symbiotic organisms with little physiological significance for genetics.

The research programmes based on cytoplasmic heredity continued to stand in direct conflict with the dominant genetic research programmes in the United States. There, the control of the cell by the nucleus continued to be a central dogma. Distinctions based on scientific *indoctrination* were used by geneticists to understand the conflict between the interpretations of genetic results. Carl Lindegren, an associate of the Morgan school, who after 1948 advocated the inheritance of acquired characteristics and believed his work on yeast supported the existence of extra-nuclear heredity, explicitly referred to the bias due to training:

> Recently in America there has been a recognition of the phenomena discovered by the German workers and much discussion of cytoplasmic inheritance especially by Sonneborn, and it is noteworthy that his early training did not involve genetic indoctrination as a member of the Morgan school.

> In yeast genetics we encountered many examples of non-Mendelian phenomena in our early work. I interpreted them as involving the hereditary transmission of autonomous entities. This was a rather difficult thing for me to do because I had been thoroughly indoctrinated in gene theory by my long association with Dr. Morgan and other members of his famous staff.[32]

In the domain of biochemical genetics, the stakes were precisely understood and the struggle between the genetic work on the cytoplasm and that on the nucleus centred largely on a struggle for control over what were considered to be the most important biochemical constituents: the cytoplasmic proteins and their specificity. Based on their biochemical genetic investigations in *Neurospora*, George Beadle and his associates proposed that a gene acts by determining the specificity of a particular enzyme (the one gene-one enzyme hypothesis) and thereby controls in a primary way enzymatic synthesis and other chemical reactions in the organism. Beadle clearly stated the issue concerning the threat of the cytoplasm in 1948 when he wrote:

> There is no convincing evidence in any single instance of the functioning of an essential gene being entirely taken over by a so-called plasmagene. On the contrary, the great bulk of available evidence bearing on the question clearly indicates that the primary control of protein specificity resides in the nuclear genes.[33]

Despite Beadle's claim, there was no direct evidence that the nuclear genes controlled protein specificity, beyond the fact that the evidence obtained was compatible with it. In order to create a place for alternative theories of protein specificity the formal methodological principles considered to be underlying biochemical genetics had to be scrutinized. In 1946 the influential physicist Max Delbruck made this attempt by showing that the experimental method applied in the *Neurospora* work predetermined the answer by restricting possible alternatives. In his view, *Neurospora* geneticists had only been pretending to be working with inductive methods, when they were actually using deductive reasoning. Delbruck's own polemic was based on what might be seen as naive falsificationism:

> To sum up: in order to make a fair appraisal of the present status of the thesis of a one-to-one correlation between genes and species of enzymes, it is necessary to begin with a discussion of methods by which the thesis could be *disproved*. If such methods are not available, then the mass of 'compatible' evidence carries no weight whatsoever in supporting the thesis.[34]

Advocates of the 'one-to-one theory' argued that it was difficult to devise an experiment that would disprove the thesis. In effect it was claimed that Delbruck was being idealistic in his proposal of the right way to do science. The *Neurospora* geneticist David Bonner defended *Neurospora* work by turning the question into one of scientific *value* and *judgement*. He saw little value in questioning the one gene-one enzyme proposal at least as a working hypothesis. Bonner argued on the basis of the rule that Auguste Comte had articulated: that one should in all cases form the simplest hypothesis consistent with the whole of the facts to be presented. The one gene-one enzyme proposal had merit in terms of its simplicity and in terms of its ability to account for the data:

> Since the data now available are accounted for on a one-one basis, it would be of little value to devise a more complex thesis of the relation of genes to enzymes.[35]

While *Neurospora* geneticists who upheld the predominant role of the nuclear genes in heredity saw no value in attempting to construct more definitive experiments or in entertaining alternative theories, the reverse was true for defenders of the cytoplasm who were pushing to make their genetic work more relevant to developmental phenomena and to the views of most embryologists. A number of alternative interpretations of the genetic control of protein specificity existed which would bring genetic control in line with cytoplasmic heredity and therefore in line with the problem of genetic regulation. In direct conflict with the claims of *Neurospora* geneticists, Ephrussi argued that there was no convincing proof of nuclear control of protein specificity and upheld the value of his results in yeast as providing convincing evidence that plasmagenes controlled protein specificity.[36]

However, during the early 1950s the validity of cytoplasmic heredity relied on its recognition by Ephrussi's and Sonneborn's chief competitors in biochemical genetics. The *Neurospora* school led by Beadle and his associates were the dominant group and they upheld the importance of their work on nuclear genes and maintained Morgan's dictum that nuclear genes played the exclusive or primary role in heredity. The most common interpretation of Ephrussi's results in the early 1950s followed the suggestion of A. H. Sturtevant that they could be explained by a chromosomal phenomenon called 'position effect', a change in a gene's function due to a change in its position in the chromosomes.

Some *Neurospora* geneticists only began to consider the evidence for non-Mendelian, cytoplasmic heredity when a member of their school reported an example. In 1952 at the California Institute of Technology, there was an unexpected report of a case of non-Mendelian inheritance in *Neurospora* called 'poky' which resembled, both biochemically and genetically, Ephrussi's non-Mendelian '*petite*' mutations in yeast. A brief glance at some of the discussion surrounding this case underlines the social processes involved in validating knowledge claims.

To some members of Beadle's school, the case of non-Mendelian inheritance reported in *Neurospora* actually verified the case in yeast reported in France. Norman Horowitz, for example, a former student of Beadle's, changed his interpretation of Ephrussi's results. He wrote to Ephrussi in 1953:

> Incidentally, I don't remember whether I mentioned it to you, but I think that the *poky* results eliminate whatever doubts there may have been concerning the cytoplasmic inheritance of *petite*.[37]

Ephrussi recognized the role played by social elements in the verification of scientific findings. In an attempt to defend his priority, he quickly wrote back to Horowitz protesting against the lack of credibility given to his results:

> . . . seeing the stand you are taking now that 'the *poky* results eliminate whatever doubts there may have been concerning the cytoplasmic inheritance of *petite*'. I hope you will take this opportunity to explain what system of logic has come to dominate the cartesian one so as to enable you to remove the doubtful elements of the results obtained with yeast by working on *Neurospora*. I am afraid, dear Norm, that it will be hard for us to agree, for, contrary to you, I am beginning to believe that Sturt [Sturtevant] is right in invoking position effect in this business: the credibility of the same results obtained in Paris and Pasadena is obviously very different.[38]

Despite his rhetorical response to Horowitz's claim, Ephrussi considered 'a case of cytoplasmic heredity discovered in Neurospora' to be 'an event of major importance'.[39] Ephrussi had placed cytoplasmic inheritance at the centre of his investigations in France, and had a considerable vested interest in its biological importance. Taking the long-awaited results in *Neurospora* into consideration with other evidence for cytoplasmic heredity, it seemed to him that cytoplasmic heredity would lead ultimately to the downfall of American Mendelian supremacy in hereditary phenomena. Immediately upon hearing the news of 'poky', he wrote to Sonneborn:

I suppose you have heard of 'poky', the presumably cytoplasmic case in *neurospora* discovered by the Mitchells. It fills me with joy: it's a dream to have this thing happen in a fortress of all exclusive Mendelism![40]

To Ephrussi, the result produced in *Neurospora* seemed not only to verify his own results, but to legitimate the general occurrence of cytoplasmic heredity in all organisms.

Despite Ephrussi's excitement, the real significance of 'poky' remained a subject of contention. To Sonneborn, for instance, the results in *Neurospora* were not seen as a discovery at all, but merely another isolated example of cytoplasmic heredity. After all, Sonneborn's own work had convinced him of the importance of cytoplasmic heredity for some time. Several years before the report of 'poky', he reviewed the evidence for cytoplasmic inheritance and wrote: 'Cytoplasmic inheritance . . . is . . . not an hypothesis, but a fact — one of the capital facts of biology'.[41] Moreover, unlike Ephrussi who had reported only one example in yeast, Sonneborn had accumulated three examples of cytoplasmic inheritance in *Paramecium* based on different interactions with nuclear genes. It was his kind of results, in his opinion, that would change the views of most American geneticists. As he wrote to Ephrussi, the verification of the existence of cytoplasmic heredity required not another case of non-Mendelian inheritance in another organism, but inversely, several cases in one organism.[42]

This flurry of method discourse over the significance of '*petites*' in yeast and 'poky' in *Neurospora* highlights the social negotiations that take place in the process of scientific discovery. We cannot understand the significance of the results independently of the power relations in the field. Each participant in this controversy struggled to occupy the dominant position by attempting to impose the greatest significance on the scientific values most closely related to him personally or institutionally. However, Ephrussi's hope for a revolution in genetics, thought to be signalled by the report of 'poky', was not realized during the 1950s, and 'poky' remained just another isolated genetic anomaly.

4. 'THE CELL AS A WHOLE' IN A HOSTILE MENDELIAN MILIEU

Despite the enormous potential importance Ephrussi and Sonneborn placed on plasmagenes in heredity, they recognized that their autonomy in the cell, like that of nuclear genes, was only relative. Like many physiologists and embryologists, they viewed the cell as an integrated unit that could be broken up only for analysis. Moreover, in the process of development plasmagenes could only be *instruments* not autonomous *agents* of cellular differentiation. They were secondary to another cytoplasmic property that played the predominant role in development. The spatial arrangement and organization of the cellular materials were also considered to be crucial parts of the genetic system. Although Ephrussi and Sonneborn recognized the integrative relations between parts of the whole cell or organism, they did not believe that these would have to be explained by unknowable forces. However, they did want to alert geneticists to the problem of organization and spatial relations in the cell and steer genetic research away from purely reductionistic approaches. If genes (nuclear and cytoplasmic) were in complete and exclusive control of heredity, they reasoned, it would have to be concluded that genes isolated under conditions that permitted their multiplication would be capable of reconstructing cells of the same kind from which they were taken. If this did not happen, then it would have to be concluded that the cell itself, including the cytoplasm, somehow served as a necessary model for the formation of new cellular material in essentially the same sense as the genes are necessary models for the formation of new genes.

However, Ephrussi and Sonneborn lacked the appropriate techniques for investigating the genetic role of the supra-molecular structure of the cell. This work would later be carried out by Sonneborn and his co-workers beginning in the 1960s, in direct conflict with the theory of self-assembly that was then gaining ground among molecular biologists. In the meantime, they reasoned on grounds of logical necessity and supported their position by citing biologists who held similar views. Sonneborn stated the issue in 1951 in the following figurative terms:

> Perhaps it will be objected that there are some self-duplicating cytoplasmic elements which the nucleus cannot make. Then suppose that these too can be cultivated in vitro. Is anyone willing to believe that, if all such self-duplicating components of the cell were thrown together in a test tube in the proper proportions with adequate food for their multiplication, a *Chilomonas* cell or any cell at all would result? Although the whole picture is admittedly imaginary, it makes the nature of the problem sharp and clear. If cells cannot be reconstituted in the way suggested, then it seems to me we are forced to admit that the molecular and particulate arrangement of the cellular materials, their arrangement into a working system is itself a part of the genetic system of the cell.[43]

The claim that the organization of the cell represented a hereditary property stood in direct conflict with the dominant doctrines of genetics and demanded that the technical capacity of Mendelian genetics be brought into question. The autonomy of the nuclear genes and their primary role in heredity was a doctrine based on conclusions from previous genetic experiments.[44] In order to challenge this doctrine one had to question in technical terms the experiments on which it was based, matters normally ignored by defenders of the nuclear monopoly. As discussed earlier, the genetic notion of heredity had become endowed with connotations that restricted its usage to phenomena studied by certain techniques. In this instance, opponents of the exclusive or primary role of the nuclear genes employed arguments focusing on the prevailing analytical techniques in an attempt to distinguish what occurs *in vitro* from what possibly occurs *in vivo*. Genetics was based on differences between organisms, not on similarities; and cross-breeding analysis presented a picture of genes as discrete units acting autonomously from the rest of the cell and dictating its activities. In 1953, Ephrussi raised the issue of technique dictating theory:

It is less obvious that the method of genetics, although it involves no 'bloodshed' is as analytical in its essence. Indeed the 'resolving power' of this method is amazing. It provides us with a picture of the cell's nuclear constitution with unequaled 'definition'. But so long as the basis of genetics is the study of differences it cannot be expected to give us an undistorted picture of the cell as a whole. The integrative character of the cell, which is its fundamental property, is bound to escape our notice most of the time.[45]

In the 1950s an acknowledgement of the hereditary properties of 'the cell as a whole' entailed a confrontation not only with the procedures of genetics, but also with the biochemical perspective on the cell. When, with the aid of the Rockefeller Foundation, geneticists turned decisively to the study of genic expression, many physicists and biochemists immigrated into biology.[46] They brought with them the techniques, theories, standards, objectives and doctrines of their disciplines. Ephrussi vigorously opposed the purely biochemical view of the cell which conflicted with his integrative physiological conception of heredity. In this case, however, technical and theoretical arguments could be drawn over disciplinary lines:

The present knowledge of the biochemical constitution of the cell was achieved largely by the use of destructive methods. Trained in the tradition of the theory of solutions, many a biochemist tends, even today, to regard the cell as a 'bag of enzymes'. However, everyone realizes now that the biochemical processes studied *in vitro* may have only a remote resemblance to the events actually occurring in the living cell.[47]

The struggle between geneticists and the 'emigrants' from the physical sciences was represented at the social level as well. For example, in 1956 when Ephrussi made a recommendation concerning a position in physiological genetics at Berkeley, he raised the issue of the disciplinary dichotomy between the objectives of physiological genetics and those of biochemistry.[48] Nor were Ephrussi and Sonneborn the only geneticists to object to the reductionist biochemical approach that threatened to dominate biology, or at least genetics, after World War II. The situation of the geneticist Sol Spiegelman as he saw it at Washington University in St. Louis in 1946 provides vivid testimony to the social and intellectual conflicts facing some geneticists with the emergence of 'big science' in the United States:

> ... affairs at this university, as far as the biochemical sciences are concerned, have deteriorated considerably over the summer. The great American 'team spirit' has hit this place with a vengeance. ... In the biological sciences they have decided to back the 'team' headed by Carl Cori. ... [He] is a great biochemist but he is no biologist and has very little understanding or sympathy with biological programs. If you can't isolate it and crystallize it you aren't doing science as far as he is concerned. Since in terms of money, our program is the biggest thing outside of Cori's department, you can guess what might result. While we were away for the summer, without our knowing it, they almost organized us out of existence. When Kamen and I returned, we put up quite a fight and managed to save the program. ... So we have a truce in which we are free to carry out our program but it is obviously no long term solution. ... Biological research as we understand it, will be stifled here for some time to come and no amount of brilliant biochemistry on cell-free extracts can substitute for it. As a result everybody is leaving.[49]

By the early 1950s, then, biologists who supported the view of 'the cell as a whole' faced a great deal of resistance. However, they did not understand the 'incorrect' views of their opponents in terms of a lack of adherence to a formal scientific methodology. Instead, they accounted for conflicting views in terms of divergent techniques and objectives in the field and in terms of scientific indoctrination. In the following section we will see from a more detailed examination how accounts based on the technique-ladenness of observations, scientific indoctrination, and power relations in science can be used by scientists in an attempt to delegitimate opposing and dominant views in the field.

5. THE TECHNIQUE-LADENNESS OF OBSERVATIONS, DISCIPLINE, AND THE VALUE OF THE CYTOPLASM

The postulation of cytoplasmic heredity represented a major effort to fill the gap between the problems of development and those of heredity. While this interpretation had great merit as a theoretical model in accounting for cellular differentiation and morphogenesis, it also had some striking weaknesses into the 1950s. It had been greeted mostly with hostility, and relatively little genetic evidence had been obtained for cytoplasmic inheritance as a general phenomenon. During the 1950s, in an attempt to counteract these two major weaknesses of their genetic theory, advocates of cytoplasmic heredity raised various methodological issues and drew them to the centre of a complex socio-political understanding of the nature of scientific development and scientific activity. In effect, Sonneborn and Ephrussi were attempting to make a new and major scientific fact out of a few genetic anomalies by placing them in a theoretical construction irreducible to genetic evidence. Within this context, arguments based on interpretations of the development of science, the social nature of scientific activity, and the technique-ladenness of observations would play instrumental roles.

If the relative frequencies of known examples of nuclear and cytoplasmic inheritance were an index of the frequencies of their occurrence in nature, then the cytoplasm would clearly play a minor role in heredity. On the other hand, Sonneborn and Ephrussi reasoned that such a measure did not provide a reliable index of the occurrence of the phenomena in question: it merely represented the extent to which cytoplasmic inheritance could be, and had been, investigated by geneticists. There was a technical incommensurability. One could not compare the results produced in favor of nuclear inheritance with those of cytoplasmic inheritance. Sonneborn summarized the major issues concisely in 1951:

> In the first place, there is a simple, familiar and highly successful method — the Mendelian method — for studying genic inheritance; no method comparable in simplicity, power, and reliability is available for the study of cytoplasmic inheritance. This results in considerable selection in the examples of heredity studied and reported in the literature: the simpler cases that yield readily to a familiar methodology are preferentially attacked and reported. Complex or less standard cases tend to be put aside or interpreted formally in terms of accepted theory. Secondly, genetic methods are designed chiefly for analysis of differences between individuals that can interbreed, hence, properties common to all individuals of an interbreeding group remain largely unanalysable by the ordinary methods of genetics except under unusually favorable conditions.[50]

This quotation contains several independent and important suggestions that were articulated throughout the 1950s by defenders of the cytoplasm, and that deserve specific attention. First and most generally, the facility of Mendelian analysis was seen to be largely responsible for deceiving geneticists and for leading many geneticists away from what actually occurs in the cell.[51] But although technical opportunism could help explain the attention focused on nuclear inheritance, this alone could not account for the attitude of most American geneticists towards alternative forms of inheritance. Since advocates of cytoplasmic heredity could not understand the attitude of nucleo-centric geneticists in rational terms (that is, the terms they claimed they themselves were using in arriving at an opposite conclusion), they turned to historical interpretations of the nature of scientific development and postulated that throughout the twentieth century, Mendelian geneticists had in part been advocating a cause and had necessarily had to ward off alternative theories in their scientific struggle. This strategy would no longer be necessary once the chromosome theory became accepted and the presumed limits of its possibilities were realized. Within this context, the argument for the final recognition of cytoplasmic heredity in the 1950s was held to rest not only on the intrinsic properties of nature, but also on the natural order of social processes. Thus, Ephrussi wrote in 1951,

> . . . the rapid rise of genetics, resulting in the ever widening recognition of the universal character of Mendelian mechanisms, has indubitably overshadowed both the recurrent claims of evidence for the occurrence of extra-nuclear heredity, and the theoretical need for postulating its existence. No doubt, to be effective, revolutionary ideas, whether political or scientific, have for a time to disguise themselves in simplified forms and pretend to the monopoly of virtues. Moreover, since science proceeds by successive approximations, a prevailing theory has of necessity to explain all the facts it can account for, before a new theory can be allowed in. It is only natural therefore that the complete exploration of Mendelian principles should have delayed the recognition of cytoplasmic heredity.[52]

Continuing his essentially political understanding of the nature of scientific development, Ephrussi concluded, ' "The gene as a basis of life" is only a revolutionary slogan which must not over-shadow the concept of "the cell as a whole" '.[53]

Another major question raised by cytoplasmic geneticists concerned the actual ability of Mendelian procedures to detect cytoplasmic inheritance when it occurred. This problem was addressed in Sonneborn's second suggestion that genetic analysis dealt mainly with *differences* between individuals of an inter-breeding group, not with *similarities*, or properties *common* to members of an interbreeding group. Implicit in this suggestion was the idea that non-Mendelian cytoplasmic heredity might be concerned especially with properties that distinguished higher taxonomic groups. This view had two consequences that could elevate the importance of cytoplasmic heredity and its researchers. On the one hand, 'macro-evolution' has recourse to cytoplasmic as well as nuclear gene mutations and, on the other hand, cytoplasmic heredity is particularly concerned with fundamental cellular functions. It was also noted that if the latter were true, cytoplasmic mutations would usually be incompatible with survival and would therefore escape detection. In effect, this latter theoretical move allowed cytoplasmic geneticists to denote the lack of reported experimental evidence for cytoplasmic heredity and turn it into an argument for the cytoplasm's crucial role in heredity.

In some instances, the loss of a cytoplasmic element was in fact shown to be incompatible with survival. The '*petite*' mutation in yeast investigated by Ephrussi's group was held to present an illuminating example.[54] Here a fundamental function, respiration, could be abolished. However, the detection of this phenomenon raised the issue of the role of the organism itself as a technological tool for investigating various types of biological mechanisms. In the case of respiratory deficiency associated with mitochondria in yeast, it was argued that the mutation could only be detected because the organism in question possessed a remarkably efficient alternative energy-yielding metabolic pathway. As Ephrussi argued:

> It is to these circumstances, hardly realized with similar perfection in many other sexually breeding organisms, that we owe the possibility of its detailed study. The cytoplasmic control of fundamental functions may escape detection precisely for this reason.[55]

These collective arguments, then, permitted Ephrussi to revitalize the notion that nuclear genes may be concerned primarily with only 'superficial' characteristics:

> The emphasis placed in the above statements to the term 'fundamental' will probably be strongly objected to by many a reader as reminiscent of the old distinction between 'general' and 'special' heredity . . . and to the controversies as to whether gene mutations concern both 'superficial' and 'fundamental' characteristics, rather than the former alone. I think that the question is today in need of serious reconsideration, and that it should not be answered by metaphors or by the usual counter-questions . . .[56]

Despite the theoretical merit of cytoplasmic heredity and the arguments of cytoplasmic geneticists for a major change in the genetic conception of heredity, victory could be achieved only on the terrain of the technical procedures of that discipline. Cytoplasmic geneticists simply lacked the necessary equipment required to effect a revolution in genetics. Nucleo-centric geneticists had perpetuated a system of norms that suited their particular interests and assured the continuing growth of the assembled data that could be treated by their technical methods. Holistic or synthetic conceptions of heredity were ignored. The legitimacy of the nuclear monopoly of the cell during this period cannot be understood independently of the social struggles in the field where participants upheld those values that gave them the competitive advantage. As we have seen, it has been difficult for us to decide whether scientific knowledge claims are accepted because those who uphold them have an interest in truth or whether they simply uphold the truth which suits their interest. This problem was recognized by Ephrussi and Sonneborn, and left them with the question of how certain and objective knowledge could ever be attained by a community of scientists. Ephrussi could only allude to the Baconian statement that 'truth was the daughter of time' rather than of authority, which in his opinion was responsible for the nuclear monopoly. Thus he wrote in the preface of his small book on nucleo-cytoplasmic relations in 1953:

> I am . . . aware that many of the suggested deductions may not stand the test of time. I know also that very few of my fellow geneticists would subscribe to the views I have expressed, but I am encouraged by the thought that 'we cannot determine the truth of hypothesis by counting the number of people who believe it' and that 'a hypothesis does not cease to be a hypothesis when a lot of people believe it'.[57]

Throughout the period since the emergence of molecular biology the conflict over the significance of nuclear and cytoplasmic heredity has continued. Advocates of cytoplasmic heredity have continued to use social accounting to help explain the attitudes of their competitors. On the one hand, historical interpretations have been employed to describe the collective behavior of scientists, and the technical advantage of one research programme over another has been used to explain how the balance was tipped in favor of Mendelian inheritance. On the other hand, the individual behaviour of those scientists who focused on the study of traits which could be easily investigated by Mendelian methods was explained by an inquiry into the nature of laboratory activity, the motivations of competitors, and publishing practices (a domain of accounting that we have not surveyed). During the late 1950s, D. L. Nanney, a former student of Sonneborn, went further and raised the question as to whether social forces acting upon individual scientists biased the evidence against cytoplasmic heredity:

We are all aware that many considerations — both conscious and unconscious — are involved in the selection of traits to study and are cognizant of the non-randomness of the published accounts. Certain types of traits are more likely to be examined and certain types of results are more likely to be reported. One may argue, moreover, that these biases systematically prejudice judgement in favor of chromosomal inheritance.[58]

These biases, Nanney argued, began with a decision as to which traits should be studied. He maintained, like Sonneborn and Ephrussi, that from the beginning of genetics attention had been focused on differences that could be readily distinguished between individuals, not on similarities. However, even after the study of a trait is initiated a further source of bias may appear. Unlike the clear operational approach to Mendelian characters, the establishment of a cytoplasmic basis was a difficult and tedious task, since many possible vagaries of chromosomal behavior had to be eliminated. Thus Nanney remarked,

It is perhaps only natural that investigations of 'messy' characteristics are discontinued before publication and that investigators move on to traits more readily analysed.[59]

If this was, in fact, a correct evaluation of the practice of geneticists, the role of the chromosomes in heredity could be seriously over-emphasized. 'On the other hand,' Nanney argued, 'one would like to believe that the rule regarding the cherishing of exceptions is not only an ideal but a widespread practice . . .'[60] In this light nuclear geneticists were simply skimming off the easy work that yielded quick rewards and leaving the difficult problems to others. Nanney concluded his account of possible sources of bias in the reported accounts with the claim that 'a completely satisfactory answer' to the significance of cytoplasmic inheritance would necessarily entail '*an analysis of the pressures to which scientists are subjected and the motivations of the investigator*'.[61] But here he drew a disciplinary distinction and claimed that such analysis was beyond his competency.

Certainly, in the struggle for recognition in science, a researcher has to weigh the time and risk involved in investigating exceptions to standard rules against the yield in symbolic profit. The Morgan school had been certified and celebrated with two Nobel Prizes, one to Morgan in 1933 and one to H. J. Muller in 1948. To geneticists trained in this school and committed to the importance of nuclear genes, the risks simply out-weighed the perceived profits. This is not to say, however, that those who focused their investigations on the exceptions to Mendelian laws and the difficult analysis of cytoplasmic inheritance did not have high hopes of success or that they were disinterested. Indeed the development of research on cytoplasmic inheritance can be understood (objectively at least) as a strategy directed toward the maximization of scientific recognition and prestige. Given the intense competition triggered off with the genetic analysis of nuclear genes, one can easily understand the departure of a fraction of the researchers towards other topics, such as the study of the role of the cytoplasm in heredity and the principles of somatic cell differentiation, where competition was less intense. The possibility that cytoplasmic genetic factors were if not more so, at least as important as nuclear genes in evolution, and that they held the key to understanding ontogenetic development, increased the potential profits.

However, it would be false to believe that a scientific field is a purely free intellectual enterprise system and that the background of training and associations is the only restraint imposed upon a scientist. There are a number of other sociological factors which persistently limit the freedom of thought and expression, and these may be briefly mentioned because some of the participants in this dispute deployed them as accounting resources. In 1966, Lindegren (quoted earlier), who continued to oppose the primary importance of nuclear genes, described a series of *disciplinary* issues that he felt had to be taken into consideration when evaluating the relative importance of nucleus and cytoplasm in heredity. His points comprise a digest of commonplaces in the sociology of science, and many of the issues raised by Lindegren in what he described as the 'Cold War in Biology' continue to be addressed to the present day by some American geneticists who attempt to defend non-Mendelian and non-Darwinian principles in heredity and evolution

(i) During the 1960s there were few attempts to oppose the primary importance of the nuclear genes in heredity. However, Lindegren did not account for this in terms of a consensus emerging from an inherent logical necessity. The strict disciplinary organization of American genetics, in his view, was largely responsible for maintaining orthodoxy;

it posed one of the most serious obstacles to opposing the primary importance of the nuclear gene.[62]

(ii) The referee system, Lindegren argued, represented another persistent factor limiting the freedom of expression of the scientist in the United States. Although it purported to prevent incompetence, scientific publications were littered with trivial papers.[63] In fact, the referee system functioned as a mechanism of control to exclude the 'lunatic fringe' rather than incompetents. It functioned to maintain orthodoxy; 'new ideas are almost invariably labeled as "lunatic" when they are first introduced'. He describes this process of control in the American scientific community as follows:

> To be accepted by most scientific journals, a scientific paper must meet the approval of two of three referees. If it is a pedestrian paper confirming conventional ideas, it generally meets with approval. If it is unorthodox, the editor can assure its rejection simply by his choice of referees, who remain anonymous as far as the author is concerned.
>
> If an editor receives a scientific communication which offends him, it is a simple matter to refer it to a hot-headed emotional colleague. The referee is protected by anonymity and the editor 'regrets' that the rules will not permit him to accept the communication . . .[64]

This form of discipline was strictly practiced in American genetics as evidenced by the fact that few, if any, papers on cytoplasmic heredity were ever published in the journal *Genetics*. Instead, the relatively few American geneticists who voiced opposition to the 'nuclear monopoly' throughout the first half of the century had made heavy use of other vehicles: the few non-refereed American journals, those of related specialities, monographs, and foreign genetics journals.

In contrast, in Germany a strict disciplinary approach to science was resisted, many investigators openly criticized the hegemonic role of the nuclear genes, and investigations and theorizing on cytoplasmic heredity remained prominent. Within the institute system of German academic science, the referee system was not used. Dogmas were maintained not through journals but by powerful professors, and there were almost as many dogmas as there were professors. It was thus difficult to draw a distinction between orthodoxy and heterodoxy in German biology:

> Conformity in scientific thought is not demanded in Germany, and the German professor is a man of such importance in the community that it would be as

inconceivable to believe that he would be afraid to express his opinions as it would be to imagine that any means of communication would be denied him.[65]

(iii) Finally, 'job control' represented a major constraint limiting the freedom of thought and expression of young scientists rely on established figures for jobs and research grants. In Lindegren's opinion, the control of the deviant in a liberal democracy, though more subtle than in a dictatorship, was 'much more effective',

... because the movement of the disciplined worker out of the field is generally voluntary, and the opportunity for employment in other fields is not impaired by the deviationist's defection in one field.[66]

These issues continue to be seen, by some American geneticists, as a major obstacle to an objective appreciation of some extra-nuclear mechanisms of inheritance.[67] This illustrates how explanatory schemas from 'meta-scientific' disciplines can be turned into accounting devices within science. In such cases topos from sociology of science can be used as tools for accounting. In a structurally analogous way philosophy of science has often provided them in the form of formal methodologies.

6. CONCLUSION

In the complex structure of a scientific field, in the midst of continuous competition and conflict, scientists have recourse to an elaborate body of rhetorical and political strategies. Amongst these must be included not only the deployment of any available 'formal' methodological doctrines as accounting devices, but also the other discursive resources used in social/cognitive accounting in the field. A comprehensive history of 'methodology' must take into consideration all these discursive strategies and their modes of deployment. In this essay, I have attempted to emphasize those which are not normally considered by methodologists and historians of method. In particular, I have stressed those which are not framed out of the materials of 'formal' method doctrines, and which seem to occur often in the context of disciplinary disputes.

First, there is the issue of specialization and the formation of the discipline through which groups of scientists attempt to establish those objectives and explanatory standards which can be embraced by their technical procedures and theory. The success of this strategy, though conditioned by institutional structure, results from the ability of organized groups to develop lines of inquiry rapidly and to define the field in a way that gives them the competitive advantage. As discussed in the present account, the nuclear monopoly of the cell resulted largely from the ability of Mendelian geneticists to form their own discipline with their own objectives, technical methods, explanatory standards, doctrines, journals and societies, and to restrict their field of inquiry to problems that could only be dealt with effectively by 'the Mendelian Method'. As a result, the controversy over the roles of the cytoplasm and the nucleus in heredity continued as a dispute understood by the participants in terms of scientific values which included the motivations of researchers and the problem of establishing priorities in scientific research.

In this context, scientists questioned the motivations of their competitors in their attempts to delegitimate conflicting results and groups of contenders. These included not only judgements about the intentions of competitors to provide room for 'false' theories (i.e. Lamarckism and/or the belief that the nucleus controlled only relatively trivial traits while the cytoplasm controlled fundamental characteristics), but also their motivations in seeking prestige and recognition, and in maintaining orthodoxy as opposed to truth. As we have seen when discussing the arguments of Sonneborn, Ephrussi, Nanney, and Lindegren, these issues became especially poignant when dissenters infiltrated the discipline and attempted to challenge dominant views. Formal methodological issues were also deployed as rhetorical devices in inter-disciplinary disputes over property rights and in individual priority disputes. These were used by competitors when attempting to make room for alternative theoretical possibilities and research problems.

On the basis of this study it can also be seen that when scientists lack the scientific equipment to effect a revolution, and when the equipment required to aid them in their struggle can only be legitimately acquired from within the scientific establishment, the technique-ladenness of observations offers another resource to be used in accounting for the 'false' beliefs of their competitors. This may include claims of both quantitative and qualitative misrepresentation of reality resulting from specific experimental procedures and materials. In the view of cytoplasmic geneticists, cross-breeding analysis was responsible for perpetuating the idea that genes were autonomous agents acting independently from the rest of the cell. The facility of the Mendelian procedures was largely responsible for the attention focused on the nucleus. The kinds of materials (i.e. organisms) employed conditioned the range of results that could be obtained.

Finally, it should be clear that there is no reason to accept the claim of Gilbert and Mulkay that social accounting for error does not occur in the formal scientific literature and that 'there is seldom if ever any published attempt to explain why false beliefs are accepted by one's colleagues'. As we have seen in this account, scientists use a variety of accounting resources in their publications. Those who upheld the 'nuclear monopoly' accounted for the 'false' views of their opponents in terms of their lack of adherence to formal methodological doctrines (such as pragmatism, empiricism, falsificationism, etc.); their 'refusal' to accept the dominant role of the nucleus; and their training in terms of a 'false' theory. Those who supported the importance of the cytoplasm in heredity accounted for the 'false' beliefs of Morganists in terms of 'defensive attitudes', 'failure to put enough effort', the technique-ladenness of observation, the socio-political nature of scientific development, 'job control' in the United States, etc. All these accounts occur in the published scientific literature. Moreover, as I have stressed throughout this esssay, social accounting, ascriptions of technique-ladenness of observation, and formal methodological doctrines all play similar roles in scientific controversy. None of these accounting resources is any less 'scientific' in any *a priori* sense, that is, any less likely to be deployed in the constitution of knowledge-making and knowledge-breaking claims.

As illustrated in the present account, social accounting feeds back into the scientific enterprise and allows competitors to adjust their activity, not only in terms of what they are doing, but also in terms of what they have to do. In the midst of the continuous conflict and competition within the scientific field whereby scientists attempt to make their opinions pass for 'scientific facts', the manner in which a person propounds his or her views may be decisive at least in terms of persuading competitors to adopt one of various alternative theories. In this context, when critically evaluating contributions, scientists necessarily have to consider the history of the development of an idea, the specific indoctrination and point of view of the contributor, the technical nature of the experimental procedure and the data presented. These issues are especially salient when attempts are made substantially to modify a dominant theory in order to reconcile it with contradictory observations and theoretical problems that are persistently ignored by members of dominant groups.

ACKNOWLEDGEMENT

I am grateful to H. E. Le Grand, W. R. Albury, D. Turnbull and especially the editors for their helpful comments and suggestions.

University of Melbourne, Australia

NOTES

[1] For discussions of the symmetrical approach see the relevant essays in Karin D. Knorr-Cetina and Michael Mulkay (eds.), *Science Observed* (London, 1983).

[2] See for example, W. B. Provine, *The Origins of Theoretical Population Genetics*, (Chicago and London, 1971); R. Olby, *The Path to the Double Helix* (London, 1974); Garland Allen, *Life Science in the Twentieth Century* (Cambridge, 1975); H. F. Judson, *The English Day of Creation. The Makers of the Revolution in Biology* (New York, 1979); E. Mayr and W. B. Provine (ed.), *The Evolutionary Synthesis. Perspectives on the Unification of Biology* (Cambridge, MA., 1980).

[3] W. Coleman, 'Bateson and Chromosomes: Conservative Thought in Science', *Centaurus* **XV**, 1970, pp. 228—314. See also Allen, *op. cit.* (Note 2), p. 57.

[4] For attempts to account for the attitude of French neo-Lamarckism in purely social terms see for example C. Limoges, 'A Second Glance at Evolutionary Biology in France', in Mayr and Provine, *op. cit.* (Note 2), pp. 322—329. See also D. Buican, *Histoire de la génétique et de l'évolutionisme en France* (Paris, 1984).

[5] This over-simplified view is predominant in Buican, *ibid.*, pp. 355—405, and *Idem., L'éternal retour de Lysenko*, (1978). For a review of the literature on Lysenkoism see R. Young, 'Getting Started on Lysenkoism', *Radical Science Journal* **VI—VII**, 1978, pp. 81—106. Those who have accounted for Lysenkoism in purely 'political' terms have overlooked the fact that many leading non-communist biologists in the West also opposed Morganist doctrines and held views of heredity that were quite similar to those advocated by Lysenkoists. See Jan Sapp, 'Cytoplasmic Inheritance and the Struggle for Authority in the Field of Heredity, 1891—1981' (Doctoral Dissertation, The University of Montreal, 1984). There is only one other account which looks at the 'scientific' merits of Lysenko's work. See Nils Roll-Hansen, 'A New Perspective on Lysenkoism' (in preparation).

[6] Pierre Bourdieu, 'The Specificity of the Scientific Field and the Social Conditions for the Progress of Reason', *Social Science Information* **VI**, 1975, pp. 19—47.

[7] Michael Mulkay and G. Nigel Gilbert, 'Accounting for Error: How Scientists Construct their Social World When they Account for Correct and Incorrect Belief',

Sociology **XVI**, 1982, pp. 165–183 (my emphasis). See also G. Nigel Gilbert and Michael Mulkay, *Opening Pandora's Box* (Cambridge, 1984).

[8] Mulkay and Gilbert, *op. cit.* (Note 7).

[9] *Ibid.* p. 179 (my emphasis).

[10] The present essay, focusing on some special aspects of method discourse, necessarily has to limit severely its account of the theoretical and empirical reasons many biologists gave for advocating cytoplasmic inheritance. For a detailed account of the conceptual and technical developments concerning cytoplasmic inheritance throughout the twentieth century, see Sapp, *op. cit.* (Note 5).

[11] See for example, D. L. Nanney, 'Microbial Precursors of Developmental Processes', *Verh. Dtsch. Zool. Ges. Bd.* **LXXVII**, 1984, pp. 24–30; Joseph Frankel, 'Development and Evolution: A Report', *Evolution* **XXXVIII**, 1984, pp. 1160–1162. This hereditary phenomenon concerns the genetic autonomy of the solid (cortical) cytoplasm which T. M. Sonneborn and his collaborators first demonstrated in sensational graft hybrids of *Paramecium* in the early 1960s. For an historical account of the development of this research see Sapp, *op. cit.* (Note 5). It might be mentioned that some investigators may also adhere to the belief that hereditary specificity also resides in the fluid part of the cytoplasm.

[12] J. Maynard Smith, 'Evolution and Development', in *Development and Evolution. Symposium of the British Society for Developmental Biology* (Cambridge, 1983), pp. 33–45, at p. 39.

[13] One exception is the recent account of Jonathan Harwood, 'The Reception of Morgan's Chromosome Theory in Germany: Inter-War Debate over Cytoplasmic Inheritance', *Medizin historisches Journal* Band **XIX**, Heft 1/2, 1984, pp. 3–32. See also the excellent paper of D. L. Nanney, 'The Ciliates and the Cytoplasm', *Journal of Heredity* **LXXIV**, 1983, pp. 163–170.

[14] Francois Jacob, *The Logic of Life. A History of Heredity*, tr. Betty E. Spillmann (New York, 1976). There is no mention of cytoplasmic heredity in Jacob's text.

[15] Ernst Mayr, *The Growth of Biological Thought. Diversity, Evolution, and Inheritance* (Cambridge, MA., 1982). Mayr only briefly mentions cytoplasmic inheritance (see pp. 786–790) in the attempt to exclude some work as being 'premature', 'disproved' or 'eventually shown to have a genic-chromosomal explanation'. He erroneously excludes the cytoplasmic control of respiratory deficiency, investigated by Ephrussi, as being due to infection. As will be discussed later, throughout the twentieth century many geneticists who upheld the predominant role of the nuclear genes in heredity used the idea that cytoplasmic inheritance was due to infectious parasites, in order to exclude its general occurrence and significance. Yet, while mitochondria genetics is one of the most rapidly growing specialities of modern genetics, Mayr can only state that mitochondria 'may' have their own DNA. Finally, in outright conflict with other evolutionists such as Maynard Smith, Mayr makes the amazing claim that the inheritance of supra-molecular structure does not fundamentally contradict the chromosome theory.

[16] For a detailed account of the problems of embryology in relation to the cytoplasm and the nucleus, see Sapp, *op. cit.* (Note 5).

[17] T. H. Morgan, 'Genetics and the Physiology of Development', *The American Naturalist* **LX**, 1926, pp. 489–515, at p. 491.

[18] For a more detailed account of the early field of heredity and the discursive and social strategies deployed by Mendelian geneticists, see Sapp, 'The Struggle for Authority in the Field of Heredity, 1900—1932: New Perspectives on the Rise of Genetics', *Journal of the History of Biology* **XVI**, 1983, pp. 311—342.

[19] L. C. Dunn, 'Nucleus and Cytoplasm as Vehicles of Heredity', *The American Naturalist* **LI**, 1917, pp. 286—300, at p. 286.

[20] Ibid., p. 296.

[21] T. H. Morgan, *The Physical Basis of Heredity* (New York, 1919), p. 241.

[22] A. F. Shull, 'Cytoplasm and Heredity', *The Ohio Journal of Science* **XVII**, 1916, pp. 1—8, at p. 6.

[23] For an account of the obstacles to the formation of new disciplines posed by the Institute structure and centralized control of the French University system see, for example, T. Clark, *Profits and Patrons: The French University and the Emergence of the Social Sciences* (Cambridge, 1973). For a comparison of the institutional structures of the German University system with that of the United States, see J. Ben-David, 'The Universities and the Growth of Science in German and the United States', *Minerva* **VII**, 1968, pp. 1—35. These institutional structures as they relate to the development of genetics, in France, Germany and the United States are discussed in Sapp, *op. cit.* (Note 5).

[24] Many biologists claimed that although Mendelian genetics had supplied a basis for understanding one (minor) aspect of evolution, it had the major deleterious effect of confining biologists' attention to the nuclear genes and therefore had led them into an impasse with respect to development. These views were also expressed by Bateson who claimed that the chromosome theory, though providing much that is certainly true and of immense value, had fallen short of the 'essential discovery'. See W. Bateson, 'Segregation: Being the Joseph Leidy Memorial Lecture of the University of Pennsylvania, 1922', *Journal of Genetics* **XVI**, 1926, pp. 201—235. Wilhelm Johannsen, who coined the central genetic terms, 'genotype', 'phenotype' and the 'gene', also recognized that genetics mostly operated with characters which were rather 'superficial' in comparison with the 'fundamental, specific or generic nature of the organism'. He came to believe in a 'central something' as yet not divisible into separate factors presumably located in the cytoplasm of the cell. See W. Johannsen, 'Some Remarks about Units in Heredity', *Hereditas*, Band **IV**, 1923, pp. 133—141, at p. 137. The point here is that when it is recognized that many leading biologists shared Bateson's attitude towards the idea of particulate genes on chromosomes as the exclusive mechanism of heredity, the question of his 'idealism' and 'conservatism' in biology loses much of its analytic value.

[25] E. Guyénot, *L'Hérédité* (Paris, 1924) p. 289 (my emphasis).

[26] See Sapp, *op. cit.* (Note 5) and Harwood, *op. cit.* (Note 13).

[27] E. M. East, 'The Nucleus-Plasma Problem', *The American Naturalist* **LXVIII**, 1934, pp. 289—303; 402—439.

[28] *Ibid.*, p. 409—410.

[29] For an account of the development of the work of Sonneborn and Ephrussi, see Sapp, *op. cit.* (Note 5).

[30] The work of Sonneborn in particular was singled out by Lysenkoists as providing proof of the downfall of the Mendelian-chromosome theory. The German work on cytoplasmic inheritance was never cited by Lysenkoists. Lysenkoists in France cited the work of L'Héritier in favour of their views. See Sapp, *op. cit.* (Note 5).

[31] Many Lysenkoists denied the existence of particulate genes. Cytoplasmic geneticists,

in contrast, denied only the autonomy of the genes from the rest of the cell.

32 Carl Lindegren, *The Yeast Cell, Its Genetics and Cytology* (St. Louis, 1949), p. 27.

33 W. Beadle, 'Genes and Biological Enigmas', *Science in Progress, 6th series*, G. Baitsell (ed.), (New Haven, 1948), pp. 232—233.

34 M. Delbrück, discussion following David Bonner, 'Biochemical Mutations in Neurospora', *Cold Spring Harbor Symposium on Quantitative Biology* **XI**, 1946, pp. 14—24, at p. 23.

35 *Ibid.*, p. 23.

36 Boris Ephrussi, *Nucleo-Cytoplasmic Relations in Micro-organisms* (Oxford, 1953), p. 121.

37 N. H. Horowitz to B. Ephrussi, 10 June 1953, Ephrussi Papers, Centre de Genetique Moleculaire du C.N.R.S., Gif-sur-Yvette, France.

38 B. Ephrussi to N. H. Horowitz, 26 June 1953, Ephrussi Papers.

39 B. Ephrussi to T. M. Sonneborn, 9 November 1952, Sonneborn Papers, Lilly Library Archives, Indiana University, Bloomington, Indiana.

40 B. Ephrussi to T. M. Sonneborn, 24 May 1952, Sonneborn Papers.

41 T. M. Sonneborn, 'The Cytoplasm in Heredity', *Heredity* **IV**, 1950, pp. 11—36, at p. 11.

42 T. M. Sonneborn to B. Ephrussi, 17 November 1952, Sonneborn Papers.

43 T. M. Sonneborn, 'The Role of Genes in Cytoplasmic Inheritance', in L. C. Dunn (ed.), *Genetics in the 20th Century* (New York, 1951), pp. 291—314, at p. 311.

44 Ephrussi and Sonneborn opposed the view of genes as 'autoreproducing particles'. Instead they viewed them as 'conservative units' of the cell. The view of genes as 'units of life' or the fundamental basis of life was dicussed most prominently by the Nobelist, H. J. Muller. See, for example, H. J. Muller, 'The Gene as the Basis of Life', *Proceedings of the International Congress of Plant Sciences* **I**, 1929, pp. 897—921; *Idem.*, 'The Development of the Gene Theory', in L. C. Dunn (ed.), *op. cit.* (Note 43), pp. 77—99.

45 Ephrussi, *op. cit.* (Note 36), p. 108. At the same time, Ephrussi was careful to distinguish his views from those who might assert a special unknowable vital force, a group sometimes referred to as the 'obscurantist biological right': 'Neither should my opinions be taken as a sign of defeatism, for they are based on the belief that ultimately cell integration will find its explanation in terms of knowable, if not known, molecular structures and forces'.

46 For discussions of the secondary literature treating the Rockefeller Foundation's role in fostering the development of molecular biology, see John A. Fuerst, 'The Definition of Molecular Biology and the Definition of Policy: The Role of the Rockefeller Foundation's Policy for Molecular Biology', *Social Studies of Science* **XIV**, 1984, pp. 225—237; Ditta Bartels, 'The Rockefeller Foundation's Funding Policy for Molecular Biology: Success or Failure?', *ibid.*, pp. 238—247; Robert Olby, 'The Sheriff and the Cowboys: Or Weaver's Support of Astbury and Pauling', *ibid.*, pp. 244—247; E. J. Yoxen, 'Scepticism about the Centrality of Technology Transfer in the Rockefeller Foundation Programme in Molecular Biology', *ibid.*, pp. 248—252; Pnina Abir-Am, 'Beyond Deterministic Sociology and Apologetic History: Reassessing the Impact of Research Policy upon New Scientific Disciplines (Reply to Fuerst, Bartels, Olby and Yoxen)', *ibid.*, pp. 252—263.

47 Ephrussi, *op. cit.* (Note 36), p. 108.

48 B. Ephrussi to J. A. Jenkins, 4 December 1956, Ephrussi papers.

49 S. Spiegelman to B. Ephrussi, 16 November 1946, Ephrussi Papers.

[50] T. M. Sonneborn, 'Beyond the Gene-Two Years Later', *Science in Progress*, 1951, 7th Series, pp. 167—203, at p. 199.
[51] As Sonneborn put it [*ibid.*, p. 201], 'During the past fifty years research necessarily had to concentrate on what it *could* do. It could analyse the architecture of the nuclear material and it did. The theory of the gene and the chromosome theory of heredity were the grand result'.
[52] B. Ephrussi, 'Remarks on Cell Heredity', in L. C. Dunn (ed.), *op. cit.* (Note 43), pp. 241—262, at p. 242.
[53] *Ibid.*, p. 260.
[54] Another example was the loss of photosynthetic ability accompanying the loss of chloroplasts in some forms of algae.
[55] Ephrussi, *op. cit.* (Note 36), p. 119.
[56] *Ibid.*, p. 119. During the 1950s, Sonneborn maintained a similar, though slightly less radical position: 'Long ago it was suggested that these so-called *fundamental* traits were controlled by the cytoplasmic part of the genetic system, the nuclear genes modifying only relatively minor superficial individual variations. This view, in the extreme form just set forth, is certainly untenable. Mutations of nuclear genes are known to alter not only individual traits, but also traits of high taxonomic importance and properties on which life itself depends, ... Yet it is also remarkable that among the traits and properties known to be controlled by cytoplasmic materials, a large proportion are undoubtedly very fundamental'. Sonneborn *op. cit.* (Note 50), p. 199.
[57] Ephrussi, *op. cit.* (Note 36), p. vi.
[58] D. L. Nanney, 'The Role of the Cytoplasm in Heredity', in W. D. McElroy and H. B. Glass (eds.), *The Chemical Basis of Heredity* (Baltimore, 1957), pp. 134—166, at pp. 137—138.
[59] *Ibid.*, p. 138.
[60] *Ibid.*, pp. 138—139.
[61] *Ibid.*, p. 139 (my emphasis).
[62] Carl C. Lindegren, *The Cold War in Biology* (Ann Arbor, 1966), p. 10.
[63] *Ibid.*, p. 9.
[64] *Ibid.*
[65] *Ibid.*
[66] *Ibid.* Lindegren also points out that when deviation assumed 'serious proportions', American geneticists were not above supressing it by direct action. He illustrates this point through the example of the Lysenko Affair when, in 1949, members of the American Genetics Society established a committee for 'Scientific Freedom' in which there was an attempt to prevent the teaching of Lysenkoism in American colleges and universities. However, after protests from some members of the society the committee remained inactive. *Ibid.*, p. 8.
[67] For example, in a recent paper presented before the German Zoological Society, Nanney, a leading expert on the genetics of ciliated protozoa, pointed out the more recent experimental evidence for a Lamarckian principle in evolution and the threat to the evolutionary synthesis represented by the inheritance of supra-molecular structure. After noting the chequered histories of the idea of the inheritance of acquired characteristics as allied with frauds, and Lysenkoism, Nanney clearly stated the risks of adhering to his views: 'Obviously it is dangerous to one's career to become too closely involved with issues of such doctrinal significance'. [*op. cit.* (Note 11), p. 29.]

W. R. ALBURY

THE ORDER OF IDEAS: CONDILLAC'S METHOD OF ANALYSIS AS A POLITICAL INSTRUMENT IN THE FRENCH REVOLUTION

1. INTRODUCTION

At the time of his death in August 1780, the abbé Etienne Bonnot de Condillac was recognised throughout France as one of the leading philosophers of his age. This reputation was based not only on Condillac's major philosophical texts dating from the middle of the century — his *Essay on the Origin of Human Knowledge* (1746), *Treatise on Systems* (1749) and *Treatise on Sensations* (1754) — but also on his later, more practically-oriented works, published in the 1770s after he had returned to France from nine years in northern Italy as tutor to the Prince of Parma. These practical works included the multi-volumed *Course of Studies* (1775) which Condillac had composed for the Prince, and an economic treatise entitled *Commerce and Government* (1776). Finally, just a few weeks after his death, Condillac's *Logic* — his last completed work — appeared in print. Originally commissioned by the government of Poland for the reform of its educational system, this text was hailed in France as a concise and compelling account of Condillac's doctrine of scientific method.[1]

Despite his renown, Condillac did not leave a school of immediate successors behind him to carry on his philosophical work. A quiet and diffident man, he had lived in retirement on his country estate since 1773. On his death, the philosophic mantle passed to Antoine de Condorcet, whose regard for the work of Condillac was lukewarm at best.[2] The feeling was mutual: it was reported that Condillac had always detested Condorcet and that he attributed his fatal illness to a cup of bad chocolate which he had been served at Condorcet's during his last brief visit to Paris.[3] At any rate, for about a decade, little was done to advance the development of Condillac's formal philosophy[4], although a number of scientists — most notably the the chemist Antoine Lavoisier — sought to apply his method in their own special fields of research during this period.[5]

In contrast to the relative philosophical eclipse of Condillac during the 1780s, the succeeding revolutionary decade saw an efflorescence of

J. A. Schuster and R. R. Yeo (eds.), The Politics and Rhetoric of Scientific Method, 203—225.
© 1986 *by D. Reidel Publishing Company.*

enthusiasm in France for the basic principles of his philosophy, especially those regarding the relationship between language and thought.[6] After the Thermidorian reaction ended the Terror in 1794 a distinctive school of philosophic enquiry arose to further the work of Condillac in developing what its advocates called Ideology, or the science of ideas. These philosophers looked upon Condillac 'as the founder of their school, even though a generation or more separated him' from their own time.[7] Nevertheless, it was not the substantive conclusions reached by Condillac, but rather his doctrine of method that united the school of Ideologists (or Ideologues, as they were sometimes called).[8] One of the chief spokesmen for this group, writing in 1802, held that

> there is perhaps not a single one [of us] who adopts without restriction the grammatical principles of Condillac, or who is fully satisfied with the manner in which he analyses our intellectual faculties, or who finds nothing to reproach in what he has said about reasoning . . . . It is not the decisions of Condillac which we regard so much but rather his method which we value most highly.[9]

In addition to their distinctive philosophical programme based on Condillac's method, the Ideologues also shared, during their period of greatest influence (1794—1803), a common political programme.

> In these years between the Thermidorian reaction and the rise of Bonaparte to dictatorial power, Ideology was the approved doctrine of the [French] government; as such it gave philosophic support to moderated republicanism. For nearly a decade the Ideologues proposed reforms based not upon a sudden and violent change of institutions but upon a reconstruction of the moral and intellectual life of the people.[10]

Given this context, the following two questions seem worth considering:

(1) What was the relationship between the political position of the Ideologues and the resurgence of philosophical interest in the work of Condillac, particularly as a guide to method?

(2) What was the relationship between the Ideologues' interest in Condillac's method and the various applications of that method attempted by French scientists both in the decade of the 1780s and in the decade following Thermidor?

Before suggesting answers to these questions, however, it will be useful first to describe the characteristic features of Condillac's method and then to outline its role in the scientific work of Lavoisier.

2. THE METHOD OF CONDILLAC

Condillac's method, which he called the method of analysis, consisted in a two-fold process of resolving the object of study into its simplest constituents and then reconstituting it according to its natural order of generation and development from these constituents. The term 'synthesis' he restricted to those methods of composition which failed either to start from the true origins of things or to proceed according to their natural order of development.

The secret of analysis for Condillac, then, was not just to resolve wholes into their parts, but to do this in a particular, correct way: the method presupposed an epistemology. Since all knowledge comes by way of the senses, Condillac held, the proper method of analysis is to trace all our ideas back to their origins in simple sensations, and then to reconstruct them according to their natural generation, studying the principle which brings about their formation and discovering the relations between the ideas formed at each stage of the development. The goal of this procedure is to create within the mind an ordered arrangement of ideas whose relations to one another intrinsically reveal the processes by which those ideas, or the phenomena they represent, were formed. As Condillac wrote:

> To analyse, then, is nothing other than to observe the qualities of an object in a successive order so as to give them, in our mind, the simultaneous order in which they exist outside it. . . . And it is in this simultaneous order that the knowledge which we have of things consists: for if we could not depict them to ourselves all together, we could never make judgements about the relations between them, and we would know them very inadequately.[11]

Thus far Condillac's methodology was explicitly built upon John Locke's *Essay Concerning Human Understanding*, which had found the 'materials' of ideas in sensation. But as many commentators have pointed out, Condillac was also indebted to René Descartes' *Rules for the Direction of the Mind*.[12] Indeed, up to a certain point, Condillac's method can be understood as the method of Descartes' *Rules* with simple sensations in place of simple natures.[13] Condillac repeatedly asserted that no idea can be more clear and distinct than that arising from the immediate datum of a single sense; and like Descartes, he claimed to have found a method of universal scope, applicable to all fields of inquiry. Finally, especially in his late works such as the *Logic*, Condillac offered algebra as a model of clear reasoning, and in so doing he exhibited the same concern as did Descartes for establishing a quasi-algebraic relationship between the known and unknown elements of any problem to be solved. But Condillac's treatment of algebra went beyond that of Descartes in important ways: not being a mathematician himself, he interpreted the workings of that art from a perspective which he derived at least partly from his reading of G. W. Leibniz.[14]

Characteristically, Condillac sought to discover the origin or principle of algebra's power as a means of analysis in mathematics. In his earliest work Condillac had stressed the necessity of language for intellectual development, because only by means of signs connected to ideas could the mind gain control over the connections and associations between ideas which constitute the motive power of thought. On the basis of this theory of language Condillac concluded that algebra owes its power and clarity to its symbolic system. Algebra, as he said in his *Logic*, is really a language, but it surpasses all other languages in the degree to which it is suited to expressing the conceptions which it represents. When we are presented with a mathematical problem we can, by means of an algebraic equation, relate the quantities sought to those which are given. Then, operating at the level of the signs alone, without reference to their content, we can easily translate that equation through a series of mathematically-equivalent expressions until we reach an expression which solves the problem by setting each unknown quantity equal to a known value.[15]

According to Condillac, our reasoning in the non-mathematical sphere can approach the analytic power and clarity which algebra exhibits in mathematics, if we carry out a systematic reform of our language. For each language, Condillac held, is essentially a method of analysis which structures our thought in its own way; and in proportion as our language represents more faithfully the true origins and relations of things, so will all reasoning carried out in that language be more exact.[16]

Linguistic reform, then, is an essential part of analysis for Condillac; and it must be accomplished even as we resolve and recompose our objects of study. This reform assures that the ordered results of our analyses are perpetuated in a precise terminology which, in turn, directs and assists the mind in making further analyses. We must eliminate words which give false or misleading ideas and replace them with others which exhibit the most important relations of the ideas designated. In an ideal Condillacian language — and we must have different languages in different fields of inquiry, because our needs are different[17] — the connections between our ideas would be manifest in our names for those ideas; and the relationship between that which is known and that which is sought in any problem could be found by the translation of the statement of that problem into a series of equivalent expressions, just as in the solution of an algebraic problem. Thus, as Condillac put it, 'the art of reasoning is reduced to a well-made language'.[18]

It was this linguistic, algebraic treatment of the classic methodology of resolution and composition that gave Condillac's *Logic* much of its initial appeal, because it seemed to hold out the promise that reasoning in the non-mathematical sphere could approach the analytic power and clarity which algebra achieves in mathematics, if only the languages of the various non-mathematical fields of study could be systematically reformed. One of the most consequential attempts to carry out such a reform according to the method of Condillac's *Logic* was to come within the field of chemistry during the decade of the 1780s, and it is to this subject that we now turn.

3. THE CHEMICAL REVOLUTION

Interest in the reform of the language of chemistry had been growing in Europe throughout the eighteenth century[19], but the new system of chemical nomenclature introduced by Lavoisier and his colleagues in 1787[20] differed from all previous proposals in three important ways: (1) it explicitly invoked Condillac's methodological theory as its procedural guide; (2) it implicitly incorporated a new substantive theory of chemical composition; and (3) it was to become widely accepted (largely because of that substantive theory). The degree to which the method of Condillac may have contributed to the actual *formation* of Lavoisier's new chemical theory is a problem that will not be addressed here[21]. For the purposes at hand it is the role of Condillac's method in the *presentation* of Lavoisier's theory that is of central importance.

Lavoisier's ambition to make chemistry a rational, quantitative science on the model of mathematical physics has been appreciated by many scholars and is evidenced by his collabortion with such figures as Gaspard Monge and Pierre-Simon de Laplace.[22] With his new theory of chemical composition Lavoisier felt that the basis for his 'chemical revolution' had been established[23], but he was not yet in a position to reduce the data of chemistry to a fully quantitative form. What he found, then, in Condillac's methodological combination of constituent analysis and linguistic algebra was a means of expressing his theory of composition in a quasi-algebraic fashion through the language of chemistry. If the science of chemistry was not yet ready to become wholly mathematical, it could at least lay claim to an approximation of the theoretical rigour and precision of mathematics in its language.

The fullest elaboration of this claim came in the preface to Lavoisier's textbook of 1789, *Elements of Chemistry*[24] – a work which, like the *Method of Chemical Nomenclature*, explicitly cited Condillac's *Logic* for its methodological rationale. With the new nomenclature, Lavoisier argued, chemistry could now enshrine the results of the analyses of substances in the names of those substances. 'In short', he wrote,

> we have advanced so far, that from the name alone may be instantly found what the combustible substance is which enters into any combination; whether that combustible substance be combined with the acidifying principle, and in what proportion; what is the state of the [resulting] acid; with what base it is united; whether the saturation be exact or whether the acid or the base be in excess.[25]

Here the analytic language of chemistry functioned as a system of resolution and composition, in keeping with the classic definition of chemistry as the science concerned with the separation and recombination of the constituents of natural substances.[26] But this function now operated according to a mathematical model. Condillac held that

> from the nature of our way of calculating, it is sufficient to have ideas of the first numbers, in order to form for ourselves ideas of every number that can possibly be determined. This is because the first signs being given, we have rules for inventing others.[27]

Similarly, the new chemical nomenclature was based upon the names of simple substances, operationally defined by Lavoisier as 'the last point which analysis is capable of reaching'.[28] Once these names were established, a few rules served to determine the names of all compound substances, these names being 'compounded in the same manner as are the substances themselves'.[29]

In addition, the nomenclature also carried out a taxonomic function. Lavoisier, for example, in discussing binary compounds, wrote that

> as the number of double combinations is already very considerable, the only method by which we could avoid confusion, was to divide them into classes. In the natural order

of ideas, the name of the class or genus is that which expresses a quality common to a great number of individuals: The name of the species, on the contrary, expresses a quality peculiar to certain individuals only.[30]

This statement Lavoisier followed with a lengthy quotation from Condillac's *Logic* on the classification of ideas according to genera and species, after which he noted: 'This is the logic of all the sciences, and is naturally applied to chemistry'.[31] Hence some terms, such as 'acid', 'oxide', and 'sulphide', served to designate what Lavoisier called 'genera', while others, such as 'nitrous' or 'nitric', indicated what Lavoisier called 'species' — despite the fact that these 'genera' and 'species' were logically quite different from the genera and species of a natural history taxonomy.[32]

Finally, the nomenclature had a mathematical character, making possible a qualitative algebra, the chemical equation, which for the first time permitted chemists, like 'mathematicians [to] obtain the solution of a problem by the mere arrangement of data, and by reducing their reasoning to such simple steps, to conclusions so very obvious, as never to lose sight of the [quality of self-] evidence which guides them'.[33] Lavoisier had in fact introduced the chemical equation prior to his reform of the nomenclature in 1787. In a draft memoir on alcoholic fermentation, Lavoisier used the equation form to express the following test reaction to determine the acid (x) and base (y) components of an unknown salt ($x + y$):

$$x + y + \text{vitriolic acid} = \text{nitrous acid} + \text{vitriolated tartar.}^{34}$$

But to proceed any further one had either to know the composition of vitriolated tartar from a prior analysis or to carry out such an analysis at that time. In the new nomenclature, however, the test reaction was expressed as follows:

$$x + y + \text{sulphuric acid} = \text{nitric acid} + \text{sulphate of potash}$$
$$(x + y + H_2SO_4\ [+ 2H_2O] = 2HNO_3 + K_2SO_4\ [+ 2H_2O]).^{35}$$

From this point one can proceed entirely at the level of signs, without further recourse to the chemical substances themselves or prior knowledge of their composition. By the linguistic rules alone one known that

$$\text{sulphate of potash} = \text{sulphuric acid} + \text{potash,}$$
$$(K_2SO_4 + [2H_2O] = H_2SO_4 + 2KOH)$$

and hence that

$$x + y + \text{sulphuric acid} = \text{nitric acid} + \text{sulphuric acid} + \text{potash}.$$
$$(x + y + H_2SO_4\ [+ 2H_2O] = 2HNO_3 + H_2SO_4 + 2KOH).$$

Furthermore, by deleting the sign 'sulphuric acid' from both sides of the equation one sees immediately that

$$x + y = \text{nitric acid} + \text{potash}$$
$$(x + y\ [+ 2H_2O] = 2HNO_3 + 2KOH).$$

Thus the components of the unknown salt are given solely by the transformation of the linguistic expression of the initial reaction. And finally the salt itself is given by the linguistic rules, which stipulate that

$$\text{nitric acid} + \text{potash} = \text{nitrate of potash}$$
$$(2HNO_3 + 2KOH = 2KNO_3\ [+ 2H_2O]).$$

It was the possibility of formulating this sort of equation, despite its lack of a quantitative dimension, that prompted the mathematician Joseph-Louis de Lagrange to say that Lavoisier's reforms had made chemistry 'as easy as algebra'.[36]

4. PRELIMINARY ANSWERS

Following this brief account of Condillac's methodology and its role in the presentation of Lavoisier's chemistry, it is now possible for us to offer preliminary answers to the two questions we raised in the first section of this paper, concerning the Ideologues' interest in the method of Condillac. As we shall see, however, these answers — which largely represent the consensus of scholarly opinion on the subject of the Ideologues — can only be accepted as a partial solution to our problem, and further considerations will be necessary before we may conclude our discussion satisfactorily.

A preliminary answer to our first question, concerning the relationship between the political position of the Ideologues and their interest in Condillac's methodology, would go as follows. Since the Ideologues were the continuers of the general tendencies of Enlightenment thought[37], and since Condillac was the most explicit and systematic philosophical writer on method in France during the eighteenth century, the Ideologues would naturally look to Condillac's work for methodological guidance when they sought a philosophical basis for the reconstruction of French society after the ending of the Terror in 1794.[38] Furthermore, Condillac's pedagogical reputation, based on his *Course of Studies* for the Prince of Parma and his *Logic* for the Polish schools, would have given his methodology additional appeal in the eyes of philosophers like the Ideologues who believed that education was the most effective means of social reform and regeneration.[39] The Ideologue D. J. Garat, for example, lecturing at the newly established *Ecole Normale* in 1795, said of Condillac that 'the works which he composed for the instruction of a child [now] prepare a revolution in the teaching of all peoples'[40]; and 1400 copies of Condillac's *Logic* were purchased for distribution to the students of that institution.[41]

A preliminary answer to our second question, concerning the relationship between the Ideologues' interest in Condillac's method and scientific applications of it, is also readily available. As we have seen, the new nomenclature of chemistry was introduced by Lavoisier and his colleagues under the aegis of Condillac's theory of method, and Lavoisier's *Elements of Chemistry* published in 1789 explicitly appealed to Condillac's *Logic* for its methodology. Because of this association,

> the progress that was made in the new chemistry after the adoption of the reformed nomenclature was commonly invoked in the 1790s to substantiate Condillac's claim that the advancement of knowledge and the perfecting of a science depended upon a complete overhaul, and perhaps a total reconstruction of language.[42]

For the Ideologues in particular, the chemistry of Lavoisier offered both a model of the application of Condillac's method and a validation of that method's claims as a guide to scientific discovery. Garat, for example, wrote that Lavoisier had 'accomplished in chemistry a revolution so striking, so fecund in truths and happy prospects, by bringing to that science his genius, his fortune, and the principles and the language of Condillac'.[43] And his fellow Ideologue Antoine Destutt de Tracy challenged any savant who denied that knowledge of the human understanding had a 'necessary influence' upon the progress of science,

> to open the works of the illustrious and unhappy Lavoisier, and to see if each page of the writings of this celebrated man, forever deserving of our regrets, does not testify that it was the analysis of ideas which led him and his colleagues to make so much progress in the analysis of bodies.[44]

And since the Ideologues were looking not simply for a philosophical basis to support their social reforms but a rigorous scientific discipline in which to ground them[45], they were naturally led by Lavoisier's success to take Condillac as their principle methodological preceptor.[46] Pierre Cabanis' vision of Ideology as 'the science of man' and Destutt de Tracy's reference to it as 'a branch of zoology', give ample evidence of the scientific ambitions which the Ideologues had in this regard.[47] In addition, Cabanis' medical orientation and Destutt's zoological reference show the continuity of the Ideologues' philosophical programme with the biomedical researches of such contemporaries as Philippe Pinel and Xavier Bichat, who also hoped to emulate Lavoisier's success in their own fields using the method of Condillac.[48]

Now while these preliminary answers may go some way toward resolving the two questions facing us, we cannot regard them as completely adequate responses because they do not take sufficient account of the fact that the Ideologues were political activists as well as intellectuals. Most of the leading Ideologues were participants at some level in the various revolutionary governments formed during the 1790s, and their policies were put forward not as academic recommendations but as elements of factional politics. For this reason we must now consider in more detail the political position of the Ideologues during the period in question.

5. POLITICS AND THE IDEOLOGUES

As reforming intellectuals, most of the Ideologues actively supported the Revolution in its early stages, naively expecting the political events then unfolding to follow the rule of reason as they understood it. With the establishment of the Republic, the Jacobin government and the Terror, however, the Ideologues became disillusioned with revolutionary democracy and began formulating a programme to recapture the position of moral and intellectual leadership which they believed they were entitled to hold.[49] Their activities during the period of Thermidor and the Directory, including their collusion with Napoleon in his coup of the 18th Brumaire, must be viewed in the light of this struggle for ideological hegemony (in both senses of 'ideological').

In the period following the Terror, the chief problem facing the Ideologues in their attempt to gain moral and intellectual leadership in France was to repudiate the excesses of the Jacobins without abandoning the gains of the revolution which met with their approval. With the ending of the Terror and the return of many *émigrés*, the Royalist party became a threat to even moderate republicanism; and the Royalist press began an energetic campaign to exploit public revulsion against the Terror by accusing the Ideologues of being responsible for it, both through their political activities and through their espousal of Enlightenment philosophy.[50]

The Ideologues' tactic in the face of this accusation was to portray the Jacobins as anti-intellectual vandals — a characterization which soon reached mythical proportions[51] — and to emphasize the fact that several eminent philosophers and scientists had died during the Terror. Here is one typical example, taken from Cabanis' *éloge* of the anatomist Félix Vicq-d'Azyr. In order to support his claim that fear of persecution by the Jacobins was responsible for Vicq-d'Azyr's death in 1794, Cabanis writes:

> He had seen his most illustrious friends, Lavoisier, Bailly and Condorcet, perish under an almost general proscription of talents and virtues — Lavoisier, one of the principal authors of perhaps the most important revolution which the natural sciences had ever experienced; Bailly, equally distinguished as a scientist [*savant*], a philosopher, a writer and a sincere friend of liberty; and Condorcet, whom the judgement of Voltaire had long placed at the head of philosophy . . . .[52]

This roll-call of scientific and philosophical martyrs, frequently and ritually invoked, functioned as a rhetorical refutation of the Royalist accusation that the Ideologues had been responsible for the Terror. Rather than being an embodiment of philosophic reason, the Terror had persecuted reason in an outburst of passion and madness quite comparable with the cases of madness which the Ideologue-physician Pinel treated at the Asylum of Bicêtre. Indeed, it was in his famous *Medico-Philosophical Treatise on Mental Alienation* of 1801 that Pinel included the following political parable in the form of a case history taken from the period of the Terror.

> At a period of the revolution, which it is to be wished could be effaced from the annals of our history, a case of mania without delirium, gave rise to an extraordinary scene at the Asylum de Bicêtre. The Brigands, after the massacre of the prisons, broke like madmen into the above hospital under pretence of emancipating certain victims of the old tyranny, whom it had endeavoured to confound with the maniacal residents of that house. They proceeded in arms from cell to cell, interrogating the prisoners, and passing such of them as were manifestly insane. A maniac, bound in chains, arrested their attention by the most bitter complaints which he preferred with apparent justice

and rationality. "Is it not shameful," said he, "that I should be bound in chains and confounded with madmen?" He defied them to accuse him of any act of impropriety or extravagance. "It is an instance of the most flagrant injustice." He conjured the strangers to put an end to such oppression, and to become his liberators. His complaints excited, amongst the armed mob, loud murmurs and imprecations against the governor of the hospital. They immediately sent for that gentleman, and, with their sabres at his breast, demanded an explanation of his conduct. When he attempted to justify himself, they imposed silence upon him. To no purpose did he adduce, from his own experience, similar instances of maniacs, who were free from delirium, but at the same time, extremely dangerous from their outrageous passions. They answered him only with abuse; and had it not been for the courage of his wife, who protected him with her own person, he would have been sacrificed to their fury. They commanded him to release the maniac, whom they led in triumph, with reiterated shouts of Vive la République. The sight of so many armed men, their loud and confused shouts, and their faces flushed with wine, roused the madman's fury. He seized, with a vigorous grasp, the sabre of his next neighbour, brandished it about with great violence, and wounded several of his liberators. Had he not been promptly mastered, he would have soon avenged the cause of outraged humanity. The savage mob then thought proper to lead him back to his cell, and, with shame and reluctance, yielded to the voice of justice and experience.[53]

The Ideologues' claim to represent the voice of justice and experience, to whose control the passions and the populace ought willingly to yield, rested on their insistence that Ideology was a rigorous science leading to a rational consensus on sociopolitical maters. The analysis of ideas, it was held, was the only basis for reaching agreement in this field and if properly carried out would infallibly produce its conclusions.[54] In this context the image of Lavoisier as a martyr of science and philosophy at the hands of the Jacobins, was added to the image of Lavoisier as a successful reformer of chemistry using the method of Condillac, to doubly determine the Ideologues' enthusiasm for Condillac's method. Not only had Lavoisier associated this method with the progress of science, but he had also given it, by his death, strongly anti-Jacobin connotations.

But why would not the figure of Condorcet have served the Ideologues' purposes just as well as that of Lavoisier? Condorcet also was martyred during the Terror. He also had been a scientific success, in mathematical fields rather than chemistry. And he also had advanced a philosophy of analysis, although one that differed from Condillac's in being more mathematically based. Why, then, could the Ideologues not have taken their contemporary Condorcet as their model rather than returning *via* Lavoisier to Condillac, the philosopher of an earlier generation?

The answer, I believe, lies in the linguistic basis of Condillac's approach as opposed to the mathematical basis of Condorcet's. As politicians, lawyers, literary men and philosophers — in short, intellectuals whose stock in trade was language — the Ideologues found in Condillac's linguistic philosophy a legitimization of their claims to moral and intellectual dominance. If the analysis and reform of language rather than mathematical technique was the key to scientific progress, then the Ideologues were the legitimate cultivators of the science of society. If, on the other hand, Condorcet's application of the mathematics of probability to social questions represented the way forward in social science, then only a few individuals such as Laplace, who had carried on Condorcet's statistical work, could make a claim to dominance in this area.

Therefore, in spite of the Ideologues' enthusiasm for Condorcet's *Outline of an Historical Tableau of the Progress of the Human Mind*, which they had printed at government expense and distributed throughout the country[55], his approach to scientific method was overshadowed in the Ideologues' eyes by the method of Condillac. Destutt de Tracy even disputed the claim that probability could be studied scientifically and specifically criticized the theories of Condorcet on this point.[56] Like the young Auguste Comte some years later, the Ideologues were able to embrace the doctrine of progress set forth in Condorcet's *Outline* while at the same time repudiating his concept of a statistical social science.[57] Although they presented Ideology as an elite, technocratic solution to the problems of French society, they were unwilling to hand over the methodological control of their own movement to an even more elite and more technocratic group of mathematicians.[58]

6. THE ORDER OF IDEAS

Lavoisier's chemistry textbook, according to its subtitle, set out the elements of that science 'in a new order and according to the modern discoveries'. Both the analytic form and the new empirical content of the chemical revolution were thereby signalled, without the relationship between the two being made explicit. For many Ideologues, however, it was the analytic order of Lavoisier's thought that had made his discoveries possible. Destutt de Tracy, for example, 'regarded Lavoisier as a great Ideologue who, by the concatenation of his ideas, had foreseen the majority of his discoveries before confirming them empirically'.[59]

Similarly, in the social sphere, a correct analysis of our ideas and a reform of our language should permit a deductive derivation of new truths about human behaviour and society. Thus Garat hailed Condillac's method as bringing the certitude of mathematics to moral and political discourse.[60] And Pierre Laromiguière, principal editor of the complete edition of Condillac's works published in 1798, claimed that the philosopher's goal at the end of his life had been to reform the language of metaphysics and morals so that 'those ideas which had seemed the most inaccessible to the human mind would have sprung by themselves and without effort from the most common notions'.[61] It was with this view of the analytic method that Destutt de Tracy, in his writings on Ideology, set out 'to prove by means of facts what Locke and Condillac have shown us by means of reasoning — that morals and politics are susceptible of demonstration'.[62]

The reform of the national education system on Ideological principles, and the establishment of the Second Class of the new *Institut de France*, devoted to Moral and Political Sciences and dominated by the Ideologues, were to be the means by which the order of ideas could be translated into a new social order. With the Terror ended, it was no longer the impassioned mob and the dangerous lunatic of Pinel's case history that offered a model of the populace to the Ideologues, but rather the feral child of Aveyron — mute, squalid and unfeeling — whom the good doctor J.-M.-G. Itard sought to rescue from animality by a special educational programme based on the precepts of Condillac.[63] Contrary to the teachings of Jean-Jacques Rousseau, favourite philosopher of the Jacobins, natural man was not a noble savage but rather a dumb beast of the field — unless he was subjected to an analytically structured process of education.[64] Likewise the populace of France, now liberated from the brutalizing oppression of the *ancien régime*, stood in need of the Ideologues' educational system in order to realize their potential. Earlier in the Revolution, reformers had seen analysis as an intellectual equivalent to political liberty: 'Analysis is to the sciences, it is to teaching, what liberty is to political institutions; both make a man feel his worth and contribute to his perfection'.[65] Now, however, the Ideologues offered their analytic method not as a parallel to liberty but as the very condition of its existence.

Meanwhile, if Pinel's madman was no longer to play a role in the Ideologues' model of the populace at large, he could still serve as a model for the deviant, the individual whose ideas and passions are 'disordered'. For these individuals a 'moral' or educational treatment – based on the analytic order, both spatial and social, of the asylum – would provide a means of re-integration into the general social order.[66] Pinel described his method as follows: 'We must examine the different species of derangement analytically, consider them separately, and thence deduce the principles of both medical treatment and hospital police'.[67] The internal order of the asylum, or 'hospital police' in Pinel's words, required that the different types of madness thus distinguished analytically should also be rigorously segregated spatially. If, therefore, in the design of the hospital

its compartments have been so arranged as not to admit of insulating in a great measure different classes of patients, so as effectually to prevent their reciprocal communication; and, at the same time, to facilitate the labours of the governor, physician or other assistance officers, it is defective in a most important circumstance.[68]

Given this spatial segregation, it was then by means of a controlled social environment for each group, rather than by drugs or by physiological therapies (such as cold baths), that Pinel sought to restore his patients to society at large. In the treatment of 'mania', for example, he considered it 'a fundamental principle . . . to watch over the impetuosities of passion, and to order such arrangements of police and moral treatment as are favourable to that degree of excitement which experience approves as conducive to recovery' in that class of patient.[69] And since the line between insanity and criminality is a fine one, a similar regime of segregation and moral treatment was proposed by Ideologue legislators for the prison system.[70]

In the 'science of man' foreshadowed by Cabanis as a medically-based science of society, in Destutt de Tracy's demonstrative science of morals and politics, in Pinel's analytic of disordered and reordered minds, and in countless proposals for educational reform, the Ideologues offered a technocratic solution to the problems of French society as they understood them in the aftermath of the revolution.[71] In the wake of the Terror and their consequent disillusionment with revolutionary democracy, they championed the method of Condillac as a means of restoring order to a society which they felt was still beset by confusion. 'When we acquire ideas by the analytic method', Condillac had written,

they are arranged in an orderly way within the mind; there they retain the order which we have given them, and we can easily reproduce them for ourselves with the same clarity as that with which we acquired them. But if, instead of acquiring ideas by this method, we accumulate them at random, they will be in a state of great confusion and they will remain that way. . . . To speak intelligibly, we must conceive and present our ideas in the analytic order . . . .[72]

According to the programme of the Ideologues, French society — or at least its more educated classes[73] — would be taught to speak not only intelligibly but also unanimously, without dissent. For as the Second Class of the *Institut* suggested in 1797, were not differences of opinion and disputes in the moral and political sciences merely an effect of the present inexactitude of our language? And was it not the role of analysis to remedy this inexactitude?[74] The irenical vision of a perfect language, which had animated reformers since the seventeenth century, would thus become in the programme of the Ideologues a projection of a monolithic, linguistically-based social order to be constructed by the technocrats of the analytic method. The feasibility of realizing such a programme in practice was still a matter of vigorous discussion within the Second Class in 1803 when that body was overtaken by events. In that year Napoleon Bonaparte, whom the Ideologues had originally supported, abolished the Second Class of the *Institut* and imposed his own form of order upon French society, employing far more tangible means than the analysis of ideas.

Deprived of their political influence as Napoleon consolidated his power, some of the Ideologues, such as Cabanis, soon withdrew from public life altogether. Others stubbornly persisted with the theoretical development of Ideology: between 1801 and 1815, for example, Destutt de Tracy completed and published the five volumes of his *Elements of Ideology*, originally intended as a textbook for the *Ecoles Centrales* which Napoleon abolished. Still others, such as Pinel, trimmed their sails to the prevailing winds of the new Napoleonic regime and maintained their positions of public importance. In keeping with Napoleon's contemptuous use of the terms 'Ideology' and 'Ideologue' as pejorative epithets suggesting fatuous intellectualizing and 'visionary speculation in contrast to the realities of statecraft'[75], Pinel added to the second edition of his treatise on insanity a new case history which could serve as an ironic comment on the aspirations and melancholy fortunes of the Ideologues' movement.

A strong capacity for application and a profound ability for drawing connections seem constantly to be the lot of the melancholic, and they would assuredly bring to his judgements a rare character of exactness and stability, if he were not unceasingly carried away by some moral affection that dominates him, that makes him one-sided and always apt to take things beyond their limits. A man with a cultivated mind and a happily gifted memory had made such an assiduous study of the works of Condillac, and had been so impressed by them, that he believed he had found therein the epitome and embryo of all the other sciences; and he claimed that it was necessary to burn, as useless, all other books on history, physics, chemistry and even mathematics. His imagination was more and more inflamed, and he believed himself to be the messenger of the Almighty, sent to propagate this doctrine everywhere upon the earth and to render it universal. Like all great men, he expected to be persecuted; and one day when someone threatened to have him sent to an institution for the insane, he seemed to rejoice in this — for him it was a subject of triumph: 'So much the better!' he said, smiling; 'Behold now how my enemies fear me and how they redouble their hateful and vindictive efforts in proportion as my principles are propagated over the terrestrial globe'.[76]

Shortly after Pinel's work was published, Destutt de Tracy unwittingly echoed Pinel's 'melancholic' in his correspondence with Thomas Jefferson, when he commented on the success he expected his ideas to have in America and the trouble they would bring him from the authorities in France. 'The destiny of truths', he wrote, 'is to spread around the world once they are known; and I can easily console myself that these will not return to Europe until after my death, for they would disturb my life'.[77]

7. CONCLUSION

The focus of this discussion has been the political dimensions of the Ideologues' enthusiasm for, and application of Condillac's method of analysis. In summary, I have argued that the intellectual prominence of Condillac's method in France during the decade following Thermidor was in large part a consequence of the political aspirations of the Ideologues. The theories of Condillac offered the Ideologues a linguistically-based method of analysis which:

(1) was associated with recent triumphs in the science of chemistry and concomitantly with the repudiation of Jacobin politics (that is, the more highly one valorized the chemistry of Lavoisier, the more one discredited the politicians who had executed him);

(2) promised to provide the foundations of a stable social order of moderate Republicanism under the guidance of an elite corps of educational technocrats; and

(3) placed Ideology at the pinnacle of the intellectual hierarchy, while relegating mathematics and the other sciences to subsidiary positions.

These characteristics of the method of Condillac made it a weapon in the hands of the Ideologues against both the Jacobin Left and the Catholic-Royalist Right, while at the same time legitimating their claim to intellectual leadership over the educated elite.

In highlighting these political considerations, however, I am not dismissing the intrinsic intellectual appeal of Condillac's method as irrelevant to the use which the Ideologues made of it. On the contrary, I would maintain that only a doctrine which had such intrinsic appeal (according to the general cognitive standards prevailing in France at the time) could have served the political ends of the Ideologues, since a doctrine without this appeal could hardly have supported their claim to intellectual leadership. But it is important to remember that Condillac's method of analysis was not the only available method doctrine which had intrinsic intellectual appeal — and it was by no means the most notable one in France during the decade leading up to Thermidor. Hence the pertinent historical question remains: why was the method of Condillac (rather than that of Condorcet, for example) given such a predominant place in French intellectual life during the post-Thermidor period? And this qustion, I suggest, is best answered by attending to the political position of the Ideologues during that particular period.

Finally, it may be worth noting that this discussion of the Ideologues illustrates, once again, the chief vices toward which intellectuals, viewed from a political perspective, appear to be drawn: namely, a tendency toward all-embracing (and ultimately totalitarian) theoretical systems, and an extreme naivety about the practical exercise of political power. These two vices constitute what might be called the darker aspect of the faith in reason; and it is perhaps fortunate that they are often found together, as in the case of the Ideologues, so that the worst political consequences of the second vice may at least mitigate those of the first.

ACKNOWLEDGEMENT

A preliminary version of this chapter was presented to the historical section of the Seventh International Congress of Logic, Methodology and Philosophy of Science, Salzburg, Austria, 11–16 July 1983.

Research undertaken for this study was supported, in part, by a grant from the American Philosophical Society (Penrose Fund), whose assistance is gratefully acknowledged. My thanks also go to the editors and referees of this volume for their helpful comments on earlier drafts of this chapter.

University of New South Wales, Australia

NOTES

[1] On Condillac's reputation and the reception of his *Logic*, see my 'Introduction' to E. de Condillac, *La Logique/Logic*, W. R. Albury (tr.), (New York, 1980), pp. 7–34.
[2] K. M. Baker, 'Un "éloge" officieux de Condorcet: sa notice historique et critique sur Condillac', *Revue de Synthèse* **LXXXVIII** [3me série], 1967, pp. 227–251.
[3] G. Baguenault de Puchesse, *Condillac: Sa Vie, Sa Philosophie, Son Influence* (Paris, 1910), pp. 23–24.
[4] C. H. Van Duzer, *Contribution of the Ideologues to French Revolutionary Thought* (Baltimore, 1935), p. 38.
[5] For interpretations of Lavoisier's use of Condillac, see, for example, C. C. Gillispie, *The Edge of Objectivity* (Princeton, 1960), Chapter 6; and M. P. Crosland, 'The Development of Chemistry in the Eighteenth Century', *Studies on Voltaire and the Eighteenth Century* **XXIV**, 1963, pp. 369–441 (on pp. 416–421).
[6] J. Knowlson, *Universal Language Schemes in England and France, 1600–1800* (Toronto, 1975), p. 166.
[7] C. H. Van Duzer, *op. cit.* (Note 4), p. 6.
[8] On the history and use of these terms, see B. W. Head, 'The Origin of "Idéologue" and "Idéologie"', *Studies on Voltaire and the Eighteenth Century* **CLXXXIII**, 1980, pp. 257–264.
[9] A. Destutt de Tracy, 'De la métaphysique de Kant', *Mémoires de l'Institut National des Sciences et Arts; Sciences Morales et Politiques* **IV** (1802), pp. 544–606 (on p. 549).
[10] C. H. Van Duzer, *op. cit.* (Note 4), pp. 6–7.
[11] E. de Condillac, *op. cit.* (Note 1), pp. 72/73 (parallel French/English text).
[12] See, for example, F. Papillon, *La Philosophie Moderne dans Ses Rapports avec la Développement des Sciences de la Nature* (Paris, 1876) **II**, pp. 55–56; F. Pelikán, 'La méthode de Condillac et Descartes', *Travaux du IXe Congrès International de Philosophie* (Paris, 1937), fasicule III, pp. 43–48; and E. Bréhier, *The History of Philosophy*, W. Baskin (tr.), (Chicago, 1963) **V**, pp. 76–77.
[13] E. de Condillac, *Cours d'Etudes. De l'Art de Penser*, in: *Oeuvres Philosophiques de Condillac*, G. Le Roy (ed.), (Paris, 1947–1951) **I**, pp. 715–776 (on p. 773a).
[14] E. de Condillac, *Les Monades*, L. L. Bongie (ed.), (Oxford, 1980); see especially part 6 of Bongie's introduction, pp. 89–107.
[15] E. de Condillac, *op. cit.* (Note 1), part II, Chapter 7, pp. 278/279–304/305.
[16] *Ibid.*, part II, Chapter 3, pp. 224/225–234/235.

[17] R. McRae, *The Problem of the Unity of the Sciences: Bacon to Kant* (Toronto, 1961), p. 95.

[18] E. de Condillac, *op. cit.* (Note 1), pp. 244/245; cf. pp. 302/303 and 332/333. With this thesis, first put forward by Condillac in his *Course of Studies* (1775), the conception of logic as an autonomous science of the laws of thought was replaced by one which subordinated the study of logic to the study of language. On this point, see S. Auroux, *L'Illuminismo Francese e la Tradizione Logica di Port-Royal* (Bologna, 1982), pp. 54—55.

[19] M. P. Crosland, *Historical Studies in the Language of Chemistry* (London, 1962), pp. 114—167.

[20] A. L. Lavoisier, L. B. Guyton de Morveau, C. L. Berthollet and A. F. Fourcroy, *Méthode de Nomenclature Chimique* (Paris, 1787).

[21] This matter will be dealt with in a monograph currently in preparation.

[22] See for example, H. Guerlac, 'Chemistry as a Branch of Physics: Laplace's Collaboration with Lavoisier', *Historical Studies in the Physical Sciences* **VII**, 1976, pp. 193—276; and M. P. Crosland, 'Chemistry and the Chemical Revolution', in: *The Ferment of Knowledge*, G. S. Rousseau and R. Porter (eds.), (Cambridge, 1980), pp. 389—416 (on p. 391).

[23] R. Siegfried and B. J. Dobbs, 'Composition, a Neglected Aspect of the Chemical Revolution', *Annals of Science* **XXIV**, 1968, pp. 275—293.

[24] A. L. Lavoisier, *Traité Elémentaire de Chimie, Présenté dans un Ordre Nouveau et d'après les Découvertes Modernes* (Paris, 1789); translated as *Elements of Chemistry, in a Systematic Order, Containing All the Modern Discoveries*, R. Kerr (tr.), (Edinburgh, 1790).

[25] A. L. Lavoisier, *op. cit.* (Note 24) **I**, p. xxv (1789); p. xxx (1790). Quotations from this work generally follow the Kerr translation, with occasional modifications for greater clarity.

[26] For eighteenth-century definitions of chemistry, see Jon Eklund, *The Incompleat Chymist, Being an Essay on the Eighteenth-Century Chemist in His Laboratory* . . . [Smithsonian Studies in History and Technology, No. 33] (Washington, 1975), pp. 2—3.

[27] E. de Condillac, *op. cit.* (Note 13) **I**, p. 731b. The statement also occurs in Condillac's *Essai sur l'Origine des Connoissances Humaines, ibid.*, pp. 1—118 (on p. 41a). Also, see E. de Condillac, *op. cit.* (Note 1), pp. 220/221.

[28] A. L. Lavoisier, *op. cit.* (Note 24) **I**, p. xvii (1789); p. xxiv (1790).

[29] *Ibid.* **I**, p. xx (1789); p. xxvi (1790).

[30] *Ibid.*

[31] *Ibid.* **I**, p. xxi (1789); p. xxvii (1790).

[32] This point will be elaborated in my forthcoming study (Note 21).

[33] A. L. Lavoisier, *op. cit.* (Note 24) **I**, pp. x—xi (1789); p. xviii (1790).

[34] A. L. Lavoisier, 'Mémoire sur la fermentation spiritueuse', in: *Oeuvres de Lavoisier*, J. B. Dumas (ed.), (Paris, 1864—1893) **III**, pp. 777—778. Although undated, the *mémoire* can be shown from internal evidence to have been drafted after 1782.

[35] The material in parentheses gives the equation in modern notation. The role of water in such reactions was not understood in Lavoisier's time.

[36] J.-B. Delambre, 'Notice sur la vie et les ouvrages de M. le Comte J.-L. Lagrange', in: *Oeuvres de Lagrange*, J.-B. Delambre (éd.), (Paris, 1867—1892), **I**, p. xxxviii.

[37] C. H. Van Duzer, *op. cit.* (Note 4), pp. 84, 94.
[38] *Ibid.*, p. 15; cf. J. Knowlson, *op. cit.* (Note 6), Chapter 7.
[39] See C. H. Van Duzer, *op. cit.* (Note 4), Chapter 3, for the educational programme of the Ideologues.
[40] D. J. Garat, 'Analyse de l'entendement — programme et première leçon', in: *Séances des Ecoles Normales Recueilles par des Sténographes* (Paris, 1800) **I**, pp. 138—169 (on p. 166).
[41] P. Dupuy, 'L'Ecole normale de l'an III', in: *La Centenaire de l'Ecole Normale, 1795—1895* (Paris, 1895), pp. 1—209 (on p. 119).
[42] J. Knowlson, *op. cit.* (Note 6), p. 173.
[43] D. J. Garat, *Mémoires Historiques sur le XVIIIe Siècle et sur M. Suard* (Paris, 1821) **II**, p. 326; cf. *ibid.*, p. 44.
[44] A. Destutt de Tracy, 'Mémoire sur la faculté de penser', *Mémoires de l'Institut National des Sciences et Arts; Sciences Morales et Politiques* **I** (1796), pp. 283—450 (on p. 287).
[45] F. Picavet, *Les Idéologues* (Paris, 1891), pp. 299, 303; E. Kennedy, *A Philosophe in the Age of Revolution: Destutt de Tracy and the Origins of "Ideology"* (Philadelphia, 1978), p. 40; and M. S. Staum, *Cabanis: Enlightenment and Medical Philosophy in the French Revolution* (Princeton, 1980), pp. 3-4.
[46]On the Ideologues' references to the example of Lavoisier's chemistry, see F. Picavet, *op. cit.* (Note 45), pp. 161—162; C. H. Van Duzer, *op. cit.* (Note 4), p. 126; E. Kennedy, *op. cit.* (Note 45), pp. 31, 37; and M. S. Staum, *op. cit.* (Note 45), pp. 156—157.
[47] 'Physiology, the analysis of ideas and the study of morals are merely the three branches of the same unique science, which can appropriately be called *the science of man*'. P. J. G. Cabanis, *Rapports du Physique et du Moral de l'Homme*, in: *Oeuvres Philosophiques de Cabanis*, C. Lehec and J. Cazeneuve (éd.), (Paris, 1956) **I**, pp. 105—631 (on p. 126). 'One has but an incomplete knowledge of an animal if one does not know its intellectual faculties. Ideology is a branch of zoology, and it is particularly in man that this branch is important and deserves to be examined thoroughly'. A. Destutt de Tracy, *Elémens d'Idéologie* (Paris, 1801—1815) **I**, p. xiii.
[48] On Bichat and Pinel, in relation to Condillac, see P. Lain Entralgo, 'Sensualism and Vitalism in Bichat's *Anatomie Générale*', *Journal of the History of Medicine* **III**, 1948, pp. 47—64; and W. Riese, 'La méthode analytique de Condillac et ses rapports avec l'oeuvre de Philippe Pinel', *Revue Philosophique* **CLVIII**, 1968, pp. 321—336. See also S. Moravia, 'Philosophie et médecine en France à la fin du XVIIIe siècle', *Studies on Voltaire and the Eighteenth Century* **LXXXIX**, 1972, pp. 1089—1151; and my forthcoming study (Note 21).
[49] S. Moravia, 'La Société d'Auteuil et la révolution', *Dix-huitième Siècle* **VI**, 1974, pp. 181—192 (on pp. 190—191).
[50] F. Picavet, *op. cit.* (Note 45), p. 89; see also the newspaper comments quoted in B. W. Head, *op. cit.* (Note 8), p. 257. For a general study of Royalist newspapers during this period, see J. D. Popkin, *The Right-Wing Press in France, 1792—1800* (Chapel Hill, 1980).
[51] R. Hahn, *The Anatomy of a Scientific Institution: The Paris Academy of Sciences, 1666—1803* (Berkeley, 1971), pp. 290—295; cf. M. S. Staum, *op. cit.* (Note 45), p. 148.

[52] P. J. G. Cabanis, 'Eloge de Vicq-d'Azyr', in: P. J. G. Cabanis, *op. cit.* (Note 47) **II**, pp. 368—386 (on p. 385). Although this *éloge* dates from around 1805, it accurately reflects the post-thermidorian rhetoric of the Ideologues.

[53] P. Pinel, *Traité Médico-Philosophique sur l'Aliénation Mentale, ou la Manie* (Paris, 1801), pp. 153—155; translation taken from the English edition, *A Treatise on Insanity*, D. D. Davis (tr.), (Sheffield, 1806), pp. 154—155.

[54] See for example, F. Picavet, *op. cit.* (Note 45), pp. 299, 303; and E. Kennedy, *op. cit.* (Note 45), pp. 39—40, 337.

[55] C. H. Van Duzer, *op. cit.* (Note 4), p. 99; K. M. Baker, *Condorcet: From Natural Philosophy to Social Mathematics* (Chicago, 1975), p. 343.

[56] F. Picavet, *op. cit.* (Note 45), pp. 300 and 374—376.

[57] K. M. Baker, *op. cit.* (Note 55), p. 378.

[58] In spite of their rejection of a mathematized social science, the attitude of the Ideologues toward mathematics and the mathematical sciences generally was one of respect and had nothing in common with the 'democratic' or 'populist' antagonism toward these areas of study which C. C. Gillispie has proposed as the characteristic feature of 'the Jacobin philosophy of science'. For the Ideologues, however, mathematics constituted only a limited, special application of the method of analysis, while Ideology dealt with analysis, and therefore with human thought, at the highest level of generality. Thus were their claims to intellectual leadership defended against the mathematicians. For Gillispie's thesis, see Gillispie, 'The *Encyclopédie* and the Jacobin Philosophy of Science: A Study in Ideas and Consequences', in: *Critical Problems in the History of Science*, M. Clagett (ed.), (Madison, 1959), pp. 255—289. For critiques of this thesis see the commentaries of Henry Bertram Hill and Henry Guerlac, *ibid.*, pp. 309—320; and, more recently, H. Moss, 'Scientists and Sans-culottes: The Spread of Scientific Literacy in the Revolutionary Year II', *Fundamenta Scientiae* **IV**, 1983, pp. 101—115.

[59] S. N. Destutt de Tracy, 'Notice sur M. Destutt de Tracy', in: *Essais Divers, Lettres et Pensées* (Paris, 1852—1855) **I**, pp. 305—404 (on p. 314); quoted in S. Moravia, *Il Pensiero degli Idéologues: Scienza e Filosofia in Francia* (*1780—1815*) (Firenze, 1974), p. 321.

[60] D. J. Garat, *Corps Législatif: Conseil des Anciens: Discours Prononcé par le C. Garat en Offrant les Oeuvres de Condillac à la Séance du 3 Fructidor An VI* (Paris, 1798), p. 3; quoted in J. Knowlson, *op. cit.* (Note 6), p. 264.

[61] P. Laromiguière *et al.*, 'Note des éditeurs', in: E. de Condillac, *op. cit.* (Note 13) **II**, p. 529; quoted in J. Knowlson, *op. cit.* (Note 6), p. 174. In defending Condillac's theory that analysis leads to new truths by a quasi-algebraic process of deduction, Laromiguière appealed, like Destutt de Tracy, to Lavoisier's discoveries. See P. Laromiguière, *Les Paradoxes de Condillac* (Paris, 1805), p. 60.

[62] MS note of Destutt de Tracy, quoted in F. Picavet, *op. cit.* (Note 45), p. 303.

[63] M. Lyons, *France Under the Directory* (Cambridge, 1975), p. 121; H. Lane, *The Wild Boy of Aveyron* (London, 1977), p. 92; and J.-M.-G. Itard, *The Wild Boy of Aveyron*, G. and M. Humphrey (tr.), (New York, 1962), *passim*, especially pp. 49 and 55.

[64] Ironically, although Itard set out to vindicate the theories of Condillac in his treatment of the 'wild boy' of Aveyron, his final report on the boy's progress, sent to the Minister of the Interior in November of 1806, gave prominence to a conception of imitation which seems to owe more to Rousseau than to Condillac. See D. E. Leary,

'Nature, Art and Imitation: The Wild Boy of Aveyron as a Pivotal Case in the History of Psychology', in: *Studies in Eighteenth-Century Culture*, O. M. Brack, Jr. (ed.), (Madison, 1984) **XIII**, pp. 155–172.

[65] *Rapport et Projet de Décret sur la Composition des Livres Elémentaires Destinée à l'Instruction Publique* (Paris, n.d.), quoted in G. Boas, *French Philosophies of the Romantic Period* (Baltimore, 1925), p. 5.

[66] M. Foucault, *Madness and Civilization: A History of Insanity in the Age of Reason*, R. Howard (tr.), (New York, 1965), Chapter 9; and J. Postel, *Genèse de la Psychiatrie: Les Premiers Ecrits de Philippe Pinel* (Paris, 1981), pp. 258–260.

[67] P. Pinel, *op. cit.* (Note 53), p. 134 (1806).

[68] *Ibid.*, p. 174.

[69] *Ibid.*, p. 99. As Foucault points out, physiological measures such as sudden cold showers were still employed in Pinel's asylum, but they were now used as punishments for misbehaviour rather than as therapeutic agents. M. Foucault, *op. cit.* (Note 66), pp. 266–268.

[70] M. Foucault, *Discipline and Punish: The Birth of the Prison*, A. Sheridan (tr.), (London, 1977), part II.

[71] Cf. M. S. Staum, 'Medical Components in Cabanis's Science of Man', *Studies in History of Biology* **II**, 1978, pp. 1–31 (on p. 25); and B. W. Head, 'The Origins of "La Science Sociale" in France, 1770–1800', *Australian Journal of French Studies* **XIX**, 1982, pp. 115–132 (on pp. 127–129).

[72] E. de Condillac, *op. cit.* (Note 1), pp. 78/79.

[73] As Higonnet points out, the post-Thermidorian educational policy of the Ideologues concentrated on the development of the elite *Ecoles Centrales* at the expense of primary education. Destutt de Tracy, for example, wrote that he considered it 'a great error for the primary schools to be connected with the *Ecoles Centrales* and to be the gateway to them. . . . There must be two complete systems of instruction which have nothing in common with one another'. P. L.-R. Higonnet, 'The Politics of Linguistic Terrorism and Grammatical Hegemony during the French Revolution', *Social History* **V**, 1980, pp. 41–69 (Destutt de Tracy quoted on p. 63).

[74] J. Knowlson, *op. cit.* (Note 6), pp. 176–177; cf. H. B. Acton, 'The Philosophy of Language in Revolutionary France', *Proceedings of the British Academy* **XLV**, 1959, pp. 199–219 (on pp. 206–207).

[75] E. Kennedy, *op. cit.* (Note 45), p. 80.

[76] P. Pinel, *Traité Médico-Philosophique sur l'Aliénation Mentale* (2e éd., Paris, 1809), p. 94.

[77] Destutt de Tracy to Jefferson, 21 October 1811, quoted in E. Kennedy, *op. cit.* (Note 45), p. 212.

DAVID PHILIP MILLER

METHOD AND THE 'MICROPOLITICS' OF SCIENCE: THE EARLY YEARS OF THE GEOLOGICAL AND ASTRONOMICAL SOCIETIES OF LONDON

1. INTRODUCTION: THE SOCIAL USES OF SCIENTIFIC METHOD

Method discourse occurs on a variety of levels. The most obvious, and most intensively studied, are statements presented as definitive discourses on method by practising scientists or philosophers. Insofar as the study of the history of scientific method has been seen as primarily a philosophical enterprise, past discourses on method by the great philosophers have been studied with a view to first establishing the 'true' meaning attached to them by their advocates, and second, with a view to showing in what ways that meaning had influence or significance in the evolution of a true or satisfactory scientific method. More recently such approaches have been criticized as Whiggish and positivistic. This criticism points out that to confine oneself to the great philosophers is to miss a considerable body of significant work on scientific method by practising scientists. In addition, the context of such past investigations has been misunderstood – prior to the twentieth century the philosophy of scientific method was closely connected with the actual practice of science. In particular, as Laudan has emphasized, statements about method cannot be understood if they are separated from the scientific context in which they were made. And very often discourses on method were stimulated and shaped by quite particularistic concerns with specific scientific theories.[1]

This emphasis on the historical context of method discourse is to be welcomed much as the Koyréan revolution in historiography of science was. But, like Koyréan historiography, it remains unabashedly intellectualist and can be criticised from the vantage point of post-Kuhnian historiography and sociology of scientific knowledge. The latter is also concerned to dispel the idea of a universal scientific method. Thus Barry Barnes has stated:

> Belief in the real existence of a universal 'scientific method' is the product of constant idealization; it cannot be sustained in the face of concrete accounts of the diversity of science. . . scientists themselves do not possess any shared single set of conventions, whether for procedure or evaluation.[2]

J. A. Schuster and R. R. Yeo (eds.), The Politics and Rhetoric of Scientific Method, 227–257.
© 1986 *by D. Reidel Publishing Company.*

At first sight this approach need not necessarily clash with a contextualist historical approach depending upon how 'contextualist' is interpreted. But it does clash in the case of Laudan because he defines the context of scientific practice as the body of scientific beliefs alone. Barnes, on the other hand, thinks it better to approach the question thus:

> . . . to regard ideas as tools with which social groups may seek to achieve their purposes in particular situations. . . . Beliefs which 'work' in one situation may be quite inappropriate in another. The connection between interests and ideas is contextually mediated.[3]

If we substitute 'ideas about/statements on scientific method' for the word 'ideas' in the above quotation, then an entirely different view of the potential significance of method discourse emerges. Method discourses then become 'tools with which social groups may seek to achieve their purposes in particular situations'. Note that no prior judgment is made here of what such purposes might be. These purposes could be quite specifically theoretical. The papers by Chalmers and Schuster in this volume question the claim that a genetic relationship, of the kind which formal method discourses literally claim for themselves, exists between the generation and use of such discourses and the detailed production and evaluation of theories. This is not to deny that method discourse – like any other discourse – can be partly constitutive of the making, or unmaking, of scientific knowledge. Though I find these arguments persuasive it is not my intention to tackle such questions in this paper. My concern is to make a plausible case that statements on method can be used to further the 'professional vested interests' or other social interests of particular groups within a scientific community or of a body of scientists within society at large. However, if one adopts a broadly 'conventionalist' view of the generation and establishment of knowledge claims, then analytically distinct uses of method discourse in 'intellectual' and 'social' contexts necessarily inosculate.[4]

The broader social uses of method discourse – one might say the public relations uses of method – have recently been explored in the case of the early Royal Society and also in mid-nineteenth Britain with particular reference to the British Association.[5] These studies of method discourse in connection with scientific institutions have tended to follow Roger Hahn's notion that 'the scientific institution is the anvil on which the often conflicting values of science and society are shaped into a viable form'.[6] They consider science as a social institution faced with the perennial task of establishing its identity and negotiating its status within the wider society. At this level of analysis, method discourse can be treated as part of a broad ideology of science, carefully tailored to mandate a particular role for science, and scientists, in society. But what is often missed as a result of such a broad treatment is the political importance of method discourse within scientific communities.[7] I will be concerned with an intermediate level of analysis – with science as a group activity. Method discourse is shaped not only by the cognitive development and situation of the individual and by the requirements of a broad consensual ideology of science but also by the 'micropolitics' of science. For example, it has been recognized for some time that methodological issues were a vital part of the debates on the wave theory of light in Britain in the 1830s and 1840s, but it now seems probable that those issues were simultaneously part of a fundamentally political struggle involved in defining the role and territory of the mathematical physicist.[8] To generalize, method discourse can be employed by actors in the political task of defining 'best' scientific practice in their own image. A great deal of what I have to say below concerns not only methodological statements *per se* (talk of fact, theory, induction, etc.) but also talk of skills, technique and competing research programmes. In this my account reflects the creative juxtaposition of these repertoires by actors. It is precisely in the creative manipulation of these repertoires and 'slippage' from one to the other, that variation of meaning results from superficially uniform methodological statements. The re-unification of such statements with the penumbra of method discourse is essential to an appreciation of its micropolitical role. Attention to this intermediate level of analysis is important to a fully integrated study of the multiple, interdependent uses of method discourse which should ultimately bring together its fine-grained relations, if any, with theory production and evaluation, its uses in small group politics and its broader ideological uses.[9]

The focus of my study will be two specialist scientific societies founded in London in the early nineteenth century – the Geological Society (f. 1807) and the Astronomical Society (f. 1820). Both institutions were publicly identified in their early years with broadly 'inductive' methodological prescriptions for work in their respective sciences.

Yet in both cases this identification was subsequently weakened, not to say repudiated. I ask how and why did this happen?

2. THE GEOLOGICAL SOCIETY OF LONDON

The Geological Society of London is among the most intensively studied of early nineteenth-century British scientific institutions. It is widely apprehended that the early years of the Society coincided with a period in which geology was finally 'made' as a scientific discipline. Nor, for most historians, was this a coincidence, since the institutionalization of a field of inquiry is usually regarded as a vital aspect of its definition and recognition. The official methodological stance associated with the Society in its early years has also been accorded much significance. What we might term the conventional wisdom runs thus: the Society was early and fervently identified with a radically empirical, 'Baconian', methodological stance. This was a reaction against the rampantly speculative theorizing about earth history characteristic of the later eighteenth century, which had exploded into a vituperative and, it was believed, unproductive struggle between the disciples of Werner and those of Hutton. The founders of the Geological Society looked to co-operative fact gathering, co-ordinated from the metropolis, as the only way to provide the new science with a firm basis. And only in the very distant future, if ever, would a sound theory of the earth be realizable. But by the 1820s and 1830s the Geological Society was dominated by men such as Adam Sedgwick, Roderick Murchison, William Buckland, W. D. Conybeare and Charles Lyell. They, in varying degrees, took a more 'enlightened' view of the role of theory in both their practice and in their public pronouncements.[10]

Judgments of the propriety of this initial and changing methodological stance have been implicit or explicit in most accounts. Martin Rudwick's early study implicitly accepted an initial radical empirical emphasis and its subsequent gradual desuetude as perfectly understandable, natural and functional. More recently Rachel Laudan has argued forcefully that the radical empiricism of the Society during its early years was both unnecessary and damaging to the progress of the science of geology in England.[11] The claim here is that more sophisticated methodologies were available to the founders of the Society – methodologies which showed their fruitfulness subsequently in the work of Charles Lyell. But these were not taken advantage of either because of ignorance or because collateral concerns with recruitment to the Society and with its public reputation 'intruded'. Thus Laudan claims that the Society lay for many years outside the mainstream of progressive geology and, insofar as useful contributions were made under its auspices, they were made in spite of, rather than because of, the Society's official methodological stance.

Laudan's general argument is that we should not assume that the setting up of formal institutions is always and inevitably to the benefit of science. We must assess the heuristic intellectual value of any methodological approach sanctioned by an institution, and that value may in certain cases be (and in the case of the Geological Society certainly was) negative. She acknowledges that there were social functions served by the Society's radical empiricist methodology. These included a public relations exercise in distancing the Society from what was perceived as rampantly speculative geology, and the exigencies of recruiting far and wide to establish the Society's numerical strength and support. An extreme inductivism was most valuable for these purposes because in Bacon's terms it presented the view that proper method could level men's wits and enable all who were interested to contribute. Laudan draws a sharp line between heuristic intellectual and social utilities of the Society's early methodological stance. The leaders of the Society failed to act 'rationally' in their choice of methodology. Their irrational act was caused by intruding social goals. There is an asymmetry (or dualism) in this explanation. A 'rational' methodological stance is treated as virtually self-explanatory, an irrational one is caused by social factors diverting the path of reason.[12]

It would be futile to pursue a judgmental counterargument involving some assessment of whether the achievements of the 1820s and 1830s would have been possible without a strong Geological Society and hence without the social utilities of empiricist methodology.[13] The task now is to *understand* rather than to judge the multiple and changing uses of method discourse associated with the Geological Society and its Fellows from its foundation into the 1830s. In so doing we will seek to establish the notion that method discourses are 'tools with which social groups may seek to achieve their purposes in particular situations' and provide a symmetrical form of explanation for their deployment.

First an important question must be answered: in what senses can a methodological stance be attributed to, or adopted by, a scientific society? Perhaps the least problematic forms of evidence for such an attribution are public statements made in the name of the Society. Such material might include general statements of purpose issued publicly at its foundation or on the occasion of Presidential addresses to anniversary meetings. Although there are obvious problems in identifying such pronouncements with the actual practice of members of the Society in question, they do at least have some stamp of official policy about them. A second form of evidence comprises public statements made by members of the Society but not readily describable as part of official policy. These might include reviews of the Society's *Transactions* or of other geological works and those geological works themselves. Potentially this category would include a great deal of published literature and published discussion of that literature. Clearly the historian has a major interpretive task here in deciding whether a given piece of material of this kind can legitimately be taken as indicating something about the Society. A third category of evidence is based on deduction from what the Society does. Thus we might infer a methodological stance from such things as the choice of papers to be published in the *Transactions* (and, if accessible, the reasons for those choices), the kinds of individuals who are admitted to the Fellowship and those who are not, those who hold offices within the Society or are awarded medals, and so on. Thus it will be granted that there are problems of attribution.

Great care must also be taken to avoid an animistic approach to institutions. Institutions do not have methodological stances, people do. It may be that 'official' policy is widely accepted and supported, or it may be acquiesced in, or it may be opposed by various individuals and sub-groups. Once again these matters should not be prejudged or elided, but must be assessed seriously in particular contexts.

There is yet another dimension of complexity which relates to the question of 'audience'.[14] If method discourse occurs in particular contexts for specific purposes, then a sense of the intended audience is vital to an understanding of what is said. Thus a disquisition in, say, the *Edinburgh Review* assessing a volume of *Transactions* of the Geological Society would be intended for a wide readership and as such anything said there about method could have its eye very much upon the public image of the science of geology in the wider scientific community and amongst the reading public. A Presidential Address delivered only to the Society, or a paper in its journal would probably have a very different status in being directed primarily at an 'in-group'. The significance and functions of any statements on method contained in it would have to be interpreted accordingly. A lack of attention to such interpretive problems has been a major weakness of previous accounts of the Geological Society and its methodological stance.

A variety of statements support the view that the Geological Society very early became associated with a radical empirical stance. In particular the *Geological Inquiries* issued by the Society in 1808 as a blueprint for its cooperative research programme is taken as being suffused with, and predicated upon, such a stance. This pamphlet defined geology as 'knowledge of the system of our earth' and as a 'sublime and difficult' science. But its difficulties could best be overcome, it was argued, by cooperative effort in which all manner of men including miners, surveyors, engineers and observant travellers could contribute observations. If such men were guided by the Society's *Inquiries* a coherent body of geological data could be gathered together. In this way the progress of geology could be assured.[15]

The *Inquiries* was produced at a time (1808) when solidarity amongst the various groups in the Society was important to the fledgling institution, and factional disputes were to be held in abeyance. Also recruitment of members was of paramount importance. Thus a 'Baconian' stance of a radical empirical kind, which glossed over distinctions later to become important, was quite appropriate. The *Inquiries* could be read as relatively non-committal about drawing distinctions between the crude empiricism of local fact gathering and 'higher' forms of geological enquiry. To emphasize such distinctions too much would have been to alienate a major potential source of support for the Society. In this respect the *Inquiries* asserted the 'democracy of the intellect'. Also apparent is a creative confusion between a Baconianism which was both inductive and utilitarian, and one which was merely inductive. In this way what was to become a major point at issue subsequently was avoided at this early stage.

We might deduce that this cooperative research programme was radically empirical and anti-theoretical. But the *Inquiries* can be read another way. Indeed in discussing the division of labour possible in geology the pamphlet states that geology is 'susceptible of division into many different departments, several of which are capable of being extended by mere observation'. But the implied corollary of this is, of course, that there were several departments which were not capable of being extended by mere observation. The Preface to the first volume of the Society's *Transactions*, published in 1811, confirms the view that its founders may have been more circumspect in their methodological stance than previous commentators have allowed. It reiterated the value of cooperative fact-gathering and placing 'the collection of facts in a repository', but the explicit reference to theory is couched in very careful terms:

> In the present state of this science, it cannot be supposed that the Society should attempt to decide upon the merits of the different theories of the earth that have been proposed. In the communications, therefore, which are now submitted to the public, every latitude has been allowed to authors, with regard to their theoretical inferences from the observations which they record; it being understood, according to the rule of Literary and Philosophical Societies in general, that the writers alone are responsible for the facts and opinions, which their papers may contain. . .[16]

This, I would suggest, is in no sense rigidly anti-theoretical as it stands. Contributors are to be granted latitude in the expression of theoretical positions. And the Society's reluctance to endorse one theory or another is couched in the same terms as those used by the Royal Society, for example, in expressing its unwillingness to pronounce publicly on the merits of papers.[17]

When we examine the contents of the first volume of *Transactions* the impression of a degree of flexibility is confirmed. To be sure, Leonard Horner took a fairly hard rhetorical line against theory in his paper on the Malvern Hills.[18] But we must not generalize from one instance. Henry Holland and J. F. Berger engaged in theorizing without the kind of apologia or strict qualification offered by Horner.[19] Thus in neither 'official' policy statements nor in the *Transactions* themselves was a consistent, or unambiguous radical empiricism advocated. The most extreme statements, significantly, are to be found in the 'Reviews'. And it is these which have been most effectively used to attribute extreme inductivism to the Society.[20] It is in this context that we find eighteenth-century theories of the earth described as a 'species of mental derangement'; it is here that William Fitton makes a positive virtue of the lack of training of many members of the Geological Society; and so on.[21] Yet, as we have noted, the context of use here is very different given the wider audience.

So what were the uses of radical empiricism? Why did it appear at its most strident in the more open public forums? Why was empiricism more restrained, but still very evident, within the circle of members of the Geological Society? How was method discourse put to use, and how did its uses change over time? To answer these questions we have to examine the circumstances of the Society's foundation and early years, paying particular attention to what Rudwick has usefully called the 'socio-cognitive topography' of the Society and its members.[22]

At its foundation and for a number of years thereafter, the Geological Society had a variety of problems in establishing itself. There was first of all the problem of establishing an independent Society in the face of opposition from Sir Joseph Banks and from the Banksian regime more generally. Although this was certainly in part a jurisdictional dispute in which Banks and his allies sought to preserve the Royal Society's claim to the promotion of all natural knowledge, it was also much more than this. Not only the constitution of the new Society but also the programme it would pursue were at issue. And there were also broader cultural dimensions to the conflict. Dissenters and others who were reacting in 'idealistic frustration at, and rejection of, corrupt late eighteenth century aristocratic polite society' were well represented among those founders of the Society who pursued the case for independence.[23]

Even when the question of the Society's independence was settled there remained divergent interests within its ranks. The most notable was the divergence between those who, with George Greenough, pursued 'geology' as the search for knowledge of the Earth on the large scale, and a significant group whose interests were simultaneously more narrowly mineralogical and more broadly utilitarian. Also, as stratigraphy grew in importance, a distinctive Wernerian stratigraphy based on mineralogical techniques and concentrating upon the Primary rocks gained ascendency. But this was challenged in the late 1810s and 1820s by a Cuvierian stratigraphy which made extensive use of paleontology and focussed upon Secondary and Tertiary formations. Clearly, to see the Geological Society as having simply a public image problem and a jurisdictional one is to miss the varied socio-cognitive interests involved with it.

The early Geological Society incorporated individuals and groups with a variety of skills, expertise and scientific programmes. In many ways this variety produced complementary work, but it also produced competing claims about such crucial matters as the nature of geology, the importance of particular types of research, and the ideal 'equipment' of the geologist. To appreciate those varied interests can take us a long way in understanding the manner in which, and the contexts in which, methodological empiricism was advocated. Without pretending to a comprehensive account of the early years of the Geological Society, we will examine three episodes in turn: first the relation between mineralogists and self-styled geologists; second an episode of 'class conflict' regarding stratigraphy; and third, tensions between proponents and practitioners of Wernerian (or mineralogical) stratigraphy and of Cuvierian (or paleontological) stratigraphy.

Paul Weindling has been largely responsible for drawing attention to our first episode in arguing that a distinctive 'mineral history' programme was important in the years immediately preceding and following the foundation of the Geological Society.[24] This programme represented one idea of the form the Society should take and of the desirable range of its activities. It contrasted with the polite, gentlemanly, non-utilitarian geology which has typically been identified with the Society in that its supporters emphasized practical, economically useful mineralogy. This was a programme in which mineralogical collectors with economic interests in the science (Charles Greville and Sir John St. Aubyn being two examples), mineralogists and chemists such as those who made up the earlier British Mineralogical Society (Arthur Aikin, William Babington, William Phillips among others), and practical surveyors (William Smith, John Farey) could cooperate. The programme was not realized, and its demise was an accomplishment of those who followed, or shared, Greenough's distinctly non-utilitarian line. In addition, whereas the mineral historians argued that mineralogy, and mineral chemistry in particular, were *the* basis of geological investigation, their opponents criticized the narrowness of interests of mineralogists and argued that geology was an independent science with its own distinct methods and concerns, notably stratigraphy *tout court.*

In reality, however, the battle lines were neither clearly nor consistently drawn. Many mineralogists extended their interests beyond the laboratory science of substance analysis and 'sought to understand the distinctions between particular rocks and minerals in terms of their different situation in the order of the strata'.[25] They engaged in fieldwork and were understandably attracted to areas with a complex lithology in which their particular expertise could be used to most advantage.[26]

Nevertheless, at the ideological level considerable sniping continued on both sides. As spokesmen for the mineralogists and chemists, Arthur Aikin and William Phillips reasserted the case for mineral chemistry as the foundation of geology.[27] Such single-mindedness was countered by, for example, Thomas Tredgold.[28] And W. D. Conybeare in his introduction to the major geological text of the 1820s and beyond argued that knowledge of mineralogy was only a 'subordinate acquisition' for the geologist.[29] William Fitton, in a different strategy, denied the historical existence of an English school of mineralogy in the course of his mythic account of the history of geology.[30] Fitton, writing for a general scientific audience, was essentially reconstructing the history of the science in a way which identified 'geology' with the research practice of a particular group in the Geological Society, namely those engaged in descriptive stratigraphy. In this way public acceptance was sought, and largely gained, for a certain type of research practice as geology and for a certain type of practitioner as the arbiter of matters geological. The 'skills' of the field stratigrapher were elevated above those of the mineral chemist.[31] The work of the latter, though admitted as a useful tool, did not constitute geology. In the struggle between mineralogists, mineral chemists and stratigraphers we have one example of the way in which attempts were made to establish hierarchies within the Geological Society. But these early tensions were not decisively resolved. Indeed their focus shifted to a conflict within stratigraphy between those who placed primary emphasis upon mineralogical stratigraphy and those who, on the contrary, emphasized paleontological stratigraphy as the essence of geology.

Before turning to such matters we will examine another, better known example of conflict involving the Geological Society – the 'minor class war' during the 1810s between the 'geological intelligentsia' (in Porter's phrase) and practical surveyors such as John Farey and William Smith. In this case the establishment of a hierarchy was in one sense more easily accomplished because of the social status and advantages of the gentlemanly geologists. But in another way it proved harder, or rather had to be accomplished by more subtle means. This was because Smith in particular pioneered stratigraphy in England.

The credit due to Smith for the first systematic use of fossil remains to establish the order of the strata, and the virtues, or otherwise, of his great geological map were contentious issues in the 1810s. The geological elite rapidly seized upon his mode of inquiry and made it their own. In this respect Smith, and other professional surveyors, were very influential in giving stratigraphy the key role in geology which it subsequently acquired. And yet Smith was shabbily treated by Greenough and other leaders of the Geological Society. The professional surveyor was at a social disadvantage in trying to make his way among the geological intelligentsia. This was due to the exigencies of making a living and of securing patronage, compounded with difficulties in communicating and publishing ideas in what was considered an appropriate form. In addition, the practical economic concerns of the surveyors' geological work allied them with the utilitarian ethos of the 'mineral history' programme, if not necessarily with its research practice. These, then were some of the issues and circumstances which underlay tensions between the surveyors and the geological intelligentsia of the Geological Society.[32]

John Farey led the assault on Greenough and the Society. For example, in 1815 Farey wrote 'in favour of Mr William Smith's irresistable claim to the discovery of the order, and to priority in the actual tracing and mapping of the surfaces of the principal part of the British series of Strata'. In so doing he described himself as supporting 'the cause of practical English Geology against the theoretical pretentions of an Anglo-German Geognosy, which has too high and proudly raised its intolerant head amongst us'.[33] There follows a diatribe against Greenough and the Geological Society in which Farey makes it clear that when it comes to facts, he has them:

> . . . on matters of fact, I rarely if ever find myself at issue with the practical Colliers, Miners, &c, but on matters of inferences, or involving their beliefs, or things not actually seen by themselves, I . . . find myself point blank at issue with them . . . I am, I hope, alike incapable of being influenced by numbers, to adopt or reject any position or deduction to be made from the study of Geological phaenomena, as I am of yielding in any such cases to authority, however academical or imposing its aspect may be. . .[34]

The import of this is clear. The Geological Society, and particularly Greenough, are only pretenders to empirical geological work. Farey's own proximity to a reliable source of factual information is emphasized as is his independence from the theoretical pretensions of the geological intelligentsia in making deductions from the phenomena. But this should not be taken as an anti-theoretical stance. Subsequently Farey gave an interesting twist to his claims in his 'Free Remarks on the

Geological Work of Mr. Greenough'.[35] He examined Greenough's *A Critical Examination of the First Principles of Geology*, a book which evinced a scepticism about theory and theory-laden terms which even Greenough's friends found exasperatingly fundamentalist. Farey did not 'deny it the merit of containing a great many original and useful local observations, and also several Inferences and Remarks, which had not previously been published', but in the final analysis he argued that the work would retard rather than advance 'the march of geological knowledge'.[36]

Farey is now saying that what is valuable in Greenough's book is limited in being mundanely empirical, and that Greenough's thoroughgoing scepticism about generalization is an unproductive way of prosecuting geological work. It seems likely that by 1819—20 Farey's campaign was increasingly directed at Greenough personally and not at the Geological Society leadership *en masse.* He is certainly siding methodologically with a number of the latter in his criticism of Greenough's book. One of Farey's chief aims throughout his campaign had been to establish the priority of William Smith's work. Some members of the Society's leadership were recognizing this more generously than ever before by about this time, but Greenough was perhaps the most reluctant to do this. In such circumstances Farey's use of method statements took the twist which we have just described.

For their part, the geological intelligentsia dominated the *Transactions* of the Society with their stratigraphical work. Porter notes that most of this work was done by a 'small cadre of London and university based members' with extensive opportunities for travel. Herein lay their strength when compared with the surveyors:

> . . . because the Society's stratigraphy was the work of a much travelled core, it easily transcended the problems of an excessively localized 'mineral geography' approach which beset many of the professional surveyors, and could assimilate Continental investigations.[37]

It was precisely in seeking to transcend local descriptive stratigraphy that the intelligentsia distinguished themselves. They might do so by stratigraphical investigations designed to compare different conceptions of earth history, or, as Buckland and Conybeare began to do around 1820, by tackling the questions of revolutions in the history of the earth and of the creation.[38] In short, systematic comparative stratigraphy and stratigraphical palaeontology (the latter inspired notably by the work of

Cuvier and Brogniart on the Paris Basin) became one of the hallmarks of the gentlemanly geologist distinguishing him from the practical geological surveyor, and distinguishing the geological devotee from the dilettante.

Recently Nicolaas Rupke has drawn a compelling portrait of what he calls the 'English School' of geology.[39] He sees this School as having its roots in Oxford in the late 1810s with the Reverend William Buckland and his circle, including W. D. Conybeare, and as taking in Adam Sedgwick and other Cambridge men as well as London-based geologists such as Roderick Murchison. Although there were certainly internal conflicts in the English School, notably about Buckland's early diluvialism, they were unified in opposition to other groups. They certainly distanced themselves from the scriptural geologists and what was seen as their biblical literalism. But they were also opposed, in varying degrees, to the geology of the 'Scottish School'. The latter was centred on Robert Jameson's Wernerian Natural History Society, Edinburgh University and the *Edinburgh Philosophical Journal.* As Rupke points out, there were a number of lines of quite self-conscious divergence between these Schools.[40] Whilst the English School would not countenance the scriptural geologists, they did take seriously the question of the relation between new geological discoveries and biblical history and they identified even more unanimously with the Paleyan natural theology of the *Bridgewater Treatises.* The Scottish School was much more circumspect in such matters. The diluvialism, and later progressivism, of the English School was quite distinctive, though the latter was shared by a number of Wernerians. But most important from our point of view was the concentration of the Scottish School upon the older Primary rocks as distinct from the English School's preoccupation with the Secondary, and also Tertiary, formations. Given the fossiliferous nature of the latter, the English School quite naturally placed considerable store by paleontological stratigraphy, whilst those north of the border relied much more heavily upon mineralogical stratigraphy. However, neither School was content with a division of labour and skill. Both were concerned to transcend local descriptive stratigraphy and establish a generalized geological schema. Each sought to identify its geological practice with 'true geology'. And it is in this context that their method discourse should be examined.

William D. Conybeare's introduction to the *Outlines of the Geology of England and Wales* has rightly been taken as incorporating a major statement on method. Superficially it might be interpreted as simply restating earlier injunctions against theory and hence as advocating a radical empiricism virtually unchanged from the earliest days of the Geological Society. But when seen in the context of the aims and aspirations of the English School it appears in a different light. Conybeare, whom Rupke aptly describes as the 'party theoretician' of the diluvialists, was concerned to assert the claims of diluvialism and of paleontological stratigraphy and to defend them against criticism from both the scriptural geologists and the Scottish School.[41]

The institutional background of Conybeare's remarks was certainly much broader than the confines of the Geological Society. Indeed in explaining the nature and course of the debates engaged in by the English School Rupke concentrates on the traditions and exigencies of the academic environments of Oxford, Cambridge and Edinburgh Universities. But capture of the official stance of the Geological Society was also a very important issue. And in the later 1810s and early 1820s there was a steady but clear changing of the guard in the Geological Society with a number of the Scottish School and their allies losing influence there and members of the English School coming to the fore.[42]

In the *Outlines* Conybeare did criticize what he referred to as the Edinburgh School for 'that excessive addiction to the theoretical speculations, which has converted the members of that school into the zealous partisans of rival hypotheses, and led them to contribute far less than they otherwise must have done, to the real progress of inductive geology'.[43] To the general reader these remarks would certainly reinforce the idea that geology should be a strenuously empirical science undiverted by speculative theories of the earth and the fulminations of the Edinburgh school. But the initiated reader, aware of the stance which Conybeare represented, would glean a more subtle message.

We have already noted that in the *Outlines* Conybeare played down the amount of mineralogical knowledge which the geologist needed to have at his fingertips. But it was distinctly otherwise in the case of organic remains. This 'important branch of geological inquiry' required 'all the resources of perfect acquaintance with many departments of Zoology'.

> . . . it is a subject which can be treated with advantage only by those who bring to it a mature and precise knowledge of the branches of natural history with which it is

connected, a remark extorted by the flippant manner in which some writers have treated conclusions, the premises of which they were incompetent to comprehend.[44]

Conybeare singled out a work on the geology of the Yorkshire coast as 'in some degree liable to this censure'. He found the author defective in his knowledge of conchology and that 'his ideas of geology are formed only from the inspection of a very limited district'. The author of the work was George Young, a Whitby Clergyman, and one of a group of biblical literalists who attacked the work of the Oxford School in the 1820s not only by directly confronting Buckland's diluvialism and his famous hyena den theory concerning the Kirkdale Cave remains, but also by questioning the utility of Cuvierian inspired comparative anatomy as applied to fossil remains.[45]

The kind of statement which Conybeare makes is a way of defining the essentials of scientific practice in one's own image, in terms of one's own skills and acquirements. Thus, whilst it is certainly possible to extract statements which Conybeare makes in the *Outlines* as evidence for an unchanging radical empiricism, when taken in context they have a quite different import.[46] To the uninitiated, Conybeare's introductory remarks would reinforce the status of geology as an empirical science not to be confused with the speculative eighteenth-century theories of the earth. To the initiated, Conybeare's pleas for more facts and the avoidance of speculative theorizing held a more specific message when taken in conjunction with his claims about the appropriate agenda for the science and the requisite equipment of the true geologist. The discovery of facts in the science depended upon an appreciation and comprehension of increasingly elaborate collateral theoretical constructs and premises. To deny the value of those premises, or to fail to comprehend them, was to exclude oneself from participation in the most important branch of geology.

Significantly, George Young had found an audience for his critique of the Oxford School at the Wernerian Natural History Society. And from that quarter, in the pages of the *Edinburgh Philosophical Journal*, and also in those of the *Edinburgh Reivew*, came a concerted attack upon the geology of the English School, upon the claims made on behalf of paleontological stratigraphy and an assertion of the rival claims of mineralogical stratigraphy.

Consider, for example, Leonard Horner's review of John Macculloch's *A Description of the Western Isles of Scotland.* Horner contended that, for the most part, geology had been in the hands of 'a minor class of philosophers' who simply accumulated facts 'without one idea of a generalization having entered into their mind'. Thus, immediately, Horner seeks to distance Macculloch and the School he was part of from dilettantism and crude empiricism. Horner then suggests that though such minor philosophers were useful because their work could be used by the generalizers, they were also a liability insofar as their sometimes inaccurate work gave comfort to 'cavillers' who would seize on any opportunity to discredit geology and geologists. Thus:

> . . . an accurate acquaintance with the minute details of the subject is of the highest importance; and it is to be regretted that some of the distinguished persons who have led the way in the higher departments of the science, have not been sufficiently familiar with its more minute branches.[47]

Such complaints, Horner argued, could not be raised against Macculloch who combined 'general views and accurate information' in a way which was rare among contemporary geologists.

Horner, I suggest, sought to accomplish essentially two things by these remarks, and in the rest of his review. First, in the wider public arena, he sought to distinguish 'sophisticated' geology from crudely empirical work. This was done by emphasizing the importance of 'general views'. Here he is engaging in essentially the same hierarchy-generating tactic which we saw in Conybeare's *Outline*. But, secondly, Horner also sought to elevate Macculloch and the style of geological work which he represented above the kinds of work engaged in by other groups and particularly above that of the English School. For Macculloch's strengths lay in mineralogy. Such skills were undoubtedly important in deciphering the complex lithology of a region such as the Western Isles. But a much broader claim was being advanced – namely that such work, rather than stratigraphical palaeontology, was the true foundation of geology. My contention is borne out by Macculloch's own assessment of his friend's review which he admired as 'a very readable and popular-looking article, calculated to make geology go glibly down among the readers of reviews'. But he also felt that the review would do good on a narrower front 'by recalling the attention of our *namby pamby cockleologists and formation men* to that which is the proper business of geologists'. Undoubtedly Horner and Macculloch had the new regime emerging at the Geological Society in mind here. Horner remarked at about this time that the Society had fallen into 'very feeble hands'.[48]

Macculloch reiterated his complaints about the dominant direction which geology had taken in his *System of Geology* published in 1831. There he suggested that very little had been accomplished in geology in the last decade.[49] A spirited response came from W. D. Conybeare:

> I have been principally induced, in the present summary of the progress of geological science, to draw a line at the close of the first series of our *Geological Transactions* in 1821, because an author. . . has asserted in a recent publication, that 'since that year geology has received scarcely any valuable additions, and not a single fundamental one' . . . I shall endeavour to give a slight sketch of the contributions which have really marked this supposed period of inanition, leaving it to your judgment how far they really deserve the above depreciating character.[50]

Conybeare's argument was essentially this. Although by 1821 the fundamentals of the geology of England were well developed, especially those of the 'secondary system', much had since been accomplished, particularly in the study of the tertiary formations. Moreover, English geologists had been very active in the study of the geology of the Continent of Europe. And this was not just a study of additional regions in isolation but a comparative enterprise. This was methodologically important 'since extensive *comparative* geology affords the only materials for obtaining the *fundamental facts* of our science'.[51] And only in this way was it possible to address those questions necessary to the construction of a viable geological theory: whether the same geological causes have acted during the same periods; whether geological deposits in different places contain similar groups of organic remains, or not. Certainly Conybeare stated that it was as yet premature to believe that enough had been accomplished to enable a geological theory to be constructed. However, he asserted that anyone aware of these problems and cognizant of the kind of evidence required to solve them, could not fail to see that the last decade had witnessed 'a very rapid accumulation of the most valuable materials for their elucidation'.[52] Conybeare goes on to discuss the work of Buckland, Sedgwick, Murchison, De la Beche, John Phillips, Charles Lyell and a number of continental geologists including, notably, Elie de Beaumont.

The point to be made here is that Conybeare is acting as the spokesman for the group of geologists (the English School) who came to dominate the Geological Society in the 1820s. They were a group of geological devotees with a well-defined communal agenda and well-developed skills in stratigraphy and paleontology (what Conybeare calls 'zoological geology'). They functioned as a group, exchanging ideas regularly by private correspondence and through the Geological Society, and undertaking joint geological expeditions both in England and on the continent of Europe.

To extract method statements from Conybeare's report as evidence of empiricism of precisely the same kind as that contained in the *Geological Inquiries* is to take them badly out of context. Although Conybeare regards the theoretical system building of the likes of Elie de Beaumont and Charles Lyell as premature, he clearly sees it as a legitimate activity by proven geologists. Though on numerous occasions he enjoins the careful search for facts, this does not amount to an indiscriminate radical empiricism. His response to Macculloch shows that the search for facts being advocated is a strategic one, conditioned by high-level generalizations or hypotheses. Taken in context then, Conybeare's method discourse was a means of legitimating and bolstering the claims of the English School to be the arbiters of matters geological and to set the agenda in the science.[53]

Enough has been said to indicate some of the complexities of the 'socio-cognitive topography' of the Geological Society's early years. The account offered here is, of course, neither comprehensive nor definitive. And it deliberately stops short of tackling the method discourse of Charles Lyell and his critics.[54] But it does allow us to draw some broad conclusions. The combination of descriptive stratigraphy and directionalism which came to constitute geology in the 1820s, and beyond, can be seen as the outcome of competing programmatic views of the science advocated by different groups in an attempt to maximize the value of their own skills and to identify those skills as *the* basis of geology.[55] In this context, statements about method were frequently employed as a tool to demarcate the 'true geologists' from the mere pretenders to the title. Clearly any blanket description of the Geological Society as being in the grip of a radical empiricism, or even for that matter gradually emancipating itself from that grip, gravely misrepresents the situation. And in the process it greatly underestimates the political utilities of method discourse within the geological community itself.

3. THE ASTRONOMICAL SOCIETY OF LONDON

There are some remarkable similarities between the early history of the Geological Society and that of the Astronomical Society of London which was founded in 1820. Both met with opposition from Sir Joseph Banks, although in the case of the Astronomical Society Banks was not associated with a particular view of the nature of the discipline involved. His concerns centred on jurisdictional matters and the broader politics of 'reform'. The Astronomical Society was from its beginning a bastion of the reform movement in British science;[56] but the Society's leadership also advocated and sought to effect a reformation of astronomy.[57] And here, as in the case of the Geological Society we see the use of both a radically inductive methodological stance and one emphasizing the role of theory and rejecting crass empiricism. However, method discourse did not loom so large in the Astronomical Society. Astronomy was not 'in the making' in the same sense that geology was. The 'higher theoretical departments' were generally recognized as such by all: they were firmly institutionalized as an exemplar elsewhere, notably amongst Continental astronomers. So the following account, whilst illustrating once again the use of method discourse in the 'micro-politics' of science, also suggests that the extent to which it is employed in such contexts depends upon the degree of 'openness' of the area to competing claims of 'best practice'.

The *Address* explaining the aims of the new institution included the following remark:

> Some vast and general views of the construction of the heavens, and the laws which may regulate the formation and motion of sidereal systems, have, it is true, been struck out; but, like the theories of the earth which have so long occupied the speculations of geologists, they remain to be supported or refuted by the slow accumulation of a mass of facts: and it is here, as in the science just alluded to, that the advantages of associated labour will appear more eminently conspicuous.[58]

The 'vast and general views of the construction of the heavens' here referred to were those of Sir William Herschel. The great astronomer's bold generalizations on the basis on his vast surveys of stars and nebulae had long been regarded with both awe and suspicion by more conventional astronomers.[59] The Astronomical Society clearly set itself a more sober agenda. And the impression given in the *Address* was that anyone prepared to learn the rudiments of astronomical observation and computation could be recruited to the cause and make a valuable contribution. To this end education in the use of instruments, 'especially such as are likely to be found in the hands of travellers, nautical men and others who may be placed in interesting situations in remote parts of the world', was vitally important.[60] The centrepiece of this astronomical fact-gathering enterprise was to be a minutely detailed co-operative survey of the heavens with members being allocated their own particular portion of the heavens to survey. The radically inductive, co-operative, 'democratic' stance taken in the official policy of the Society had the desired effect on recruitment. Numerous practical astronomers from the ranks of London's 'Mathematical Practitioners' swelled the membership list as did a significant number of nautical men.

From the very beginning, however, there were doubts among the more proficient astronomers, particularly those versed in the complex mathematical study of physical astronomy, about the official course which the Society had set itself. The *Address* had been written jointly by Francis Baily and John Herschel. Baily was a retired stockbroker with a long-standing interest in astronomy, and a seemingly infinite capacity for sustained, tedious work on astronomical catalogues and tables. But he had no serious pretensions as either a practical or theoretical (i.e. mathematical) astronomer.[61] Herschel, by contrast, combined expertise of a high order as both a practical and a theoretical astronomer.[62] Correspondence between Baily and Herschel reveals that Baily was the author of the idea for the ambitious systematic survey of the heavens. Herschel was sceptical but the passage remained. A year or so later, when they were engaged in producing the first annual report of the Society, Herschel suggested to Charles Babbage, in mocking tone, that something ought to be done about the progress of the great survey of the heavens which had featured so prominently, '*ad captandum vulgos*', in the *Address*.[63]

In fact the survey had made no progress and was never to get under way. Many found the work produced by the Astronomical Society during its first decade rather disappointing. The star catalogues produced by Baily – essentially reductions of previous observations to the current epoch – were certainly valuable. But the early *Memoirs* of the Society were in large part a farrago of observations of comets, eclipses, and occultations, reports on determinations of latitude and longitude, and papers on instrumental improvements and errors. There was a distinct lack of theoretical astronomy in the *Memoirs*. And this despite the fact that the Council had repeatedly offered the Gold Medal of the Society for analytical studies on topics such as the theory of the satellites of Saturn, of the new planets, or of the lunar motions.

Such complaints about the lack of attention to important but abstruse consequences of the theory of gravitation were made, not surprisingly, by a group well versed in modern analysis as applied to physical astronomy. Unlike most mathematical practitioners and nautical men, they were products of academic institutions where they had reached the forefront of modern analysis. Cambridge men were prominent, particularly men like George Airy and John Lubbock who were heirs to the changes wrought in Cambridge mathematics by John Herschel, Charles Babbage, George Peacock and William Whewell.[64] Those Cantabrigians who had formed an important part of the founding group of the Astronomical Society (notably Herschel and Babbage) had co-operated in seeking a broad base of membership in the Society and, despite occasional scepticism, acceded to the 'democratic' empiricism of the initial programmatic announcement. They wanted a strong Astronomical Society not only to promote astronomy but also for broader political reasons relating to the reform movement in British science more generally. But once the Society was firmly established, the stirrings of an emergent elite were felt and legitimations for the existence of such an elite, couched in methodological terms, began to appear.

Consider George Airy's report on astronomy to the British Association.[65] After an extensive survey of the development of astronomy from the beginning of the nineteenth century, Airy turned to the question of British accomplishments in the science. He asserted that 'our principal progress has been made in the instrumental and mechanical parts, and in the lower parts of Astronomy; while to the higher branches of the science we have not added anything'.[66] Airy admitted to exceptions to this general assertion, citing work by John Herschel, James Ivory, John Lubbock and, implicitly, himself. According to Airy, the highest form of astronomical work was 'the comparison of theory with observation, and its immediate consequences, the reducing of complicated phaenomena to simple laws, or the showing that new supplementary laws are necessary. . .'[67] In areas where such activity was possible – the lunar and planetary theories, the theory of the new planets and periodical comets – nothing had been done in Britain. On one of the major reasons for this it is worth quoting Airy at length:

In England, an observer conceives that he has done everything when he has made an observation. He thinks that merely noting the passage of a star over one wire and its bisection by another, is all that can be expected from him; and that the use of a Table of logarithms, or anything beyond the very first stage of reduction, ought to be left to others. In the foreign observatories, on the contrary, an observation is considered as a lump of ore, requiring for its production, when the proper machinery is provided, nothing more than the commonest labour, and without value till it has been smelted. In them, the exhibition of results and the comparison of results with theory, are considered as deserving much more of an astronomer's attention, and demanding greater exercise of his intellect, than the mere observation of a body on the wire of a telescope.[68]

Although Airy does not mention the Astronomical Society by name, it would have been eminently clear to anyone acquainted with its character that such a statement amounted to an indictment of the activities of the bulk of its membership.

Thomas Galloway made very similar remarks in reviewing the Astronomical Society's *Memoirs* for the *Edinburgh Review*. He considered practical observational astronomy to be well cultivated, but for this the only mathematical knowledge necessary was 'confined to a few elementary theorems of Geometry, the simple rules of Algebra, and Spherical Trigonometry'.[69] An even more general interest, it seemed to him, had been taken recently in the observation of double stars and nebulae for which one need only to be able to read angles. Even though mathematics had undergone a considerable improvement in Britain over the last twenty or thirty years 'the *Memoirs of the Astronomical Society* exhibit no traces of its renovation. Within the compass of the three volumes yet published, we find only a single paper that can be properly said to be devoted to the improvement of the analytical theory of gravitation, and of this the author is a foreigner'.[70]

Both Airy and Galloway were complaining of a lack of attention to theoretical astronomy. But obviously they were not advocating the kind of speculative *cosmology* which the Astronomical Society had foresworn at its institution. Rather they were concerned that attention be given to astronomical problems within the purview of the analytical theory of gravitation and, of paramount importance, that empirical investigation of such problems be constantly guided by reference to that theory. This was their methodological prescription. And its import was that only the lower departments of astronomy were open to the kind of co-operative, democratic, inductive programme which the Astronomical Society had set for itself. And even here very careful attention to the standardization of observational techniques and the reduction of observations was necessary. The progress of the higher departments of the science depended primarily on the activities of a mathematically sophisticated elite.[71] So, once again, we see those with particular skills defining astronomy 'proper' in their own image by methodological prescription. In fact this was part of a much broader effort on the part of the Cambridge group and their associates to so redefine the essentials of physical science more generally during the 1820s and 1830s.[72]

There were however important differences between the cases of the Geological and Astronomical Societies. As already noted, astronomy was a more mature science than geology, and a widely recognized model of 'best practice' was readily available in the sophisticated mathematical astronomy prosecuted on the Continent. As a result the emergent elite in the Astronomical Society did not face competition from alternative elites and rival conceptions of 'best practice'. In this their situation differed from that in which the emergent elite of the Oxford/English School of geologists found themselves. And I suggest that as a result of this method discourse was a political resource of more marginal utility in the astronomical world than in the geological. Also, the Geological Society could claim quite legitimately to be *the* major institution devoted to the science of geology in Britain.[73] The institutional structure of astronomy was more complex. Public observatories with the Royal Observatory at the apex, university observatories, the Board of Longitude before its demise in 1828, and the Office of the Superintendant of the *Nautical Almanac*, were all crucial institutions for anyone anxious to redefine practice and power relations in astronomy.[74] For this reason the Astronomical Society was less exclusive a forum for making methodological prescriptions and affirmations designed to consolidate a hierarchy within the science than the Geological Society. In the world of public institutions the 'politics of place' was vital. Crucial appointments, such as that of Airy to the Astronomer Royalship in 1836, clearly depended upon much more than the politics of hierarchy within voluntary institutions like the Astronomical Society. But by the 1830s, more so than in previous times, governments did look for the highest expertise as defined within the scientific community.[75] For this reason the establishment of hierarchies within that community, affirmed by methodological prescriptions and the expertise they implied for the elite, had ramifications beyond the internal politics of voluntary scientific societies.

4. CONCLUSION

We began with Larry Laudan's valuable reminder that method discourse is not restricted to formal tracts on scientific method. It is an integral part of scientific practice often concerned with the advocacy, or legitimation, of particular theories or approaches within a science. This means that to understand method discourse we must take careful account of the scientific context in which it is embedded historically. But to do this, I have suggested, is only one step in the right direction. There is now a considerable body of historical work which demonstrates the social and political uses to which scientific ideas have been put and indicates quite clearly that those ideas are actively shaped by their context of use.[76] Statements on method are no exception to this. I have shown by examining two specialist scientific societies and various groups within them, that method discourse is to be understood not only in terms of the cognitive development and 'situation' of the individual and of the requirements of a broad consensual ideology of science but also as a vital aspect of the 'micropolitics' of science.

The case studies which have been offered here are more indications of what might be done than fully-formed studies. But they do demonstrate a number of important points. First it is clear that to cull method statements from a variety of sources and to ascribe them to an institution as the methodological stance of that body is a dangerous procedure. This does not mean that generalizations cannot be made. In the case of the Geological Society it is quite valid to attribute a stance of radical empiricism as represented in the *Geological Inquiries* to the Society as a body in its very earliest years. This gave a public face to the science which both encouraged recruitment and dissociated the Society from 'unsound' speculative theorizing. And no doubt the latter message continued to be directed at a general audience for some time through method statements. But, equally, the very same statements which would serve that function before one audience served quite different functions before others.

The supposed uniformity of method discourse associated with the early Geological Society dissolves when we place it in both literary and socio-political context. What emerges instead is a continuing struggle between various sub-groups in which statements about method of a *prima facie* similar character acquire substantially different meanings. An empiricist exhortation to a search for 'facts' can mean quite different things depending on the conception of what constitutes a 'fact' in the science. And we saw how that term did mean quite different things to Greenough, Farey, Macculloch and Conybeare, to Baily and Airy, depending upon their perception of the agenda and the crucial problems to be tackled in their favoured science. Those problems in their turn tended to be defined in terms of group skills and acquirements so that method discourse functioned *inter alia* to create and sustain hierarchies within specialist scientific communities.

Much remains to be done. My characterization of various geological and astronomical programmes is rather crude, necessarily so given the scope of the present paper. And much individual variety in methodological stance is therefore obscured.[77] The main import of my paper, however, is to indicate one important way in which the social history of scientific institutions can began to incorporate programmatic changes in scientific practice. To do this effectively would involve not only valuable contributions to the study of the politics and rhetoric of method but also to the historical sociology of scientific knowledge more generally.

ACKNOWLEDGEMENT

For their valuable comments on an early draft of this chapter I would like to thank David Oldroyd, John Schuster and Richard Yeo.

University of New South Wales, Australia.

NOTES

[1] See L. Laudan, *Science and Hypothesis: Historical Essays on Scientific Methodology* (Dordrecht, Boston & London, 1981), pp. 6–9. Whilst Laudan does allow for a legitimating function for statements about methodology he sees this in the context of arguments about scientific theories, not in a socio-political sense. See L. Laudan, 'Theories of Scientific Method from Plato to Mach', *History of Science* **VII**, 1968, at pp. 8–9.

[2] B. Barnes, *Scientific Knowledge and Sociological Theory* (London, 1974), pp. 45–46.

[3] *Ibid.*, p. 116.

[4] See B. Barnes, 'On the Conventional Character of Knowledge and Cognition', *Philosophy of the Social Sciences* **XI**, 1981, pp. 303–333.

[5] P. B. Wood, 'Methodology and Apologetics: Thomas Sprat's *History of the Royal Society*', *The British Journal for the History of Science* **XIII**, 1980, p. 1–26; R. Yeo, 'Scientific Method and the Image of Science, 1831–1819', in: *The Parliament of Science: The British Association for the Advancement of Science 1831–1981*, R.

MacLeod and P. Collins (eds.), (London, 1981), pp. 65–88. See also the chapter by Richard Yeo in this volume.

[6] R. Hahn, *The Anatomy of a Scientific Institution: The Paris Academy of Sciences, 1666–1803* (Berkeley, 1971), p. x.

[7] Wood hints at one such use in the final part of his paper. Though perhaps subsidiary to its broad apologetic function, Baconianism allowed Fellows without technical abilities to take part in the natural history programme and hence in the Society's activities. That methodological stance could be used to 'attract support for the Society from virtuosi in London and the provinces'. P. B. Wood, *op. cit.* (Note 5), pp. 19–20.

[8] See G. N. Cantor, 'The Reception of the Wave Theory of Light in Britain: A Case Study Illustrating the Role of Methodology in Scientific Debate', *Historical Studies in the Physical Sciences* **VI**, 1975, pp. 109–132; *idem, Optics after Newton: Theories of Light in Britain and Ireland, 1704–1840* (Manchester, 1983), pp. 177–185. And compare J. Morrell and A. Thackray, *Gentlemen of Science: Early Years of the British Association for the Advancement of Science* (Oxford & New York, 1981), pp. 466–472; D. P. Miller, 'The Revival of the Physical Sciences in Britain, 1815–1840', *Osiris* **II**, 1986.

[9] An excellent study which achieves such integration is S. Shapin, 'The Politics of Observation: Cerebral Anatomy and Social Interests in the Edinburgh Phrenology Disputes', in: *On the Margins of Science: The Social Construction of Rejected Knowledge*, ed. R. Wallis, *Sociological Review Monograph* **XXVII**, (Keele, 1979), pp. 139–178.

[10] M. J. S. Rudwick, 'The Foundation of the Geological Society of London: Its Scheme for Co-operative Research and its Struggle for Independence', *The British Journal for the History of Science* **I**, 1963, pp. 325–355. In subsequent work Rudwick has developed a much more sophisticated view of the Society than that presented in this pioneering study.

[11] M. J. S. Rudwick, *op. cit.* (Note 10); R. Laudan, 'Ideas and Organizations in British Geology: A Case Study in Institutional History', *Isis* **LXVIII**, 1977, pp. 527–538. Though much of what I say will be highly critical of the interpretation given in her paper, I share and applaud Laudan's concern with exploring the 'interface between institutional and intellectual history'.

[12] For a criticism of this position and delineation of a symmetrical (or monistic) approach see B. Barnes, 'Natural Rationality: A Neglected Concept in the Social Sciences', *Philosophy of the Social Sciences* **VI**, 1976, pp. 115–126; and 'On the Causal Explanation of Scientific Judgment', *Social Science Information* **XIX**, 1980, pp. 685–695.

[13] R. Porter, *The Making of Geology* (Cambridge, 1977), p. 131, for one, makes such a judgment.

[14] For a discussion of the concept and importance of 'audience' see S. Shapin, 'The Audience for Science in Eighteenth-Century Edinburgh', *History of Science* **XII**, 1974, pp. 95–121. Shapin is concerned to show how the interests of the audience shaped the distribution of effort between various areas of inquiry.

[15] The *Geological Inquiries* was originally published as a pamphlet. I have used the reprinted version in *Philosophical Magazine* **XLIX**, 1817, pp. 421–429.

[16] 'Preface', *Transactions of the Geological Society of London*, 1st series, **I**, 1811, pp. viii–ix.

[17] Such a statement was inserted routinely in the *Philosophical Transactions*.

[18] L. Horner, 'On the Mineralogy of the Malvern Hills', *Transactions of the Geological Society of London*, 1st series, **I**, 1811, pp. 281–321, at p. 321.

[19] H. Holland, 'A Sketch of the Natural History of the Cheshire Rock-Salt District', *Transactions of the Geological Society of London*, 1st series, **I**, 1811, pp. 59–61; J. F. Berger, 'Observations on the Physical Structure of Devon and Cornwall', *idem*, pp. 93–184.

[20] R. Laudan, *op. cit.* (Note 11), pp. 529–531.

[21] [W. H. Fitton], 'Transactions of the Geological Society, Vol. I', *Edinburgh Review* **XIX**, 1811–12, pp. 207–229, at p. 207; [W. H. Fitton]. 'Transactions of the Geological Society, Vol. III', *Edinburgh Review* **XXIX**, 1817, pp. 70–94, at p. 70.

[22] M. J. S. Rudwick, 'Charles Darwin in London: The Integration of Public and Private Science', *Isis* **LXXIII**, 1982, pp. 186–206, at pp. 190–194.

[23] R. Porter, *op. cit.* (Note 13), pp. 140–141. See also D. P. Miller, 'Between Hostile Camps: Sir Humphry Davy's Presidency of the Royal Society of London, 1820–1827', *The British Journal for the History of Science* **XVI**, 1983, pp. 1–47, at pp. 5–9. Rudwick, *op. cit.* (Note 10) and R. Laudan, *op. cit.* (Note 11) consider only the jurisdictional dispute.

[24] P. Weindling, 'Geological Controversy and its Historiography: The Prehistory of the Geological Society of London', in: *Images of the Earth: Essays in the History of the Environmental Sciences*, L. J. Jordanova and R. Porter (eds.), (Chalfont St. Giles, 1979), pp. 248–271. See also P. Weindling, 'The British Mineralogical Society: A Case Study in Science and Social Improvement', in: *Metropolis and Province: Science in British Culture 1780–1850*, I. Inkster and J. Morrell (eds.), (London, 1983), pp. 120–150. The attempted establishment of a mineralogical collection at the Royal Institution in 1804 was another closely related utilitarian venture with a view to serving mining and landowning interests. See M. Berman, *Social Change and Scientific Organization: The Royal Institution, 1799–1844* (Ithaca, 1978), pp. 88–92.

[25] R. Porter, *op. cit.* (Note 13), p. 172.

[26] *Ibid.* pp. 173, 180.

[27] A. Aikin, *A Manual of Mineralogy* (London, 1814), p. vii; W. Phillips, *A Selection of Facts from the Best Authorities, Arranged so as to form an Outline of the Geology of England and Wales* (London, 1818), p. 9. It would be a mistake to attribute an unchanging narrow mineralogical approach to Arthur Aikin who paid increasing attention to fossils in his fieldwork. He was undoubtedly more generous in his attitude toward paleontological stratigraphy than was Macculloch, on whom more below. See H. S. Torrens, 'Arthur Aikin's Mineralogical Survey of Shropshire, 1796–1816 and the Contemporary Audience for Geological Publications', *The British Journal for the History of Science* **XVI**, 1983, pp. 111–153.

[28] T. Tredgold, 'Remarks on the Geological Principles of Werner and those of Mr. Smith', *Philosophical Magazine* **LI**, 1818, pp. 36–38.

[29] W. D. Conybeare and W. Phillips, *Outlines of the Geology of England and Wales* (London, 1822), p. ii, footnote.

[30] [W. H. Fitton], 'Notes on the History of English Geology', *Philosophical Magazine*, 3rd series, **II**, 1833, pp. 37–57.

[31] 'Skills' as used here should be taken to include a number of characteristics more akin to opportunity than superior technique, notably the leisure and means to engage in extended field excursions.

[32] R. Porter, *op. cit.* (Note 13), p. 143.
[33] J. Farey, 'Observations on the Priority of Mr. Smith's Investigation of the Strata of England', *Philosophical Magazine* **XLV**, 1815, pp. 333–344, at pp. 333–334.
[34] *Ibid.* pp. 343–344.
[35] J. Farey, 'Free Remarks on the Geological Work of Mr. Greenough', *Philosophical Magazine* **LIV**, 1819, pp. 127–132. See also J. Farey, 'Free Remarks on Mr. Greenough's Geological Map', *Philosophical Magazine* **LV**, 1820, pp. 379–383.
[36] J. Farey, *op. cit.* (Note 35, 1819), p. 127.
[37] R. Porter, *op. cit.* (Note 13), p. 182.
[38] *Ibid.* p. 183.
[39] N. Rupke, *The Great Chain of History: William Buckland and the English School of Geology (1814–1849)* (Oxford, 1983).
[40] *Ibid.*, pp. 16–18.
[41] *Ibid.*, p. 86 on Conybeare as party 'theoretician'.
[42] An examination of the Officers and Council Members of the Geological Society reveals that by the later 1810s and early 1820s only a handful of the early leadership were left (Arthur Aikin, G. B. Greenough, W. H. Pepys)
[43] Conybeare and Phillips, *op. cit.* (Note 29), p. xlviii.
[44] *Ibid.*, pp. xiii–xiv.
[45] *Ibid.* note, p. xiv. The work referred to is G. Young, *A Geological Survey of the Yorkshire Coast* (Whitby, 1822). On Young and other biblical literalists see Rupke, *op. cit.* (Note 39), pp. 42–50.
[46] See R. Laudan, *op. cit.* (Note 11), p. 529.
[47] [L. Horner], 'J. Macculloch, A description of the Western Isles of Scotland, 1819', *Edinburgh Review* **XXXIII**, 1820, pp. 442–470, at p. 443. On Macculloch and this work see D. A. Cumming, '*A Description of the Western Isles of Scotland*: John Macculloch's Successful Failure', *Journal of the Society for the Bibliography of Natural History* **VIII**, 1977, pp. 270–285.
[48] J. Macculloch to L. Horner, 2 June 1820 and L. Horner to A. Marcet, 10 April 1821 in: *Memoir of Leonard Horner*, K. M. Lyell (ed.), (London 1890), pp. 173–174 and 192.
[49] J. Macculloch, *A System of Geology* (London, 1831).
[50] W. D. Conybeare, 'Report on the Progress, Actual State, and Ulterior Prospects of Geological Science', *Report of the British Association for the Advancement of Science, Oxford, 1832* (London, 1835), pp. 365–414, at p. 375.
[51] *Ibid.* p. 375.
[52] *Ibid.* p. 376. For an equally forceful criticism of Macculloch by Roderick Murchison from the Chair of the Geological Society in 1832 see *Proceedings of the Geological Society of London* **I**, 1834, pp. 376–377.
[53] This they succeeded in doing through the Geological Society. See M. J. Rudwick, *op. cit.* (Note 22), pp. 190–192. Also compare Rudwick's remarks (pp. 194–198) on the availability of a social role for the geological theorist in the early 1830s with S. Herbert, 'The Place of Man in the Development of Darwin's Theory of Transmutation, Part II', *Journal of the History of Biology* **X**, 1977, pp. 155–227, on pp. 157–178. See especially Herbert's use of Conybeare's supposed methodological stance on pp. 167–168 and n. 25.
[54] The extent to which the history of geology in this period has been ahistorically dominated by the shadow of Charles Lyell is a contentious question. Any study of Lyell

and method along the lines sketched in this paper would have to deal with a vast literature. But see particularly R. Laudan, 'The Role of Methodology in Lyell's Science', *Studies in History and Philosophy of Science* **XIII**, 1982, pp. 215–249; P. Corsi, 'The Importance of French Transformist Ideas for the Second Volume of Lyell's Principles of Geology', *The British Journal for the History of Science* **XI**, 1978, pp. 221–244; R. Porter, 'Charles Lyell and the Principles of the History of Geology', *The British Journal for the History of Science* **IX**, 1976, pp. 91–103.

55 On the dominance of descriptive stratigraphy and a 'directionalist' view of earth history in the 1820s see M. J. S. Rudwick, 'Uniformity and Progression: Reflections on the Structure of Geological Theory in the Age of Lyell', in: *Perspectives in the History of Science and Technology*, ed. D. H. D. Roller (Norman, Oklahoma, 1971), pp. 209–227.

56 D. P. Miller, *op. cit.* (Note 23), pp. 9–13, 18–19.

57 The following draws on material to be published in D. P. Miller, *op. cit.* (Note 8).

58 *Regulations of the Astronomical Society of London: Established February 8, 1820. To Which is Prefixed an Address Explanatory of their Views and Objects* (London, 1820), pp. vi–vii.

59 See M. Hoskin, *William Herschel and the Construction of the Heavens* (London, 1963), pp. 188–191, and S. Schaffer, 'Herschel in Bedlam: Natural History and Stellar Astronomy', *The British Journal for the History of Science* **XIII**, 1980, pp. 211–239.

60 *Op. cit.* (Note 58), pp. viii–ix.

61 On Baily see G. J. Whitrow, 'Some Prominent Personalities and Events in the Early History of the Royal Astronomical Society', *Quarterly Journal of the Royal Astronomical Society* **XI**, 1970, pp. 89–104.

62 G. Buttman, *The Shadow of the Telescope: A Biography of John Herschel* (New York, 1970).

63 J. L. E. Dreyer and H. H. Turner, ed. *History of the Royal Astronomical Society 1820–1920* (London, 1923), pp. 3–6; J. Herschel to C. Babbage, 19 December 1820, Royal Society of London, Herschel Papers, 20.106.

64 P. C. Enros, 'The Analytical Society: Mathematics at Cambridge University in the Early Nineteenth Century', (Ph. D. dissertation, University of Toronto, 1979); H. W. Becher, 'William Whewell and Cambridge Mathematics', *Historical Studies in the Physical Sciences* **XI**, 1980, pp. 1–48.

65 G. Airy, 'Report on the Progress of Astronomy during the Present Century', *Report of the British Association for the Advancement of Science, Oxford 1832* (London, 1835), pp. 125–189.

66 *Ibid.* p. 181.

67 *Ibid.* p. 183.

68 *Ibid.* p. 184.

69 [Thomas Galloway], 'Memoirs of the Astronomical Society of London', *Edinburgh Review* **LI**, 1830, pp. 81–114 at p. 105.

70 *Ibid.* p. 104.

71 There remained, however, a basic dilemma for members of the elite, particularly those like Herschel and Whewell who became major public spokesmen for science. In public, or broad scientific, forums an unreconstructed Baconian inductivism promoted an image of science as an accessible part of general culture. Yet herein also lay a danger, from the elite's point of view, of excessive egalitarianism which might form the

basis of a challenge to their scientific authority. This was guarded against by qualification of the notion of scientific method suggesting that skills were needed even in observation and that facts could not be collected in the absence of a guiding theoretical framework. In this way the elite attempted, with some success, to have their cake and to eat it too by promoting an image of science which rendered it popular and encouraged participation whilst qualifying that image to legitimate a situation in which research practice and its institutional supports remained firmly under their control. For an extended discussion of this issue see the chapter by Richard Yeo in this volume.

[72] D. P. Miller, *op. cit.* (Note 8).

[73] There were, to be sure, other societies devoted exclusively or partly to geology, but the Geological Society of London was regarded as essentially synonymous with the science. By the 1830s its leadership had become so attached to that state of affairs that they were ambivalent, to say the least, about the setting up of the national Geological Survey. See J. B. Morrell, 'London Institutions and Lyell's Career: 1820–41', *The British Journal for the History of Science* **IX**, 1976, pp. 132–146, at pp. 141–142.

[74] D. P. Miller, *op. cit.* (Note 23), pp. 32–33.

[75] D. P. Miller, 'The Royal Society of London 1800–1835: A Study in the Cultural Politics of Scientific Organization', (Ph. D. dissertation, University of Pennsylvania, 1981), pp. 313–328.

[76] S. Shapin, 'History of Science and Its Sociological Reconstructions', *History of Science* **XX**, 1982, pp. 157–211.

[77] See M. J. S. Rudwick, 'Cognitive Styles in Geology', in: *Essays in the Sociology of Perception* ed. M. Douglas (London, 1982), pp. 219–241. Rudwick's account of four geological styles exemplified in the work of various individuals of the period is very interesting. Method is only one element of a seven-fold polythetic classification of geological style which he constructs. Rudwick then attempts, with some success, to correlate the four geological styles with the social environments of the individuals he studies. And he finds a generally good agreement between those correlations and the relationships predicted by grid/group theory. It would be an enormous task to study a scientific institution like the Geological Society in this way, but it could, and should, be done.

RICHARD R. YEO

SCIENTIFIC METHOD AND THE RHETORIC OF SCIENCE IN BRITAIN, 1830–1917

1. INTRODUCTION

The *Tribune* of October 1945 carried a contribution from George Orwell entitled 'What is Science'? While noting that this question could be answered by referring to the body of knowledge about the natural world, Orwell perceived that science had also come to denote a 'method of thought'. Alluding to the contemporary debate about curriculum content, he suggested that the demand for more science education involved 'the claim that if one has been scientifically trained one's approach to *all* subjects will be more intelligent than if one had no such training.[1] At this point, Orwell identified an important theme in the public rhetoric of science – the appeal to what Beatrice Webb called the 'cult of the scientific method'.[2] This phenomenon was enshrined in Karl Pearson's *Grammar of Science*, published in 1892, in which he declared that the 'scientific habit of mind', apart from the intellectual triumphs it had accomplished since the seventeenth century, was now called upon to play a vital historical role in a period of rapid social and political change.

> I assert that the encouragement of scientific investigation and the spread of scientific knowledge by largely inculcating scientific habits of mind will lead to more efficient citizenship and so to increased social stability. Minds trained to scientific methods are less likely to be led by mere appeal to the passions or by blind emotional excitement to sanction acts which in the end may lead to social disaster.[3]

In short, Pearson claimed that scientific method was essential for good citizenship in a democracy.

The distinction between the content and method of science has been a significant element in the cultural legitimation of the scientific enterprise. In the case of the examples above, it is possible to see this theme as part of what F. M. Turner has called 'public science' – that body of rhetoric, argument and polemic by which the scientific community advanced its interest.[4] However, the appeal to scientific method in this campaign was only one aspect of the connection between methodology

J. A. Schuster and R. R. Yeo (eds.), The Politics and Rhetoric of Scientific Method, 259–297.
© 1986 *by D. Reidel Publishing Company.*

and the social relations of science during the nineteenth century. The aim of this chapter is to reveal the rhetorical dimension of methodological discourse, arguing that certain assumptions and statements, in both systematic works and less formal texts, were associated with debates about the nature of science, its public image, and the relationship between its various disciplines.

Discourses on methodology were important rhetorical resources for the early scientific community. At one level, it is arguable that appeals to the efficacy of experimental, inductive method were central to the formulation of distinctions between scientific and non-scientific knowledge. From different perspectives, Ian Hacking and Barbara Shapiro have shown how the classical distinction betwen knowledge (*scientia*) and opinion (*opinio*) was blurred during the seventeenth century as axiomatic, demonstrative certainty ceased to be the criterion of scientific knowledge. Although Bacon and Descartes had spoken of *certain* knowledge as the goal of science, the major practitioners of the new experimental natural philosophy were willing to accept the *probable* status of its theories about the empirical world. However, it was argued that this form of knowledge was far more secure than alternative sources of information about nature because it derived from the application of rigorous, publically repeatable, procedures which gave a *moral* certainty to its discoveries.[5] In this context, statements about the efficacy of a single scientific method became intimately associated with the definition of natural science. Furthermore, claims about this method entered into the apologetics by which men of science sought to promote the political, social and cultural value of their activites.

Joseph Ben-David has suggested that references to Bacon's inductive method provided a point of consensus for the nascent scientific community in the midst of theoretical and metaphysical disputes about science at the time of the Restoration.[6] More recently, Paul Wood has interpreted Thomas Sprat's *History of the Royal Society*, which appeared in 1667, as an apologetic text that eploited aspects of Baconian methodology in the service of a particular image of science: for example, the idea of collaborative work symbolizing the non-sectarian character of the Society.[7] It is also worth remarking that Sprat employed a dichotomy between scientific method and the 'specious *Tropes* and *Figures*' which thwarted the progress of knowledge; the clear procedures and plain style of science were set against the verbiage of other discourses. But although Sprat contrasted the 'solid Practice' of science with the 'glorious pomp of Words', and distinguished its method from the colours of '*Rhetorick*', it is arguable that the emphasis on method was itself a crucial element in the public rhetoric of the scientific community.[8]

The term 'rhetoric' is fraught with ambiguity. One definition – perhaps the most common – is epitomized by the phrase 'mere rhetoric', which refers to the persuasive use of words or literary devices. This sense of the term often carries a perjorative connotation – the implication that nothing of substance is being conveyed. There are now some approaches to this issue which challenge the rigidity of the distinction between literal and figural language in science, arguing that imagery and metaphor can have a constitutive, and not merely an expressive, role in scientific discourse.[9] The following discussion does not pursue this issue, but it does suggest that rhetoric can be understood in a more general sense as referring to what Richard Whately, a nineteenth-century authority on the subject, called 'Argumentative Composition'.[10] Recent work in both the philosophy and sociology of science views scientific activity as a process of argumentation and negotiation, rather than solely one of logical proof or justification, and this has led one writer to claim that the concept of 'rhetorical transaction' is crucial to an understanding of science.[11] Furthermore, there is evidence to indicate that apart from the putative heuristic role of scientific method, theories of methodology have been significant elements in this process of scientific debate.[12] This chapter extends the inquiry about the role of methodology by showing how it could function as a rhetorical resource for the presentation of arguments and contentions about the social relations of science. It does so by investigating how certain assumptions and statements about method were deployed in debates about the nature of the scientific enterprise in nineteenth-century Britain.

By the early nineteenth century, the discussion of scientific methodology had crystallized from the more general epistemological concerns of British philosophers. Larry Laudan has remarked that from this period 'entire books rather than prefaces or chapters were devoted exclusively to the subject.'[13] Indeed, between 1830 and 1843, the publication of major works by John Herschel, William Whewell and John Stuart Mill established the parameters of future debate in the philosophy of science. But until the later decades of the century, discussions of methodology were not confined to professional philosophers; rather, they were closely related to substantive scientific issues and, as this chapter will argue, to the social relations of science. In the first half of the century, references to epistemological and methodological issues occurred at several levels: in systematic treatises on the philosophy of science, in periodical reviews, and in the proceedings and transactions of scientific institutions and societies.[14] On the one hand, it is necessary to distinguish between these different levels, recognizing that statements about method could be shaped by perceived audience and context. On the other hand, however, when considering the social dimensions of method, it is useful to note that these levels were not entirely discrete.

John Herschel's *Preliminary Discourse on the Study of Natural Philosophy* – seen by contemporaries as a foundational text for the philosophy of science and a spur to scientific activity – was published in 1830 as a volume in Dionysius Lardner's *Cabinet Cyclopaedia*. While writing his *System of Logic*, Mill told Herschel that this *Discourse*, together with Whewell's works, was an invaluable aid in his 'attempt to methodize the process of investigating truth'.[15] But as well as contributing to formal discussions on the philosophy of science, Herschel aimed to describe the nature and scope of science for a wide audience and, in doing so, gave much attention to the character of scientific method, the hallmark, in his view, of the scientific enterprise. This emphasis on method was also a continuing theme in the addresses and speeches of the British Association for the Advancement of Science, founded in 1831, where it was linked with attempts to legitimate science, to defend it from conservative religious criticism, and to affirm its broad cultural importance. Certain themes in Herschel's *Discourse*, and in other formal writings on science, can be seen as part of this wider deployment of statements about methodology.[16]

This chapter suggests that there were three assumptions about method which were conditioned by, and served as rhetorical resources in, the social relations of science. At various levels of debate, scientific method was represented as *accessible*, *single* and *transferable*. These three characterizations respectively claimed that the method of science could be understood and practised by a large number of people; that there was a single method common to all branches of science; and that this method could be extrapolated from natural science to other subjects. Statements deriving from these assumptions were designed to promote the cause of science by presenting it as founded upon a well-defined and successful method, accepted by all practising researchers, accessible to laymen, and capable of being extended beyond the study of nature to the study of society. The chapter considers each of these assumptions in turn, showing how it was deployed in specific contexts: the idea of an accessible method in claims about the participation of amateurs and laymen in scientific inquiry; the notion of a single method in debates about the unity of science, the effects of specialization and the relationship between different disciplines; and the concept of a transferable method in the campaign to expand the place of natural science in education.

These assumptions about scientific method derived, to a large extent, from a Baconian framework, and it is important to recognize the influence of Bacon's work as the starting point for the philosophy of science in the early nineteenth century. His theories were subject to critical scrutiny in this period, yet Bacon was still acknowledged as the individual who most clearly comprehended the new scientific spirit and most successfully advertised its salient features. Herschel promoted this view in his *Discourse*, although he also placed an emphasis on the value of hypotheses and mathematics which was foreign to Bacon.[17] Similarly, Whewell criticized significant features of Bacon's thought but remained convinced about his historical importance.

> Bacon stands far above the herd of loose and visionary speculators who, before and about his time, spoke of the establishment of new philosophies. If we must select some one philosopher as the Hero of the revolution in scientific method, beyond all doubt Francis Bacon must occupy the place of honour.[18]

Even when his methodology was abandoned as an account of scientific thinking, various features of the Baconian programme continued to exert effects. His notion of an accessible set of rules or procedures for the investigation of nature had considerable currency in debates about the educational rôle of science and the organization of scientific research.[19] Although certainly qualified, his idea of a single, efficacious method capable of producing certain knowledge continued in the work of Herschel and Mill; and Whewell, although rejecting the Baconian theory of induction, retained a commitment to the project of a philosophy of discovery which he saw as a continuation of Bacon's enterprise. Furthermore, all these writers on the philosophy of science were attracted to his vision of an extension of the method of science to the study of politics and society.[20] The connections between theories of method and politico-moral concerns, which informed Bacon's writings, were also visible in the ninteenth-century discourse, even in writers credited with the establishment of a distinct, academic approach to the subject. Although they clashed over epistemology and metaphysics, Whewell and Mill agreed about the wider social importance of proper conceptions of method. Whewell once spoke of the need for a 'reforming philosophy' and hoped to 'get *the people* into a right way of thinking about induction'[21]; he also entertained the idea of establishing a periodical as the organ of an inductive school embracing natural science, political economy and moral philosophy.[22] Similarly, Mill's *System of Logic* was written on the premise that 'no great improvements in the lot of mankind are possible, until a great change takes place in the fundamental constitution of their modes of thought'.[23]

But the following discussion of the place of methodology in the social relations of science should not be seen as a simple reflection of Baconianism. This chapter focuses on assumptions about the accessibility, singularity and transferability of scientific method – assumptions which betray a Baconian descent – but some of the debates investigated indicate the decline of Baconian conceptions of the nature of scientific thinking. The chapter is also concerned with the tensions and contradictions in this rhetoric, with the manner in which it could conflict, in certain situations, with the interests of particular groups within the scientific community. Thus, the following sections show how statements about methods were also implicated in disputes between experts and amateurs, between physical and organic disciplines, and between the educational claims of the natural and social sciences.

2. METHOD AND THE ACCESSIBILITY OF SCIENCE

During the early nineteenth century, individuals and institutions seeking to advance the social status of natural knowledge found it useful to present science as a cooperative pursuit embracing all classes and talents. Statements about method occupied a crucial place in this campaign: in particular, the Baconian notion of a method which rendered 'all wits and understandings . . . nearly on the same level'[24] allowed an image of science as open to wide public involvement. The idea of an accessible inductive method, emphasizing factual observation rather than theoretical knowledge, appeared in the public programmes of institutions such as the Geological Society and the British Association for the Advancement of Science. William Vernon Harcourt, the first Secretary of the Association, laid much stress upon the need for massive collections of data, and indicated the scope available to the participation of amateur and part-time cultivators.[25]

In some contexts, this rhetoric of an accessible method was founded upon a contrast between observation and theory. For example, in the early years of the Geological Society, the gradual accumulation of empirical observations was presented as the guarantee of sound progress in the science; theory and hypothesis, on the other hand, were said to be dangerous. This position served the aims of the founding members of the Society who sought to distance the subject from the controversial cosmologies which had been current in the eighteenth century.[26] It also allowed them to argue that members who lacked previous knowledge were not disadvantaged, because they were not biased by a commitment to theory.[27] In this way, the concept of an egalitarian, inductive method could be linked to hostile attacks upon theory, and scientific progress construed as dependent upon collective effort rather than upon the achievements of gifted individuals.[28]

In this rhetoric, there was also a dichotomy between method and genius. In a biography of Kepler published in the *Library of Useful Knowledge*, readers were warned about 'the danger of attempting to follow his method in pursuit of truth'.[29] Kepler's undeniable achievements were seen to derive from his good fortune in 'seizing truths across the wildest and most absurd theories', and for this reason they did not provide a proper model of methodological practice.[30] In another text, also stemming from the popular education movement, the author characterized Newton as a careful observer rather than a speculative genius, and reported that the great natural philosopher had said that 'if there was any mental habit or endowment in which he excelled the generality of men, it was that of patience in the examination of the facts and phenomena of his subject'.[31] Such an attitude was compatible with the philosophy of self-help proselytized by Samuel Smiles: his redefinition of genius as 'common sense intensified' reduced all achievement, both intellectual and economic, to concepts such as hardwork, perseverance and humility. From this perspective, appropriate method was more important than inherent mental capacity.[32]

To some extent, this egalitarian view of science was sanctioned by influential general works on the scientific enterprise. In his *Reflections on the Decline of Science*, Charles Babbage suggested that the 'division of labour and of cooperation was as applicable to science' as it was to the improvement of manufactures because large areas of science involved ordinary skills of observation possessed by 'any gentleman of liberal education'.[33] Herschel supported this idea in his *Discourse*, arguing that knowledge was improved and refined when submitted to 'the pressure of minds of all descriptions'. In an explicit endorsement of the democratic image of science he urged that it 'should be divested, as far as possible, of artifical difficulties, and stripped of all such technicalities as tend to place it in the light of a craft and a mystery, inaccessible without a kind of apprenticeship.' Although science necessarily possessed a peculiar terminology, this should never give 'an appearance of superiority in its professors over the rest of mankind'. If such a situation occurred, science would be rejecting the 'light which the natural unencumbered good sense of mankind is capable of throwing on every subject, even in the elucidation of principles'.[34] Herschel returned to his theme later in the text, claiming that

> There is scarcely any well-informed person, who, if he has but the will, has not also the power to add something essential to the general stock of knowledge, if he will only observe regularly and methodically some particular class of facts which may most excite his attention. . .[35]

It is not surprising that this passage was cited by those who sought to show that science was open to broad participation.[36]

The belief in the value of data collected by part-time observers was not confined to popular sciences such as geology. In 1849 the British Admiralty published a *Manual of Scientific Inquiry* which encouraged naval officers to record observations in fields such as astronomy, mineralogy, magnetism, meterology and zoology. The book carried a preface by Herschel and included chapters by eminent scientists describing appropriate methods of observation in these fields. The Memorandum from the Admiralty, directing composition of the work, said that the instructions should be 'generally plain, so that men merely of good intelligence and fair acquirement may be able to act upon them.'[37]

The assumption of an accessible scientific method also had significance outside the context of scientific institutions. Thus, in schemes of social improvement involving popular education, the controversial notion of extending knowledge to the lower classes could be justified, on utilitarian grounds, by appeals to Baconian philosophy. In his book

On the Improvement of Society, the evangelical writer, Thomas Dick, argued that

as all science is founded on facts, and as every person possessed of the common organization of human nature is capable of observing facts . . . so there is every reason to conclude that . . . the increase of scientific observers, would ensure the rapid advancement of the different sciences, by an increase of the facts in relation to them which would thus be discovered.[38]

The contention that wide public involvement in science could actually facilitate scientific discovery also informed Henry Brougham's Society for the Diffusion of Useful Knowledge. Apart from aiming to teach science to artisans, this programme contained the notion that every man had 'a chance . . . of becoming an improver of the art he works at, and even a discoverer in the sciences connected with it'.[39]

There was, however, a danger that this emphasis on the accessibility of inductive method could be construed to imply that scientific method was simply a matter of common sense. According to some commentators, it was precisely this confusion which prevented an adequate appreciation of the systematic nature of scientific inquiry. When reviewing Herschel's *Discourse*, Mill hoped that it would succeed in demonstrating 'the superiority of science over empiricism under the name of *common sense*'.[40] Similarly, in the debate about the status of political economy, advocates of the subject found it necessary to insist that the questions it addressed could not be left to common sense, but required the adoption of systematic scientific methods.[41] Macaulay exacerbated this problem by claiming, in his essay on Bacon, that the inductive method was merely an instinctive mode of thinking, 'practised ever since the beginning of the world be every human being'.[42] This equation continued, at some levels, into the second half of the century. In a lecture at the Royal Institution in 1867, Charles Kingsley declared that scientific method needed no elaborate definition:

It is simply the exercise of common sense. It is not a peculiar, unique, professional, or mysterious process of the understanding: but the same which all men employ, from the cradle to the grave, in forming correct conclusions.[43]

But as early as the 1830s there were signs of a reaction against the more extreme espousals of an accessible scientific method. This can be illustrated with reference to the concept of the division of labour, which had been central to eighteenth-century optimism about the promise of collaborative, inductive enterprises. As noted earlier, Herschel emphasized the importance of contributions from a wide range of observers, referring to the 'immense mass of valuable information [which might] be collected from those who, in their various lines of life, at home or abroad, stationary or in travel, would gladly avail themselves of opportunities of being useful'.[44] But throughout the *Discourse* there was a tension between this theme and the need to confirm the role of theoretical experts. Herschel appeared to be searching for a formulation of the division of scientific labour which would allow space for amateur observers without weakening the status of theory and theorists. Statements about the contributions of part-time cultivators were qualified in several ways. Firstly, Herschel argued that in the advanced branches of science, such as astronomy, 'the superior departments of theory are completely disjoined from the routine of practical observation'.[45] This relegated the amateur to a minor role which was further complicated by Herschel's stress on the need for observers to possess skills and knowledge of the various branches of science.[46] He did speak enthusiastically about the participation of the public in subjects such as meteorology and geology, but also recommended that such involvement be carefully directed by experts, suggesting the 'circulation of printed skeleton forms' which would guide the more casual observers.[47]

At the third meeting of the British Association in 1833, Whewell made a further qualification to the idea of 'combined labour' by attacking the assumption that observation and theorizing could be easily separated. He argued that facts only assume significance when viewed from a certain intellectual framework and that, therefore, 'every labourer in the field a science, however humble, must direct his labours by some theoretical views, original or adopted'.[48] This requirement, which he also defined as 'a combination of theory with facts, of general views with experimental industry',[49] severely weakened the assumptions about an accessible inductive method. And by the 1840s, even in subjects usually associated with the idea of wide involvement, the importance of theoretical experts was being reasserted. Thus, the article on geology in the seventh edition of the *Encyclopaedia Britannica* claimed that a new phase in the history of the discipline had arrived: the wild speculation of the past had been overcome, and many facts had been collected; it was now time 'to direct the labours of observation into right and fertile channels'; and those best acquainted with theory were seen as the appropriate leaders.[50]

There was, therefore, a tension between the different rhetorical uses of method. At one level, deference to Baconian induction served to distinguish science from speculative and controversial activities and, by stressing the accessibility of method, encouraged an image of natural knowledge as part of general culture. At another level, statements about method were more carefully specified so as to present an account of scientific research which justified the theoretical guidance of a governing elite. To some extent, it is possible to relate this tension to the different forms and contexts in which the rhetoric of method was employed. Thus, while appeals to an accessible, inductive method continued to be made in the second half of the century, these came to be more confined to popular texts and public lectures and addresses. However, at another level of debate which embraced both formal philosophical works on science, such as those of J. S. Mill and W. S. Jevons, and the reflections of practising scientists such as David Brewster and Justus von Liebig, the emphasis was decidedly anti-Baconian.[51] In these works, there was revision of the earlier methodological canons derived from Bacon and Newton, leading to the establishment of what Laudan has called the 'method of hypothesis', or the hypothetico-deductive method, as the new orthodoxy.[52] On this account, the *Encyclopaedia Britannica* was recording a consensus when it stated, in 1875, that:

> The inductive formation of axioms by a gradually ascending scale is a route which no science has ever followed, and by which no science could ever make progress. The true scientific procedure is by hypothesis followed up and tested by verification; the most powerful instrument is the deductive method, which Bacon can hardly be said to have recognized.[53]

From the perspective of this chapter, it is important that this rejection of Baconianism should not be divorced from its intellectual and social context. The impetus for this shift in methodology cannot be seen simply in terms of an internal development within an intellectual tradition constituted by works on the nature of scientific thought from Bacon to Newton and Locke through to the Scottish philosophers such as Reid and Stewart. The qualification of Bacon's philosophy can, in some respects, be seen as an attempt to legitimize the status of the hypothetical and theoretical aspects of science in a way which countered the attitudes and expectations associated with the rhetoric of an accessible method. This can be seen in the writings of Whewell, which represent the earliest, extensive revision of Bacon's methodology.

As shown above, Whewell's public intervention in methodological debates was an attempt to qualify what he saw an an excessive emphasis on the value of empirical observation in the programme of the British Association. His subsequent major works on the history and philosophy of science were a continuation of his earlier defence of the claims of theory, and can be read as an extended vindication of the role of hypotheses and imagination in scientific thought. In his articulation of this position, Whewell had to confront assumptions about the character of inductive method which, as shown above, had currency in the rhetoric of both scientific institutions and popular societies during the early nineteenth century. Thus, in the chapter on Kepler in his *History of the Inductive Sciences*, Whewell noted that several persons,

> seem to have been alarmed at the *Moral* that their readers might draw, from the tale of a Quest of Knowledge, in which the Hero, though fantastical and self-willed, and violating in his conduct, as they conceived, all right rule and sound philosophy, is rewarded with the most signal triumphs.[54]

In making this remark, he referred to the *History of Astronomy*, published by the Library of Useful Knowledge, which had difficulty in aligning the work of Kepler with its conception of 'experiment and rigorous induction' as the true method of science. Whewell used the case of Kepler to argue, as he did throughout the *Philosophy of the Inductive Sciences*, that the process which led to major scientific discovery was not 'cautious' and 'rigorous' in the sense of eschewing imaginative guesses, conjectures or metaphysical assumptions.[55]

Whewell therefore dissented from the Baconian notion of a mechanical method which levels all intellects and produces scientific truth merely through careful observation and arrangement of facts.[56] He insisted that induction only took place when observed facts were informed by an appropriate idea or conception; such ideas did not simply emerge from the facts but had to be introduced from the mind of the scientist. The knowledge established by this process could give rise to the recognition of necessary truths, a status which, in his view, was warranted by the intuition of scientific experts.[57] At this point, Whewell admitted that scientific progress depended upon the 'peculiar Sagacity which belongs to the genius of a Discoverer';[58] there was no simple rule for the acquisition of appropriate ideas and, to this extent, scientific discovery required 'some fortunate cast of intellect, rising above all rules'.[59] Thus, although he hoped to establish a philosophy of discovery, founded upon historical investigation, which would reveal the process by which science had advanced, he did not conceive this as an art of discovery or a set of mechanical rules. Whewell warned that there was no '*Popular Road*' to truth and his philosophy of science constituted a major check on any wider involvement in science.[60]

Whewell's philosophy of science also had implications for the notion of a division of labour as it appeared in arguments supporting broad participation in science. In the third edition of the *Philosophy*, he acknowledged the importance of what he called the '*social* machinery for the advancement of science', namely, the role of societies and associations in facilitating research by 'dividing the offices of observer and generalizer'.[61] But he insisted that such a segregation did not imply the separation of observation and theory. Thus, when James Spedding commended Bacon's scheme for the involvement of large numbers of people in the collection of facts, Whewell developed the comments of his earlier speech to the British Association about the dangers associated with such a division of labour. In responding to Spedding's references to the example of the *Admiralty Manual of Scientific Inquiry*, he indicated that the areas covered by this volume were those in which a workable theory already existed, thus allowing 'subordinate observers' to confirm or extend it.[62] In the case of scientific fields in which a definite theory had not emerged – such as parts of geology and biology – Whewell insisted that there could be no separation of the observer and the theorist. In this sense, he was arguing that the areas of science most commonly presented as inviting a division of labour were precisely those in which it was inappropriate.[63]

Although Whewell continued to defend the significance of Bacon's writings on science, the basic thrust of his criticism anticipated the more aggressive attacks of Brewster and von Liebig, both of whom regarded Baconian methodology as an obstacle to scientific progress. From about the 1850s, it was commonplace for scientific commentators to stress the role of genius in science in contrast with the earlier emphasis on careful empirical observation. Thus, in breaking the link between Bacon and Newton, Brewster declared that 'the impatience of genius spurns the restraints of mechanical rules, and never will submit to the plodding drudgery of inductive discipline'.[64] In different ways, writers such as John Tyndall, William Carpenter and Karl Pearson presented science as a complex intellectual construction dependent upon hypotheses, unobservable entities and theoretical assumptions. Scientific thinking was said to involve speculative leaps and intuition: Carpenter told the British Association meeting of 1872 that science was akin to poetry, and that interpretations of science were matters of individual personal judgement. Tyndall claimed that major discoveries do not 'depend upon method, but upon that genius of the investigator'.[65] However, these pronouncements about the role of imagination invariably carried the warning that this faculty could only be effective when coupled with the mental discipline which derived from scientific training.[66]

This conception of the nature of scientific thought clearly reversed the emphasis of the earlier rhetoric, supported by Baconian assumptions, which promoted the notion of an accessible method. But this shift, apparent in both formal works and public statements, should not be conceived as a uniform rejection of induction. The major systematic works on the philosophy of science, such as those of Mill and Jevons, were inductivist accounts, but they were theories about methods of verification rather than discovery. Laudan explains this shift by reference to changes in substantive scientific practice: the increasing use of hypothetical and mathematical constructs, especially in the physical sciences, undermined the Baconian concern with mechanical heuristics of discovery, and methodology became preoccupied with the elaboration of criteria for the assessment of hypotheses.[67] Furthermore, when the genesis of scientific ideas came to be viewed as the realm of unpredictable factors such as intuition and imagination, the project of formulating the conditions and mode of their action was largely abandoned. This meant that the method of science was no longer perceived as an engine or art of discovery open to the use of most people.

It is also possible that this development reflects changes in the social conditions under which science was pursued. As science became more securely institutionalized, some of the rhetorical resources provided by Baconianism became redundant. There was less need to stress the possibility of participation by amateurs or the lay public, and the pressure to indicate the connections between science and other forms of culture was no longer so intensive. This consolidation of the scientific community also meant that it did not have to be so defensive about the cultural implications of its activities. The deference to Baconianism in the early decades of the century, related, in part, to the need to advertise the sober, empirical character of scientific inquiry and to distance it from speculation and hypotheses which had controversial political or theological overtones. In contrast, the increasing emphasis on the conjectural and conceptual features of scientific thinking in the later period arguably reflected the comparative independence from religious control, and the greater confidence of a more professionalized community.[68] In this sense, methodology continued to be available as a rhetorical resource. Whereas the Baconian philosophy promoted a careful, non-conjectural inductive method, accessible to all, the emphasis on imagination combined with rigorous verification was associated with a conception of scientific method as the mark of the trained scientific expert pursuing an exciting intellectual quest which lay beyond the reach of the public.[69]

3. THE UNITY OF SCIENCE AND THE UNITY OF METHOD

In his Presidential Address to the British Association in 1850, Prince Albert referred to the increasing specialization of knowledge and feared that it would affect the 'consciousness of its unity which must pervade the whole of Science'.[70] This concern about the relationship between the various branches of science was apparent at the time of the foundation of the Association, a body conceived as a national forum for the coordination of research. Harcourt, the first secretary, claimed that the advance of science would be retarded if different fields of investigation were insulated from each other.[71] In this context, it is significant that when Whewell coined the word 'scientist' in 1834, he did so in a review of Mary Somerville's book *On the Connexion of the Physical Sciences*, a work which attempted to indicate the unity of science by illustrating the general and universal character of natural laws.[72] In searching for a generic term to describe all men of science, Whewell was not only seeking to distinguish 'scientists' from 'artists', but was also attempting to underline the common enterprise in which astronomers, chemists, geologists, and botanists were engaged. Whewell warned that excessive specialization might vitiate the unity of science so that it might come to resemble 'a great empire falling to pieces'.[73] Harcourt saw the British Association as a means of holding this 'commonwealth of Science' together, but the very fact that the Association was divided into sections, many of which were becoming increasingly specialized, suggested that a *substantive* unity of science would be difficult to demonstrate.[74]

In looking back on Somerville's work, James Clerk Maxwell remarked that 'the unity shadowed forth in . . . her book is . . . a unity of the method of science not a unity of the processes of nature'.[75] This was a salient observation, because from the 1830s there was an emphasis on method as the unifying feature of scientific inquiry. In Herschel's *Discourse*, it is possible to see this focus arising from the recognition that the increasingly mathematical character of some fields was a major factor in the growing specialization of science. This connection was tangibly expressed in the order of the sections within the British Association in which Section A, the most influential, comprised astronomy, optics, mechanics and dynamics, subjects which, as Herschel noted, were 'almost exclusively under the dominion of mathematics'.[76] In discussing the value of natural philosophy for general education, he acknowledged that a thorough comprehension of this area demanded a 'degree of knowledge of mathematics and geometry altogether unattainable by the generality of mankind'. However, he then suggested that any person of 'good ordinary understanding' could grasp 'the general train of reasoning by which any of the great truths of physics are deduced', and he indicated that there were many other areas of science in which mathematics was not crucial.[77]

Herschel's discussion of scientific method – the largest part of his *Discourse* – was presented as a general account of the kind of procedures and principles common to all scientific inquiry. This avoided the questions about the role of mathematics and the differences between disciplines; indeed, some points were illustrated by references to that 'deservedly popular science' of geology, which did not require mathematical sophistication.[78] At the most general level, scientific method was presented as the inductive philosophy outlined by Bacon. This was the 'the alpha and omega of science' which, in Herschel's interpretation, did not really exist before this method was consciously employed.[79] Most significantly, he presented method as the defining feature of the scientific endeavour, the common bond relating the cultivators of its various branches:

> In the foregoing pages we have endeavoured to explain the spirit of the methods to which, since the revival of philosophy, natural science has been indebted for the great and splendid advances it has made. What we have all along most earnestly desired to impress on the student is, that natural philosophy is essentially united in all its departments, through all which one spirit reigns and one method of enquiry applies.[80]

This conception of method as the essential feature of science was a prominent theme in the rhetoric of the scientific community throughout the nineteenth century. For example, this assumption made it possible to argue that, in spite of theoretical disputes, the continuity of science was guaranteed by the consensus about method. Herschel claimed that the current opposition between Newton's corpuscular optics and the undulatory theory of light would be resolved by the proper application of scientific method.[81] In this sense, methodology was presented as an area above scientific controversy, an area in which at least general agreement could be expected. It is arguable that Herschel was sensitive to this role when, in his Presidential Address to the British Association in 1845, he sought to separate questions of method from contemporary disputes about epistemology. Alluding to the confrontation between empiricism and idealism, in which Mill and Whewell were major protagonists, Herschel emphasized what he took to be their basic agreement on 'the most essential features' of 'the inductive philosophy'.[82] In this way, method could be presented as a neutral and effective instrument of discovery.

Clearly, however, the scientific community was not unanimous on all points of methodology. For example, the debates about the wave theory of light involved significant methodological argument, with Brewster and Brougham confronting Whewell and Herschel on the role of hypotheses and unobservable entities.[83] But here it is necessary to recognise different levels of discourse about method. It was possible for detailed methodological disputes, conducted mainly in transactions and papers of limited circulation, to coexist with a consensus about the general nature of method in statements of a more public kind. This rhetoric about method was largely constituted by Baconian themes which were exploited to form both negative and positive images. Thus, the method of science was opposed to wild anticipations and conjectures, to dogmatic authority, to abstract deductive systems and to the various idols of the mind. Defined in positive terms, it was based on careful observation of nature, well constructed experiments, moving slowly and cautiously to generalizations. As such, scientific method was said to require and encourage certain attitudes and habits: humility, patience, caution and hard work. The negative image was often associated with Descartes; the positive with Newton.[84] Thomas Chalmers, the Scottish evangelical theologian, expressed the moral aspects of this contrast between methods when he stated that:

In the ethereal whirlpools of Descartes, we see a transgression against the humility of the philosophical character. It is the presumption of knowledge on a subject, where the total want of observation should have confined him to the modesty of ignorance. In the Newtonian system of the world, we see both humility and hardihood.[85]

Although these representations of method lacked specificity, the dichotomies they constructed were significant elements in public legitimations of science. For example, in his presidential Address to the British Association in 1839, Harcourt defended geology against conservative clerical attacks by stating that the evidence which it employed had been 'examined by all the various lights of science, and by undeniable methods of investigation'.[86] This was very much an argument from authority, based on the contention that the natural sciences possessed a method of the most reliable kind. Secondly, the notion of a single, scientific method entered into disputes involving attempts to demarcate acceptable science from unacceptable speculation. Thus, in the controversy surrounding the anonymous *Vestiges of the Natural History of Creation*, attacks upon the author's alleged ignorance of proper method served to characterize the work as pseudoscientific.[87] Even as late as 1859, when Darwin's *Origin of Species* appeared, Baconian rhetoric, similar to that of the early years of the Geological Society, was marshalled in an effort to condemn the work as speculative and conjectural. Sedgwick reviewed the book in *The Spectator* and told his readers that Darwin's theory resembled 'a vast pyramid resting on its apex'; it had not been logically derived from the facts; in short, Darwin's work had 'deserted the inductive track – the only track that leads to physical truth'. He also insinuated that this book, like the *Vestiges*, would appeal to those untrained minds who 'conclude that what is (apparently) *original*, must be a production of original *genius*'.[88] Richard Owen carried this charge further, drawing upon the Baconian notion of scientific procedure as open and accessible, and dismissing novel or controversial theories as the product of individual speculation.[89] Thus, the Baconian conception of inductive method, although increasingly rejected by the major writers on the philosophy of science, continued to function as a means of dealing with controversial theories. This emphasis on method also allowed critics to claim that their objections to a particular doctrine were based 'solely on scientific grounds' and not religious or moral considerations.[90] By attacking at the level of methodology, theories which had unwelcome ideological implications could be labelled as 'unscientific'.

These appeals to method depended upon the circulation of general and non-specific references to observation, experiment and induction. This allowed the idea of a single method, accepted by the scientific community, and capable of being appreciated by the public. However, at another level of discourse within the scientific community, the conception of the unity of method was a point of dispute in debates about the relationships between different disciplines.

The idea of a common method was seized by William Swainson, the author of the *Preliminary Discourse on the Study of Natural History*, the volume in Lardner's series which complemented Herschel's introduction to natural philosophy. Swainson wanted to ensure that the natural history disciplines were accorded scientific status. These subjects lacked the mathematical and experimental character of the physical sciences treated in Herschel's *Discourse*; and those such as zoology and botany, which Swainson discussed, belonged to a popular leisure tradition. Furthermore, they had been placed in a relatively low position in the well-defined hierarchy of sections and disciplines which crystallized in the British Association. This order was defined by the characteristics of the dominant section – the mathematically-based sciences of astronomy, mechanics and optics – and under the influence of Whewell and Herschel these disciplines, which began to assume a monopoly over the term *physical* science, were presented as the model to which all sciences should aspire.[91]

In order to defend the intellectual importance of his own subject, Swainson had to resist this tendency. He did so by asserting the links between natural history and other sciences and, perhaps in response to the hierarchical structure of the Association, referred to its position in 'the *circle* of human knowledge'.[92] 'We address ourselves', he said, 'to those who have been instructed to form more enlarged conceptions of the physical sciences; who view natural history but as a part, and consider that that part must be studied upon the same principles as any other'. He was able to argue that natural history, in spite of its non-experimental basis, shared, with natural philosophy, an important place among the inductive sciences.[93] In this sense, appeals to the idea of a common scientific method played a significant rhetorical function in the negotiations between scientific disciplines, allowing the less established subjects to claim full 'scientific' status.

But there were other situations in which this assumption of a single scientific method was more closely contested. Indeed, even at the time of Herschel's *Discourse* there was a latent tension beneath the surface of general statements about the unity of method. Although Herschel's remarks about method were taken to cover all science, his illustrations of scientific procedure were drawn largely from the advanced physical sciences. In reviewing Herschel's *Discourse*, Whewell agreed that all scientific thought consisted in a combination of induction and deduction, but emphasized that different sciences were at different stages of the 'complete cycle of ascent and descent'.[94] Thus, when compared with astronomy, chemistry was still in its infancy. In Whewell's view, a mature science was one in which the data could be expressed in the form of a 'few general and simple laws'.[95] Although he believed that, in principle, all sciences could hope to reach this point, he argued that their present stage should determine the method necessary to facilitate further progress. Thus, in some disciplines the 'Methods of Observation' would be dominant, in others, the 'Methods of Acquiring Clear Scientific Ideas'.[96] In this way, rather than acting as a counterweight to specialization, scientific method could become a focal point in disputes about the relationships between different disciplines, especially those between the physical and the organic sciences. This can be illustrated, firstly by looking at the response of Baden Powell and William Carpenter to Whewell's philosophy of science and, secondly, by investigating the connection between T. H. Huxley's methodological statements and his defence of the natural history sciences.

Whewell's philosophy of science was a search for the general features of a single philosophy of discovery. He hoped to show that 'the progress of moral, and political, and philological, and other knowledge, is governed by the same laws as that of physical science'.[97] This goal was supported by the claim that the progress of knowledge, in all these areas, depended upon a similar interaction between facts and ideas, between empirical observation and theory. But the concept of a general philosophy of knowledge was qualified by his own theory about the different fundamental ideas appropriate to various fields of science, and this stress upon the distinctiveness of scientific disciplines had implications for the notion of a single scientific method. Whewell believed that the various branches of scientific inquiry went through successive stages, characterized by different relationships between facts and ideas. However, these histories of development were not contemporaneous, and this meant that the present methods of physical astronomy, for example, could not be simply extrapolated to physiology or geology.

Whewell therefore warned that 'different sciences may be expected to advance by different modes of procedure, according to their present condition'; that different methods were appropriate at different stages of each science.[98]

When he illustrated these views in his major works, Whewell's discussion of the new sciences such as physiology and geology implied that these subjects should be conducted on principles at variance with those prescribed for the more advanced physical sciences. For example, he argued that teleological notions could play no part in the investigations of physics or astronomy, but claimed that the concept of final cause was the fundamental idea appropriate to the study of physiology; the major advances in this field had come when researchers allowed teleological assumptions to guide their thinking.[99] In criticizing this aspect of Whewell's work, Powell and Carpenter argued that his suggestion that research in physiology should be directed by speculation about purpose or final cause was tantamount to the exclusion of the subject from the domain of inductive science. In his review of Whewell's *History*, Carpenter said that if taken up, Whewell's views would retard the progress of physiology by disrupting its connections with the techniques and concepts of the physical sciences, such as chemistry. Carpenter asserted that the complexity of physiology required the training which physical science could provide and that, as an inductive science, it should be pursued by methods similar to those of physical science.[100]

Baden Powell strongly advocated the unity of science and resisted any suggestions concerning qualitative differences between the various disciplines. In 1838 he drew attention to the way in which opponents of the new science of geology attacked its conclusions by arguing that its methods of research were distinct from, and 'less strictly inductive', than those of other sciences.[101] He tried to show how the theories of the less controversial disciplines were passively accepted as certain, even though the concepts and reasoning from which they derived were more obscure than those of geology.[102] Like Carpenter, he also criticized the notion that the life sciences involved '*a new class and order of ideas* of so peculiar a kind that they must stand out as entirely exceptional cases to the general *unity of the sciences*'.[103] This attitude justified the use of concepts and methods of reasoning, in areas such as physiology and comparative anatomy, which would not be countenanced in other departments of science. Indeed, Powell's commitment to the ideal of a uniformity of method and principle across the sciences was related to his attempt to bring the species question within the province of normal scientific investigation. He believed that debate on this question was being confused by the intrusion of teleogical reasoning into scientific discourse, a situation justified by Whewell's philosophy of science. Powell criticized Whewell's approach as one which isolated 'each department of science . . . by a theory which would assign to each branch certain real differences of principle and peculiar fundamental ideas essentially characterizing it'.[104] At this level, therefore, statements about method did not express consensus within the scientific community, but were closely implicated in controversial areas of scientific debate. In this case, the emphasis on the unity of scientific method was part of a strategy to avoid the admission of qualitative differences between scientific disciplines.

During the second half of the century, the question of the relationship between physical and organic sciences was a matter of explicit contention. For example, representatives of the natural history disciplines felt it necessary to resist the dominance of the physical and experimental sciences, especially in science eduction.[105] The issue of method was often central to these confrontations, because distinctions between mathematical, experimental and observational modes of inquiry were associated with claims about the relative position of disciplines in the hierarchy of knowledge. This was the context in which T. H. Huxley equated scientific method with 'trained and organized common sense'.[106] By drawing this equation, Huxley tried to resist the assertion that the method of biological disciplines was inferior to that of the physical sciences.

> If . . . there be no real difference between the methods of science and those of common life, it would seem, on the face of the matter, highly improbable that there should be any difference between the methods of the different sciences; nevertheless, it is constantly taken for granted that there is a very wide difference between the Physiological and other sciences in point of method.[107]

Huxley thus attempted to strengthen the status of organic sciences by claiming that they shared the method of the 'Physico-chemical and Mathematical Sciences'.[108] In this view, there was only one method of scientific investigation, which consisted of both observation and experimentation.

The context in which these claims were made illustrates the way in which the question of method had become connected with conflicts between disciplines. For example, although Carpenter agreed with Huxley's defence of physiology, he found it useful, in the case of a particular controversy, to stress the *difference* between the two kinds of methods. Writing in 1871 about the scientific attitude to psychic phenomena, he dismissed one researcher by arguing that his previous experience in observational astronomy did not prepare him for this new area of investigation:

> In particular, we believe that his devotion to a branch of research which tasks the keenest powers of *observation*, has prevented him from training himself in the strict methods of *experimental* inquiry . . .[109]

Thus, in seeking to establish the priority of physiology and its techniques in the areas of psychic research, Carpenter exploited the tension between observational and experimental approaches – a tension which Huxley, for different reasons, sought to conceal.

In a number of scientific debates during the second half of the century, disputes about method, kinds of evidence, and modes of proof were central points of contention. As noted previously, the controversy over Darwin's work involved charges about his departure from sound induction and 'the true Baconian philosophy'.[110] But another aspect of these critiques illustrates the way in which the notion of a single scientific method – which Powell, Huxley and Carpenter saw as essential for the protection of the organic disciplines – could be deployed against Darwin's theory. The most forceful use of this strategy occurred in the review by William Hopkins, the Cambridge mathematician. Although he carefully avoided any moral or metaphysical objections to the *Origin of Species*, Hopkins assessed the book from the perspective of a logician rather than from that of a biologist or geologist. He lamented that 'however well the abstract philosophical rules of theorizing in physical subjects may be known, many people will still theorize as if Bacon and Newton had never lived'.[111] But for Hopkins, these methodological canons were best examplified in the most successful physical sciences, such as astronomy and optics, and his review systematically interrogated Darwin's work against criteria derived from the most established fields of science. He was happy to grant that the 'same principles of research' should apply in all sciences but insisted that the naturalist must therefore accept the standard of evidence demanded of the physicist. 'It is impossible', he said, 'to admit laxity of reasoning to the naturalist, while we insist on rigorous proof in the physicist. He who appeals to Caesar must be judged by Caesar's laws'.[112]

These disputes about methodology within the scientific community were further highlighted by the controversy between Lord Kelvin and the Darwinians over the age of the earth. The tension between the physical and organic disciplines was a central issue here, since much of the discussion exposed disagreement over modes of argument, kinds of evidence and standards of verification.[113] In these late-nineteenth-century debates therefore, methodological conflict was closely linked with clashes between increasingly specialized disciplines and rival professional interests. This made it more difficult for method to be promoted as the point of consensus in science and the means by which the layman could understand, or appreciate, scientific research. Writing in 1877, William Gladstone remarked that:

> The results, in pure science, are accepted by us as facts; but on the methods by which they are reached, the mass, even of intelligent and cultivated men, are not completely informed.[114]

4. THE TRANSFERABILITY OF SCIENTIFIC METHOD

The assumption that there waş a single, definable scientific method allowed claims about the possibility of extending it to other subjects. In his *Grammar of Science*, Pearson asserted that '*the unity of all science consists alone in its method, not in its material*'[115]; and he used this distinction to reinforce the claim for a scientific approach to social problems. There was, of course, a Baconian warrant for such a programme, and it was a significant theme in early-nineteenth-century writing on the philosophy of science.[116] In his *Discourse*, Herschel suggested that the success of proper methods of inquiry in the area of natural knowledge might lead to a situation in which legislation and politics were 'regarded as experimental sciences'.[117] Not unexpectedly, J. S. Mill was an early supporter of this idea, arguing that the first contribution of physical science to society would come not 'by the truths which it discloses, but by the process by which it attains them'.[118] The present weakness of moral (or social) science was due to a 'lack of proper method', and this could be taken from the natural sciences. Herschel returned to this topic in his Presidential Address to the British Association in 1845, urging that attention to scientific method was relevant to the understanding and progress of society, and suggesting that the work of Whewell and Mill on scientific thought had implications for '*the principles of all general reasoning*'.[119] These were early expressions of the claim to intellectual and cultural authority, beyond the realm of natural knowledge, which later become a major element in the rhetoric of public science.

Nevertheless, in the 1830s and 1840s, the extension of the method of natural science to social and political questions was still a contentious issue. Even amongst writers who defended the notion of inductive moral or social sciences, there was dispute about the precise way in which the method of physical science could be extrapolated. Herschel would have been aware that Whewell, although not unfavourable towards the scientific pretensions of political economy, was critical of hasty transfers of concepts and methods from physical to moral subjects.[120] By the end of the century, however, the analogies between natural and social inquiries were pursued with enthusiasm.

Representatives of the natural sciences generally lent their support to the wider application of scientific procedures. Indeed, the exponents of scientific naturalism such as Huxley, Spencer and Galton campaigned in favour of this kind of transfer. Pearson repeated this aspiration, stating that 'the material of science is coextensive with the whole life, physical and mental, of the universe'.[121] In 1910, this unlimited field was formally annexed by the British Association when it established a section for Education Science which, apart from objectives in the area of pedagogic technique, was inspired by the desire to take 'a scientific view of things in every department of life. . .'.[122] This was one of the contexts in which sociology, anthropology, psychology and archaeology claimed the title of social sciences and cited the use of empirical and experimental methods as evidence of their 'scientific' status.

In addition to supporting the claim that social inquiry could be modelled on the natural sciences, assumptions about the transferability of scientific method occurred in educational debates. From the early decades of the nineteenth century, those who sought to increase the representation of natural science in university education found it necessary to conform with prevailing conceptions of liberal education. In the two ancient English universities, this curriculum was founded upon the classics and mathematics, and supported values rigidly opposed to the intrusion of 'useful' or professional knowledge. In this context, an emphasis on the method of science rather than its content allowed natural science to be presented as a form of mental discipline. Thus, when addressing the readers of the Tory *Quarterly Review*, Charles Lyell agreed 'with those who maintain that the most important part of education consists, not so much in the things taught, as in the moral and intellectual habits instilled'.[123] He was then able to argue that the physical sciences trained the mind to reason about probable evidence rather than demonstrative theorems, thus providing the best preparation for the intellectual demands of practical life.[124] As late as 1867, J. S. Mill recommended instruction in 'the Methods of the Sciences' as the best example of 'the art of thinking'.[125] In this way, a rhetorical emphasis on method made it possible to argue for the inclusion of natural science in university teaching without challenging the non-professional ethos of liberal education.

The distinction between scientific method and scientific information also entered debate about the popularization of science. Important men of science such as Powell, Lyell and Herschel pointed to the danger of an unequal distribution of knowledge throughout society and to the problem of a small scientific class appearing as the cultivators of an esoteric language.[126] However, the appropriate form of popular science education was difficult to define. There was dissatisfaction with most popular accounts of science because they stressed factual information and practical applications rather than the intellectual processes by which results were attained. But, as scientific knowledge expanded and became more complex, the possibility of explaining it to a non-scientific public became unrealistic. Out of this situation, the conception of scientific method as an educational bridge between science and the public emerged as an influential solution.[127] G. C. Lewis, for example, who was concerned about the gap between experts and laymen, recommended that:

> every person should, so far as his means extend, make himself master of the methods of scientific investigation, so as to be able to judge whether in the treatment of any question, a sound and correct method has been observed.[128]

Throughout most of the century, the British Association exemplified this investment in the educational value of scientific method. At the Glasgow meeting in 1855, for example, the Duke of Argyll condemned 'cramming of facts from manuals', and suggested that 'what we want in the teaching of the young, is, not so much the mere results, as the *methods*, and, above all, the *history* of science'.[129] Pearson endorsed this theme in 1892, stating that 'it is not the facts themselves which form science, but the method in which they are dealt with'.[130] This dichotomy between the content and method of science also became the focus of a doctrine known as 'heurism', promoted by the chemist, Henry Armstrong. In the 1880s, Armstrong began to argue that current science teaching was too dogmatic, too concerned with imparting facts and theories, often by means of rote learning.[131] He advocated a shift of emphasis from content to method and outlined pedagogic techniques designed to place students 'so far as possible in the attitude of the discoverer'.[132]

Heuristic teaching attracted substantial support; but by the second decade of the twentieth century it was meeting with determined opposition. Critics complained that the system was impractical, expensive, not fully examinable and, that where implemented, it led to a situation in which science courses and students were dangerously free of content. In short, the stress on method had been excessive. This attack reached a climax in the 1917 Report of the British Association Committee on Science in Secondary Schools, headed by Richard Gregory, the editor of *Nature*. Arguing against the tradition which had culminated in Armstrong's heurism, the committee asserted that the method of scientific investigation was not separable from its matter or content:

> In other words, it is not strictly true that scientific method is one and the same wherever it is employed. The physical method and the historical method, for example, have common fundamental features, but cannot be simply identified the one with the other. In short, scientific method is an abstraction which does not exist apart from its concrete embodiments; and the person who desires adequate knowledge of it must study it in all its typical manifestations.[133]

This was a virtual recantation of the assumptions which had supported the earlier rhetoric about the role of method as an educational bridge between science and the public, and the possibility of its extension to a wide range of subjects. In order to understand this reversal it is necessary to appreciate the way in which the stress on method became a strategic disadvantage in educational debates.

By the early twentieth century, the precise meaning of 'science' and 'scientific' often required clarification. Thus, when T. P. Nunn set out to write a book on *The Aim and Achievements of Scientific Method* in

1907, he felt it necessary to begin with a preface on semantics. After explaining that 'science' acquired its restricted meaning of physical or natural science with the establishment of the British Association in 1831, he warned that the word 'scientific' had now assumed a much broader denotation.

> Thus, while there can be no doubt that the word "Science" refers most commonly to a certain body of knowledge, "scientific" refers more frequently to the method or procedure by which this knowledge has been established. By an easy extension of its meaning the adjective is applied to all processes by which knowledge is reached having the security which is thought to characterize "science".[134]

Because of this ambiguity, a Conference on the Humanities and Social Sciences held in 1917 was able to argue that the disciplines it represented employed scientific methods and therefore provided the mental training which the natural scientists were emphasizing. In response to such claims, the Committee on the Neglect of Science was forced to state that 'scientific method can be used in the study of literary, historical and other subjects; but applied in this way it can never be accepted as providing the same means of training as direct observation and experimental work'.[135] This was another attack upon the priority which Armstrong's heurism accorded to method over content and information; but it also derived from the recognition that this dichotomy could be used to argue that scientific method was no longer the exclusive possession of the scientific community.

> If, however, it is admitted (1) that the chief business of the science teacher is to train in scientific method, and (2) that scientific method is the characteristic not of science only but of every properly conducted intellectual inquiry, the science teacher is perilously near to the surrender of his special claim to existence.[136]

The debate over science education illustrates the way in which the implications of this rhetoric could, in specific situations, conflict with the interests of the scientific community. In 1917, when various committees reported on the place of science in the curriculum, there was a shift away from the heavy emphasis on method which had emerged in the second half of the nineteenth century. The rise of the social sciences and their claim to possess a 'scientific' approach made the earlier rhetoric more problematic. But furthermore, some scientists believed that the stess on method had produced a technical and instrumental image which divorced the natural sciences too sharply from humanistic studies. Given the war and the increased awareness of the technological applications of science, there were fears in some sections of the scientific community that future funding could be too closely influenced by judgements about utility.[137]

During the twentieth century, an emphasis on method continued to be significant in definitions of science and in claims about its educational and social value. In 1917, one of the first collaborative publications in the history and philosophy of science, Charles Singer's *Studies in the History and Method of Science*, gave the concept of scientific method a foundational role, explaining that 'the need for a comprehensive study of the methods of science is now widely recognized', and that the historian could teach 'the method by which knowledge has been gained'.[138] A similar focus on method was also discernible in the later attempts of J. B. Conant and I. B. Cohen to explain science to educated, non-scientific audiences. Conant's *Harvard Case Histories in Experimental Science*, published in 1957, attempted to encourage an understanding of 'the methods of experimental science' as a way of helping the lawyer, teacher or public servant to evaluate the work of scientific experts.[139] The more extreme claims concerning the political value of scientific thought also continued to depend upon statements about method. In December 1940, for example, *Nature* carried an editorial on the 'Cultural Significance of Scientific Method' which announced that 'the fate of democracy is bound up with the spread of the scientific attitude'.[140]

5. CONCLUSION

Most recent historiography of scientific method has been concerned with the influence of particular methodologies in the work of individual scientists, or with the role of methodological doctrines in substantive scientific debates.[141] This chapter has offered a different perspective by focusing on the rhetorical aspect of discourses about method in the wider social relations of science. Three key assumptions about method were delineated and it was argued that each of these was deployed, in specific contexts, to advance certain images of science. However, as a counterpoint theme, the analysis revealed ambiguities and contradictions within this rhetoric and investigated the ways in which these related to points of tension within the scientific community. Thus the emphasis on the accessibility of method entailed arguments about the relation between theory and observation and the role of experts and amateurs; the idea of a single scientific method was involved in conflicts between disciplines; and the concept of transferability led to situations in which the dominance of the natural sciences was questioned.

The approach taken here also suggests that the historical and sociological perspectives which have been applied to substantive scientific texts should be extended to works on methodology. References to method occurred at various levels, and it must be recognized that the social and strategic aspects of public lectures and addresses were more obvious than those of systematic works on the philosophy of science. For example, the nature of methodological comments at a meeting of the British Association could be expected to differ from those in works such as Herschel's *Discourse* or Whewell's *Philosophy*. Nevertheless, while these works responded to issues and questions which have occurred in a variety of historical situations, they were also written in specific social and intellectual contexts, and it should not be assumed that they conveyed only an abstract discourse on methodology.

Herschel's book lent credence to assumptions about method which were associated with an account of scientific research as accessible, unified and potentially applicable to social problems. As such, it was both a contribution to the contemporary debate about the place of science in early Victorian culture and a resource for other participants in this debate. Whewell's work, on the other hand can, in some respects, be read as a reaction against these public representations of scientific method. He severely qualified the notion of an accessible, inductive procedure and was cautious about the concept of a single and transferable mode of scientific thinking. But in taking this position, Whewell was also employing methodology to advance his preferred definition of science – one which stressed the importance of theoretical experts, restricted the division of scientific labour, and saw the mathematico-physical sciences as the highest form of natural knowledge. Whewell's work illustrates the way in which certain topics in the philosophy of scientific method could involve social and institutional considerations. As we have seen, the concept of a single method of science was central to debates about the relationship between physical and biological disciplines. In this case, writers of general philosophical works on science, such as Whewell and Powell, and practising scientists such as Carpenter and Huxley, were not simply formulating descriptive or prescriptive statements about scientific research: they were deploying method as a rhetorical weapon in negotiations about the place of different disciplines in the hierarchy of scientific knowledge.

This investigation does not support a simplistic notion of methodological discourse as an ideology which uniformly promoted the interests of the scientific community.[142] Instead, it suggests that the appeal to method in the apologetics of science was not identified with any single interest. The three assumptions about method which have been discussed did not indicate points of unqualified or unshifting consensus, but rather areas of debate in which statements and counter-statements were made. In fact, such a picture is to be expected if we recognize that the 'scientific community' embraced various groups which perceived different priorities for both the intellectual agenda and the social relations of science. The Baconian notion of accessibility, although influential in certain contexts and for particular audiences, was qualified by those who wanted to underline the exclusive character of scientific inquiry. The idea of a unity of method, which was promoted as a counterweight to the fragmentation of science, played a different and more complex role in disputes between disciplines. Thus, while there were recurring themes in the rhetorical use of method, a proper grasp of its significance also demands attention to the various points of emphasis and tensions, and to the ways in which these related to different contexts and preoccupations. The subject of methodology was an argumentative resource which could be exploited in a variety of ways in debates about the definition of science and its cultural significance.

ACKNOWLEDGEMENT

I would like to thank Geoffrey Cantor, Jonathan Hodge, Menachem Fisch and John Schuster for their helpful comments on earlier drafts of this chapter. The research was undertaken as part of study leave supported by Griffith University.

Griffith University, Australia.

NOTES

[1] G. Orwell, 'What is Science?' in *Collected Essays, Journalism and Letters* (4 vols, Harmondsworth, 1970), vol. 4, pp. 26–30, at p. 27.

[2] B. Webb, *My Apprenticeship* (London, 2nd ed. nd.), p. 113. See also p. 118 for the distinction between the 'subject matter of natural science' and the 'value of physical science as a training in scientific method'.

[3] K. Pearson, *The Grammar of Science* (London, 1892) pp. 10–11. See also N. Lockyer, 'Scientific Administration', *Nature*, 6 October 1870, pp. 449; M. Cohen and E. Nagel, *An Introduction to Logic and Scientific Method* (London, 1934), p. 403.
[4] See F. M. Turner, 'Public Science in Britain, 1880–1919', *ISIS* **LXXI**, 1980, pp. 589–608. For a comparative analysis, see P. Buck, 'Order and Control: The Scientific Method in China and the United States', *Social Studies of Science* **V**, 1975, pp. 237–67. See also Y. Ezrahi, 'Science and the Problem of Authority in Democracy', in *Science and Social Structure*, R. K. Merton (ed.), *New York Academy of Sciences* **xxxix**, (1980), pp. 43–60.
[5] I. Hacking, *The Emergence of Probability* (Cambridge, 1975), pp. 179–180; B. Shapiro, *Probability and Certainty in Seventeenth-Century England* (Princeton, 1983), pp. 15–73.
[6] J. Ben-David, *The Scientist's Role in Society: A Comparative Study* (New York, 1971), pp. 72–74.
[7] P. Wood, 'Methodology and Apologetics: Thomas Sprat's History of the Royal Society', *British Journal for the History of Science* **XIII**, 1981, pp. 1–26.
[8] T. Sprat, *The History of the Royal Society*, J. I. Cope and H. W. Jones (eds.), (London, 1959), pp. 62, 112. Sprat claimed that the members of the Royal Society were pledged to recovering the 'primitive purity' of language in which 'men delivered so many *things*, almost in an equal number of *words*'. (p. 113) For the connection between science and the concept of a 'plain style'. see R. F. Foster, *The Seventeenth Century: Studies in the History of English Thought and Literature from Bacon to Pope* (Stanford, 1951); W. S. Howell, *Eighteenth-Century British Logic and Rhetoric* (Princeton, 1971), pp. 441–502.
[9] See A. Benjamin, G. N. Cantor, J. R. R. Christie (eds), *The Figural and the Literal: Problems of Language in the History of Philosophy, Science and Literature, 1650–1800* (Manchester, forthcoming). For earlier references, see N. Frye, *Anatomy of Criticism* (Princeton, 1957), pp. 326–337; M. Hesse, 'The Explanatory Function of Metaphor', in *International Congress for Logic, Methodology and Philosophy of Science* (Amsterdam, 1964); R. M. Young, 'Darwin's Metaphor: Does Nature Select?', *Monist* **LV**, 1971, pp. 442–503.
[10] R. Whately, *Elements of Rhetoric* (5th ed., London, 1836), p. 6.
[11] W. B. Weimar, 'Science as Rhetorical Transaction: Toward a Nonjustificational Conception of Rhetoric', *Philosophy and Rhetoric* **X**, 1977, pp. 1–29. Weimar believes that the rhetorical aspect of scientific discourse is implied by Kuhn's work (p. 24).
[12] For references to the works of Cantor, Laudan and Ruse which deal with the role of methodology in nineteenth-century scientific debates, see the editors' introduction to this volume.
[13] L. Laudan, 'Theories of Scientific Method from Plato to Mach: A Bibliographic Review', *History of Science* **VII**, 1968, p. 29.
[14] See the comment about the concern of Victorian scientists with method in G. Basalla, W. Coleman and R. Kargon (eds.), *Victorian Science: A Self Portrait* (New York, 1970), p. 399 and the examples from British Association addresses. For the importance of the Scottish contribution, see R. Olson, *Scottish Philosophy and British Physics 1750–1880* (Princeton, 1975).
[15] J. S. Mill to J. Herschel, May 1843, Herschel Papers, Royal Society, London; see J. S. Mill, 'Herschel's *Discourse*', *Examiner* 20 March 1831, pp. 179–180; [W. Whewell],

'Modern Science, Inductive Philosophy', *Quarterly Review* **XLV**, 1831, pp. 374–407, at pp. 377, 398 for its philosophical importance, and M. Faraday to J. Herschel, 10 November 1832, in *The Selected Correspondence of Michael Faraday*, L. Pearce Williams (ed.), (2 vols, Cambridge, 1971), vol. 1, p. 235, for its inspirational value. For an exposition and analysis of Herschel's philosophy of science, see C. J. Ducasse, in *Theories of Scientific Method: Renaissance through the Nineteenth Century*, R. Blake, C. Ducasse, and E. Madden (eds.), (Seattle, 1960), pp. 153–182.

[16] For attempts to contextualize some of the themes in the *Disourse*, see W. F. Cannon, 'John Herschel and the Idea of Science', *Journal of the History of Ideas* **XXII**, 1961, pp. 215–239; J. Agassi, 'Sir John Herschel's Philosophy of Success', *Historical Studies in the Fhysical Sciences* **I**, 1971, pp. 1–36; R. Yeo, 'Scientific Method and the Image of Science 1831–1891', in *The Parliament of Science: The British Association for the Advancement of Science, 1831–1981*, R. M. MacLeod and P. Collins (eds.), (Northwood, Middlesex, 1981), pp. 65–88. I will also be referring to the works of William Whewell and to a lesser extent to the views of Baden Powell, William Carpenter and T. H. Huxley.

[17] J. Herschel, *A Preliminary Discourse on the Study of Natural Philosophy* (London, 1830), pp. 104–114, 204. There is some confusion about the publication date because one of the title pages carries an 1831 date.

[18] W. Whewell, *The Philosophy of the Inductive Sciences* (2 vols. London, 1847), vol. 2, p. 230. See also B. Powell, *History of Natural Philosophy from the Earliest Periods to the Present Time* (London, 1830), pp. 196–211. Like Herschel's *Discourse*, this work was published in Lardner's *Cyclopaedia*. For the various levels of the debate on Baconianism see R. R. Yeo 'An Idol of the Marketplace: Baconianism in Ninteenth-Century Britain', *History of Science* **XXIII**, 1985, pp. 251–298.

[19] See for example, R. Laudan, 'Ideas and Organization in British Geology', *Isis* **LXVIII**, 1977, pp. 527–538.

[20] For the expectation of certainty from inductive procedures in the early nineteenth century, see D. Hull (ed.), *Darwin and his Critics* (Cambridge, Mass., 1973), p. 17. Bacon claimed that his inductive method 'extends not only to natural but to all sciences'. See F. Bacon, *Novum Organum*, Book 1, Aphorism 127.

[21] W. Whewell to R. Jones, February 1831, in I. Todhunter, *William Whewell D. D. An Account of his Writings* (2 vols, London, 1876), vol. 2, p. 115.

[22] W. Whewell to R. Jones, 24 April 1831, in *ibid*, p. 118.

[23] J. S. Mill, *Autobiography*, J. Sillinger (ed.), (London, Oxford, 1971), p. 142. For further connections between method and political values, see [S. Bailey], *Essays on the Pursuit of Truth* (London, 1829); [T. P. Thompson], 'Pursuit of Truth', *Westminster Review* **XI**, 1829, pp. 478–487. For the relationship between proper method and moral duty within a religious framework, see I. Watts, *The Improvement of the Mind* (London, 1741); J. Abercrombie, *Inquiries concerning the Intellectual Powers and the Investigation of Truth* (Edinburgh, 1831); and M. Faraday, 'Observations on the Inertia of Mind', in B. Jones, *The Life and Letters of Michael Faraday* (2 vols, London, 1870), vol. 1, pp. 261–279.

[24] F. Bacon, *Novum Organum*, in *The Works of Francis Bacon*, J. Spedding, R. Ellis, and D. Heath (eds.), (14 vols, London, 1857–74), vol. 3, p. 63.

[25] W. V. Harcourt, *British Association Reports* (York and Oxford, 1831–2), pp. 23–31. Further references to this collection are cited as *B. A. R.* See also pp. 10–14

for an extract from a report of the Yorkshire Philosophical Society which stresses the role of provincial observers. For the Baconian rhetoric of cooperation, see J. B. Morrell and A. Thackray, *Gentlemen of Science: Early Years of the British Association for the Advancement of Science* (Oxford, 1981), pp. 267–9.

[26] See R. Laudan, *op. cit.* (Note 19), pp. 527–638; R. Porter, *The Making of Geology: Earth Science in Britain, 1660–1815* (Cambridge, 1977), pp. 204–208. See [W. H. Fitton], 'Transactions of the Geological Society', *Edinburgh Review* **XIX**, 1811–12, pp. 207–229, at p. 207, for the description of eighteenth-century geology as 'a species of mental derangement'.

[27] [W. H. Fitton], 'Transactions of the Geological Society', *Edinburgh Review* **XXIX**, 1817–18, pp. 70–94, at p. 70.

[28] T. Dick, *On the Improvement of Society* (London, 1833), pp. 105–110.

[29] J. Drinkwater, *Lives of Eminent Persons* (London, 1833), p. 54.

[30] *Ibid.* p. 15.

[31] G. L. Craik, *The Pursuit of Knowledge Under Difficulties* (2. vols, London, 1846), vol. 1, p. 2. See also H. Martineau, *Miscellanies* (2 vols, Boston, 1836), vol. 1, pp. 72, 100, 103.

[32] S. Smiles, *Self Help* (new edition, London, 1860), pp. 46–47.

[33] C. Babbage, *Reflections on the Decline of Science in England* (London, 1830), pp. 40, 170.

[34] J. Herschel, *op. cit.* (Note 17), pp. 69–70.

[35] *Ibid.* p. 133.

[36] See H. de la Beche, *How to Observe – Geology* (London, 1833), p. 83; also J. Phillips, *A Treatise on Geology* (2 vols, London, 1839), vol. 2, pp. 279–280 for an emphasis on the 'plainness and accessibility' of geology.

[37] J. Herschel (ed.), *The Admiralty Manual of Scientific Enquiry* (London, 1849) p. iii.

[38] T. Dick, *op. cit.* (Note 28) pp. 82–83. See also *Recreations in Science* (London, 1830) for defence of empirical fact as opposed to systems.

[39] H. Brougham, quoted in J. W. Hays 'Science and Brougham's Society', *Annals of Science* **XX**, 1964, p. 231.

[40] J. S. Mill, *op. cit.* (Note 15), p. 179.

[41] R. Whately, *Introductory Lectures on Political Economy* (London, 3rd ed., 1847), pp. 55–64.

[42] [T. Macaulay], 'Francis Bacon', *Edinburgh Review* **LXV**, 1837, pp. 1–104, at p. 87.

[43] C. Kingsley, 'Science', in *Scientific Lectures and Essays* (London, 1890), p. 210.

[44] J. Herschel, *op. cit.* (Note 17), pp. 133–134.

[45] *Ibid.* p. 132.

[46] *Ibid.* Herschel also implied that new areas of reasearch relied on 'individual penetration' and required 'a union of many branches of knowledge in one person'; it was the mature areas which could benefit from division of labour, although this was often skilled labour (pp. 102–103, 131). For a similar, but more explicit formulation, see the discussion of Whewell below.

[47] *Ibid.* p. 134.

[48] W. Whewell, *B.A.R.* (Cambridge, 1833), p. xx.

[49] *Ibid.* See also pp. xxi-xxii where he indicates the danger of dogmatic theory. See also J. B. Morrell and A. Thackray, *op. cit.* (Note 25), pp. 270–271.

[50] *Encyclopaedia Britannica* (7th ed., Edinburgh, 1830–42), vol. 15, p. 173. In fact, in the public statements of the Geological Society, which contained a strong emphasis on

Baconian empiricism, it is possible to detect another methodological message which provided a definite role for the theorist. Thus, references to the importance of empirical observation 'by persons attached to various pursuits and occupations' could be coupled with the recognition that the task of generalizing and systematizing required a knowledge of other sciences and could be 'performed only by philosophers'. See *Philosophical Magazine* **XLIX**, 1817, pp. 421–429, at pp. 421–422. Rachel Laudan's analysis of methodological pronouncements in the Geological Society overlooks the way in which the Baconian tradition could be interpreted to provide this space for theorists, as distinct from the army of observers. See R. Laudan, *op. cit.* (Note 19) and the chapter by David Miller in this volume.

[51] W. S. Jevons, *The Principles of Science: A treatise on Logic and Scientific Method* (London, 2nd ed., 1877), p. 507; D. Brewster, *Memoirs of the Life, Writings and Discoveries of Sir Isaac Newton* (2 vols, Edinburgh, 1855), vol. 2, pp. 400–403; J. von Liebig, 'Lord Bacon as Natural Philosopher', *MacMillan's Magazine* **VIII**, 1863, pp. 237–249; 257–265.

[52] L. Laudan, *Science and Hypothesis: Historical Essays on Scientific Methodology* (Dordrecht and Boston, 1981), pp. 4–5. Laudan claims that the 'method of hypothesis' had been popular in the seventeenth century but was ostracized by eighteenth-century inductivism.

[53] R. Adamson, 'Bacon', *Encyclopaedia Britannica* (9th ed., Edinburgh, 1875–89), vol. 3, p. 216. For a discussion of the reactions against Baconianism, see R. Yeo. *op. cit.* (Note 18).

[54] W. Whewell, *History of the Inductive Sciences* (3 vols, London, 1837), vol. 1, pp. 410–411. For his response to the image of Newton as merely a patient investigator, see vol. 2, p. 185.

[55] W. Whewell, *op. cit.* (Note 18), vol. 2, pp. 41, 46–59.

[56] *Ibid.* vol. 2, p. 240.

[57] W. Whewell, *On the Philosophy of Discovery* (London, 1860), pp. 336–340, 347–348.

[58] W. Whewell, *op. cit.* (Note 18), vol. 2, p. 40.

[59] *Ibid.* pp. 20–21.

[60] *Ibid.* p. 366. There was a tension in Whewell's work on the question of what could be learnt about appropriate method from a study of the history of science. Compare his *History of the Inductive Sciences* (3 vols, 2nd ed., London, 1857), vol. 1, pp. viii, 4, and *Novum Organon Renovatum* (London, 1858), pp. iii–v, 141–142.

[61] *Ibid.* pp. x–xi.

[62] [W. Whewell], 'Spedding's Complete Edition of the Works of Bacon', *Edinburgh Review* **CVI**, 1857, pp. 287–322, at p. 314.

[63] W. Whewell, *op. cit.* (Note 57), pp. 154–156. He also remarked that a 'considerable degree of scientific education is needed even for the subordinate labourers in science'. *ibid.* p. 156.

[64] D. Brewster, *op. cit.* (Note 51), vol. 2, p. 405.

[65] W. B. Carpenter, *B.A.R.* (Brighton, 1872), pp. lxix–lxxxiv, at pp. lxxi–lxxiii. Carpenter was Professor of Forensic Medicine, University of London, from 1856–1879; J. Tyndall, 'Scientific Materialism', in *Fragments of Science* (2 vols, New York, 1897), vol. 2, p. 77.

[66] K. Pearson, *op. cit.* (Note 3), p. 31. J. von Liebig, 'Induction and Deduction', *Cornhill Magazine* **XII**, 1865, pp. 296–303.

[67] L. Laudan, *op. cit.* (Note 52), pp. 181–191. Nevertheless, there were still some attempts to retain a method of discovery. See G. Gore, *The Art of Discovery* (London, 1878).

[68] See F. M. Turner, 'The Victorian Conflict between Science and Religion: a professional dimension', *Isis* **LXIX**, 1978, pp. 356–376.

[69] From 1811, the *Transactions of the Geological Society* carried as a motto a quote from Bacon's *Novum Organum*; after 1860, the last phrase of the extract, which spoke of the inner shrine of science as open to all, was deleted. For a reference to this deletion, without comment on its significance, see H. B. Woodward, *The History of the Geological Society of London* (London, 1907), p. 44.

[70] Prince Albert, *B.A.R.* (Aberdeen, 1859), p. lxiii.

[71] W. V. Harcourt, *op. cit.* (Note 25), p. 28.

[72] For Somerville's conviction about the unity of the sciences, see M. Somerville, *On the Connexion of the Physical Sciences* (London, 1834), preface and p. 413. See also E. C. Patterson, *Mary Somerville and the Cultivation of Science, 1815–1840* (Boston, The Hague, 1983).

[73] [W. Whewell], 'On the Connexion of the Physical Sciences' *Quarterly Review* **LI**, 1834, pp. 54–68, at pp. 59–60.

[74] Whewell's concern about the unity of science did not prevent him from defending the sectional divisions of the British Association. See J. B. Morrell and A. Thackray, *op. cit.* (Note 25), p. 455.

[75] J. Clerk Maxwell, *Scientific Papers* W. D. Niven (ed.), (2 vols, Cambridge, 1890), vol. 2, p. 402. For an earlier reference to the need for general rules of inquiry in the context of diverse fields of knowledge, see M. Faraday, *Some Observations on the Means of Obtaining Knowledge* (London, 1817), pp. 8–9.

[76] J. Herschel, *op. cit.* (Note 17), p. 26.

[77] *Ibid.* pp. 25–26. But in his introductory text on astronomy, intended for a large audience, Herschel warned that it assumed a basic mathematical knowledge. See his *Treatise on Astronomy* (London, 1833), p. 5. For the importance of mathematics in physical science from the 1830s, see G. N. Cantor, *Optics after Newton: Theories of Light in Britain 1704–1840* (Manchester, 1983), pp. 147–150.

[78] J. Herschel, *op. cit.* (Note 17), p. 145; also pp. 281–289.

[79] *Ibid.* pp. 104–105, 114.

[80] *Ibid.* p. 219.

[81] *Ibid.* pp. 253–254, 263–264. For an extended argument on this, see J. Agassi, *op. cit.* (Note 16), pp. 1–36.

[82] J. Herschel, *B.A.R.* (Cambridge, 1845), p. xi. Whewell himself made some attempt to acknowledge some common ground between his views on method and those of Mill. See W. Whewell, *On Induction* (London, 1849), p. 5.

[83] G. N. Cantor, *op. cit.* (Note 77), pp. 177–85 and his earlier articles cited in the Introduction to this volume.

[84] For views about the moral effects of induction and deduction, see W. Whewell, *Astronomy and General Physics Considered with Reference to Natural Theology* (London, 2nd ed., 1834), pp. 323–342. For eighteenth-century views of Newton and Descartes, see p. M. Rattansi, 'Voltaire and the Enlightenment Image of Newton', in *History and Imagination: Essays in Honour of H. R. Trevor-Roper* (London, 1981), pp. 218–231.

[85] T. Chalmers, *Evidence and Authority of the Christian Revelation* (2nd ed., Edinburgh, 1815), p. 200.

86 W. Harcourt, *B.A.R.* (Birmingham, 1839), p. 18.
87 See R. R. Yeo, 'Science and Intellectual Authority in Mid-Nineteenth Century Britain: Robert Chambers and *Vestiges of the Natural History of Creation*', *Victorian Studies* **XXVIII**, (1984), pp. 5–31.
88 A. Sedgwick, 'Darwin's Origin of Species', *Spectator*, 7 April 1860, pp. 334–335. This review was a revised version of one which appeared on 24 March 1860.
89 [R. Owen], 'Darwin on the Origin of Species', *Edinburgh Review* **CXI**, pp. 487–532, at pp. 495–496.
90 [S. Wilberforce], 'Darwin's *Origin of Species*', *Quarterly Review* **CVIII**, 1860, pp. 225–264, at p. 256. For an account of the reaction to Darwin, see A. Ellegard, 'The Darwinian Theory and Nineteenth-Century Philosophies of Science', *Journal of the History of Ideas* **XVIII**, 1957, pp. 362–393 and D. Hull (ed.), *op. cit.* (Note 20). For a contemporary defence of Darwin's method, see H. Fawcett, 'On the Method of Mr. Darwin', *B.A.R.* (Manchester, 1861), pp. 141–143.
91 See J. B. Morrell and A. Thackray, *op. cit.* (Note 25), pp. 271–275, 482–484. For a contemporary observation on the restricted meaning of 'science', see 'The British Association', *Oxford University Magazine* **I**, 1834, pp. 401–412. It should however be noted that Whewell did not suggest that non-mathematical disciplines were non-scientific. See for example his defence of natural history and the inexactness of its classifications. Whewell, *op. cit.* (Note 18), Vol. 1, pp. 493–494.
92 W. Swainson, *A Preliminary Discourse on the Study of Natural History* (London, 1834), p. 163. (my emphasis)
93 *Ibid.* pp. 163–164; also p. 150. See p. 201 where he quotes Herschel's *Discourse* in support of the concept of a single method in all the inductive sciences.
94 W. Whewell, *op. cit.* (Note 15), p. 390.
95 [W. Whewell], 'Jones on the Distribution of Wealth and Taxation', *British Critic X*, 1831, pp. 41–61, at p. 55.
96 W. Whewell, *op. cit.* (Note 18), vol. 2, pp. 336–337, 358.
97 *Ibid.* Vol. 1, p. 7. For the moral and theological dimensions of Whewell's work, see R. Yeo. 'William Whewell, Natural Theology and the Philosophy of Science in Mid-Nineteenth-Century Britain', *Annals of Science* **XXXVI**, 1979, pp. 493–516.
98 W. Whewell, *op. cit.* (Note 18), vol. 1, p. viii and *Novum Organon Renovatum* (London, 1858), p. vi.
99 W. Whewell, *op. cit.* (Note 54), vol. 3, pp. 377–381, 400–401, 467–472. It could be argued that Whewell was not being obsurantist, since teleology played a significant role in contemporary German biology. See T. Lenoir, *The Strategy of Life: Teleology and Mechanics in Nineteenth-Century German Biology* (Dordrecht, Boston, 1982).
100 [W. B. Carpenter], 'Physiology an Inductive Science', *British and Foreign Medical Review* **V**, 1838, pp. 317–342, at pp. 318, 321. See also his *Principles of General and Comparative Physiology* (2nd ed., London, 1841), p. 2.
101 B. Powell, *The Connexion of Natural and Divine Truth* (London, 1838), p. 60. Also p. 61 for the point that the non-mathematical character of geology did not render it less 'inductive'. For a similar appeal to the unity of science in the defence of geology see [R. Grant], 'Observations on the Nature and Importance of Geology', *Edinburgh New Philosophical Journal* **1**, 1826, pp. 293–302, at pp. 293–294.
102 B. Powell, *op. cit.* (Note 101), p. 68.
103 B. Powell, *Essays on the Spirit of the Inductive Philosophy, the Unity of Worlds, and the Philosophy of Creation* (London, 1855), p. 63.

[104] *Ibid.* p. 44. Powell did not name Whewell as the target here, but in his preface he foreshadowed disagreement with Whewell (p. vi).
[105] See D. Layton, 'The Schooling of Science in England, 1854–1939', in R. Macleod and P. Collins, *op. cit.* (Note 16), p. 191. For a defence of the life sciences in this context, see W. B. Carpenter, 'The Universities and Scientific Education', *Westminster Review* **XIX**, ns, 1861, pp. 381–402, at pp. 388–389.
[106] T. H. Huxley, 'On the Educational Value of the Natural History Science', in *Science and Education* (London, 1893), p. 45. Although Huxley often employed the notion of an accessible method, he did not accept the Baconianism which was usually linked with that notion in the first half of the century. In the situation discussed here, his major concern was to reject any qualitative distinctions between disciplines. For his attacks on Bacon, see C. Bibby (ed.), *The Essence of T. H. Huxley* (London, 1967), pp. 42–47.
[107] T. H. Huxley, *op. cit.* (Note 106), pp. 45–46.
[108] *Ibid.* See also 'On the Study of Biology', in *Scientific Memoirs* (4 vols, London, 1902), vol. 4, p. 258 and 'The Progress of Science' in *Collected Essays* (9 vols, London, 1893–95), vol. 1, p. 60.
[109] [W. B. Carpenter], 'Spiritualism and its Recent Converts', *Quarterly Review* **CXXXI**, 1871, pp. 301–353, at p. 342.
[110] W. Wilberforce, *op. cit.* (Note 90), p. 249.
[111] W. Hopkins, 'Physical Theories of the Phenomena of Life Part I', *Fraser's Magazine* **LXI**, 1860, pp. 739–752, at p. 741.
[112] *Ibid.* pp. 739–740; also pp. 741–744.
[113] See J. D. Burchfield, *Lord Kelvin and the Age of the Earth* (New Haven, 1975), Chapter 3.
[114] W. E. Gladstone, 'On the Influence of Authority in Matters of Opinion', *Nineteenth Century* **I**, 1877, pp. 2–22, at p. 20. This was a review of a book by G. C. Lewis. See (Note 128) below. In 1900, the philosopher, Henry Sidgwick, spoke of 'the actual impossibility of finding a satisfactory scientific explanation of the development of scientific knowledge'. Quoted in F. M. Turner, *Between Science and Religion: The Reaction to Scientific Naturalism in Victorian England* (New Haven, 1974), p. 20.
[115] K. Pearson, *op. cit.* (Note 3), p. 15.
[116] See (Note 20) above. For the eighteenth-century European notions of a science of society, see K. Kumar, *Prophecy and Progress* (Harmondsworth, 1981) pp. 13–44.
[117] J. Herschel, *op. cit.* (Note 17), p. 73. See also Herschel, *op. cit.* (Note 82), p. xi and 'Views on Public Education', *Philosophical Magazine* **VIII**, 1836, pp. 432–438, at p. 434.
[118] J. S. Mill, *op. cit.* (Note 15), p. 179.
[119] J. Herschel, *op. cit.* (Note 82), p. xl.
[120] W. Whewell, *op. cit.* (Note 95), pp. 52–57 and *op. cit.* (Note 84), p. 369. See also S. Hollander, 'William Whewell and J. S. Mill on the Methodology of Political Economy', *Studies in History and Philosophy of Science* **XIV**, 1983, pp. 113–126 and L. Goldman, 'The Origins of British "Social Science": Political Economy, Natural Science and Statistics, 1830–1835', *The Historical Journal* **XXVI**, 1983, pp. 587–616.
[121] K. Pearson, *op. cit.* (Note 3), p. 18.
[122] *B.A.R.* (Liverpool, 1901), p. 863. See also D. Layton, *op. cit.* (Note 105), p. 198 and P. Collins, 'The Origins of the British Association's Education Section', *British Journal of Educational Studies* **XXVII**, 1979, pp. 232–34.

[123] [C. Lyell], 'State of the Universities', *Quarterly Review* **XXXVI**, 1827, pp. 216–268, at p. 221.
[124] *Ibid.* p. 222. See also (C. Lyell), 'Transactions of the Geological Society of London', *Quarterly Review* **XXXIV**, 1826, pp. 507–540, at pp. 508–509.
[125] J. S. Mill, *Inaugural Address at St. Andrews* (London, 1867), p. 8. See also M. Pattison, *Suggestions on Academic Organization* (Oxford, 1868), pp. 266–280.
[126] C. Lyell, *op. cit.* (Note 123), p. 223; B. Powell, *The Present State and Future Prospects of Mathematical and Physical Studies in the University of Oxford* (Oxford 1832), p. 27; J. Herschel, *op. cit.* (Note 17), p. 70. See also [J. Bowring], 'Scientific Eduction of the Upper Classes', *Westminster Review* **IX**, 1828, pp. 328–373.
[127] For the background to this, see R. Yeo, *op. cit.* (Note 16), pp. 72–80.
[128] G. C. Lewis, *An Essay on the Influence of Authority in Matters of Opinion* (London, 1849), pp. 112–113. See R. Yeo, *op. cit.* (Note 87), pp. 7–8, 26 for some reference to this text.
[129] J. D. Campbell (Ducke of Argyll), *B.A.R.* (Glasgow, 1855), p. lxxxi. See also *B.A. R.* (Dundee, 1867), pp. xl–xlii.
[130] K. Pearson, *op. cit.* (Note 3), p. 15 John Dewey saw this distinction as central to better science education in America. See his 'Science as Subject-Matter and as Method', *Science* **XXXI**, 1910, pp. 121–127.
[131] On Armstrong, see W. H. Brock (ed.), *H. E. Armstrong and the Teaching of Science, 1880–1930* (Cambridge, 1973) and E. W. Jenkins, *From Armstrong to Nuffield: Studies in Twentieth Century Science Education in England and Wales* (London, 1979), pp. 41–57.
[132] *B.A.R.* (Newcastle, 1889), p. 229.
[133] *B.A.R.* (Newcastle, 1917), p. 126.
[134] T. P. Nunn, *The Aim and Achievement of Scientific Method* (London, 1907), p. 2.
[135] *Committee on the Neglect of Science: Report for the Year 1916–17* (London, 1917), p. 14. For another publication in this debate, see A. C. Seward (ed.), *Science and the Nation* (Cambridge, 1917).
[136] *B.A.R. op. cit.* (Note 128), p. 133. For a later perception of this implication, see G. Orwell, *op. cit.* (Note 1), p. 12.
[137] See D. Layton, *op. cit.* (Note 105), p. 201. For an attempt to restore the links between science and humanist values, see R. Gregory, *Discovery, or the Spirit and Service of Science* (London, 1916).
[138] C. Singer (ed.), *Studies in the History and Method of Science* (2 vols, London, 1917, 1921), vol. 1, p. vi. See also the references to this issue un the Introduction to this volume.
[139] J. B. Conant (ed.), *Harvard Case Histories in Experimental Science* (Cambridge, Mass., 1957), p. vii; also his *On Understanding Science: An Historical Approach* (New Haven, 1947), pp. 5–10 and *Science and Common Sense* (London, 1951), pp. 42–48. See also I. B. Cohen, *Science, Servant of Man: A Layman's Primer for the Age of Science* (London, 1948), pp. 13–15.
[140] *Nature*, 28 December 1940, p. 817.
[141] See the Introduction to this volume.
[142] This appears to be implied by Feyerabend's notion of a mythology of method. See P. Feyerabend, *Against Method* (London, 1975), pp. 220, 300–309.

NOTES ON CONTRIBUTORS

ALAN CHALMERS has a Ph.D. in History and Philosophy of Science from the University of London and is currently a Senior Lecturer in the Department of General Philosophy at the University of Sydney. He is the author of *What is this thing called science*? published by University of Queensland Press and Open University Press and of some twenty journal articles on contemporary philosophy of science and the history of physics

HOMER LE GRAND is a Senior Lecturer in the Department of History and Philosophy of Science at the University of Melbourne. His research interests include the history of chemistry, scientific change, and the social basis of science but, more recently, have extended to the development of twentieth-century geology. He is engaged in an extended study of the development and reception of "continental drift" theories.

T. D. STOKES. After undergraduate studies at Macquarie University, Sydney, T. D. Stokes completed a doctorate at the University of Melbourne. He taught social studies of science at Deakin University, Geelong, until taking up his present postdoctoral fellowship there. He is one of a group undertaking an extended sociological/ethnomethodological study of a leading Australian biomedical research institute.

JAN SAPP lectures in the History and Philosophy of Science at the University of Melbourne. His book *Cytoplasmic Inheritance and the Struggle for Authority in the Field of Heredity* is to be published by Oxford University Press. He is presently writing a book on Franz Moewus and the use by scientists of claims of fraud as an accounting device in the development of modern genetics.

W. R. ALBURY is Associate Professor and Head of the School of History and Philosophy of Science at the University of New South Wales. A native of the United States, he received his Ph.D. in the history of science from the Johns Hopkins University in Baltimore,

J. A. Schuster and R. R. Yeo (eds.), The Politics and Rhetoric of Scientific Method, 299—300.
© 1986 *by D. Reidel Publishing Company.*

before moving to Sydney in 1973. His papers have been published in leading journals of the history of science and medicine, and his annotated translation of Etienne de Condillac's *Logic* (1780) appeared in 1980. He is also the author of *The Politics of Objectivity*, published by Deakin University Press in 1983.

DAVID PHILIP MILLER studied Science and Politics at the University of Manchester and received his doctoral training in History and Sociology of Science at the University of Pennsylvania. Since 1981 he has been a Lecturer in History and Philosophy of Science at the University of New South Wales, having taught previously at the Universities of Pennsylvania and Edinburgh. His publications are chiefly concerned with the social history of British science in the late eighteenth and early nineteenth centuries.

The Editors:

JOHN A. SCHUSTER obtained his Ph.D. in the history of science from Princeton University and is a Senior Lecturer in the Department of History and Philosophy of Science at the University of Wollongong where his is also a Fellow of its Centre for Technology and Social Change. He has previously taught at Princeton, Leeds and Cambridge. Currently, he is completing a book on the interaction of methodology, natural philosophy and physico-mathematical sciences in the early career of Descartes. He is President of the Australasian Association for the History, Philosophy and Social Studies of Science.

RICHARD R. YEO studied History and Psychology at the University of Sydney where he completed a doctorate on natural theology and the philosophy of science in nineteenth-century Britain. He has published on such issues in the social and intellectual history of nineteenth-century science as natural theology, Baconianism, scientific authority, and on the work of William Whewell. Having taught previously at the Universities of Sydney and New South Wales, he is currently a Lecturer in History in the School of Humanities, Griffith University, Brisbane, Australia.

INDEX OF NAMES

GPSR Compliance

The European Union's (EU) General Product Safety Regulation (GPSR) is a set of rules that requires consumer products to be safe and our obligations to ensure this.

If you have any concerns about our products, you can contact us on ProductSafety@springernature.com

In case Publisher is established outside the EU, the EU authorized representative is:

Springer Nature Customer Service Center GmbH
Europaplatz 3
69115 Heidelberg, Germany

Batch number: 10830937

Printed by Printforce, the Netherlands